Hedges

BRITISH WILDLIFE COLLECTION 13

Hedges

Robert Wolton

B L O O M S B U R Y W I L D L I F E
LONDON • OXFORD • NEW YORK • NEW DELHI • SYDNEY

I dedicate this book with love to my wife Paula.

BLOOMSBURY WILDLIFE
Bloomsbury Publishing Plc
50 Bedford Square, London, WC1B 3DP, UK
29 Earlsfort Terrace, Dublin 2, Ireland

BLOOMSBURY, BLOOMSBURY WILDLIFE and the Diana logo are trademarks of Bloomsbury Publishing Plc

First published in the United Kingdom 2024

A catalogue record for this book is available from the British Library

ISBN: HB: 978-1-3994-1171-4; ePDF: 978-1-3994-1174-5; ePub: 978-1-3994-1172-1

2 4 6 8 10 9 7 5 3 1

Design by Susan McIntyre
Jacket artwork by Carry Akroyd

Printed and bound in India by Replika Press Pvt. Ltd.

MIX
Paper from responsible sources
FSC www.fsc.org FSC® C016779

To find out more about our authors and books visit www.bloomsbury.com and sign up for our newsletters

HALF TITLE: Recently-laid Field Maple and Hazel hedge.
FRONTISPIECE: Hedges white with May (Hawthorn) blossom are a familiar and striking feature in the spring.

Contents

Preface

It's difficult to imagine Britain and Ireland without hedges. Arriving here by plane, as one looks down, the most striking feature of our countryside is the patchwork of fields, nearly all of which are strongly edged by hedges. Green lines, brown in winter, tracing an intricate web across the land surface, irregular in parts, ordered in others. An iconic treasure, stretching from horizon to horizon.

To those of us who spent our childhoods in rural parts, the view through that plane window is a sure sign of coming home. It's not that other countries don't have hedges – many do – but we have far more than most and, crucially, they've been a defining part of our landscapes for a very long time, for centuries, in places millennia. Hedges have seeped deep into our national ethos. Without them, these islands would look and feel very different and would surely be much the poorer – not just aesthetically and culturally, but in many other ways too. It's said, and with good reason, that they are our biggest nature reserve, and by a long way. This, despite losing a great many over the last 80 years or so; we still have a richness of hedges that is the envy of our European neighbours.

For here in Britain and Ireland hedges remain everywhere, not just across our farmland but in our gardens, towns and cities too. Only in the uplands, some exposed western lands, chalk hills and fens are they absent, being replaced by stone walls, ditches or simply by tracts of open moorland. As a young man, before my interest in hedges had developed, I very nearly emigrated to New Zealand. There I saw first-hand how denuded landscapes look when stripped bare of native bushes and trees, the fields separated only by wire fences. The farmland I saw was for the most part raw, stark and quite frankly boring, a blank canvas. Here in Britain and Ireland we have cleared even more of our native woodland, but we did so far longer ago, giving time for rich cultural landscapes to develop in their stead,

OPPOSITE PAGE:
Hedges are a quintessential part of the British and Irish countryside. Iddesleigh, west Devon.

with hedges at their core. Of course, it helps that barbed wire is a relatively recent invention. The lack of hedges was not the reason why I did not end up in New Zealand, but looking back on it, I'm glad I stayed here. Studying and enjoying hedges has been infinitely rewarding.

My interest in hedges really began when we took on a small farm in Devon in the early 1990s. Until then, like many others I suppose, I had taken hedges for granted. The farm, where we raise beef cattle and sheep, consists of many small grass fields, some wonderfully full of flowers and all surrounded by thick hedges, rich in different shrubs. As I walked around our land, checking on the livestock, fixing water supplies, mending gates and carrying out the many other necessary chores, I gradually became more and more aware of the abundance of wildlife in the farm's hedges. I'd had a strong interest in birds since childhood, so naturally started to make a note of the species that were nesting in the hedges, but when I came across spherical nests of woven grass and leaves with no obvious entrance hole I was perplexed. One day I gently disturbed one of these nests and out popped a Hazel Dormouse. I think it was that moment more than anything else that hooked me on hedges.

At the time I was working for the Nature Conservancy Council (very soon to become English Nature and now Natural England), the government agency responsible for nature conservation, and I persuaded my bosses to allow me to devote some of my time to hedges. This gave me the opportunity to really get to know them. The rest is history, as they say. Although I've now retired from Natural England, we are still on the farm, and I still cannot help myself from looking for dormouse nests each autumn as I walk beside our hedges.

OPPOSITE: Dawn mist picks out the glorious latticework of hedges that defines so much of the farmland of Britain and Ireland. This shot shows the view from Pilsdon Pen in Dorset.

Acknowledgements

I have been immensely fortunate that so many people have shared their knowledge and thoughts about hedges with me over the years, and I am hugely grateful to you all. I would particularly like to thank Don Gaskins, George Pidgeon, Jo Staley, John Whetman, Jon Stokes, Lisa Norton, Nigel Adams, Ross Dickinson and Tom Hynes for their help and friendship.

I have enjoyed many a field trip, meeting and pub session with Hedgelink members, many of whom have been good friends, including Ann Davies, Clare Bowen, Clare Burrows, Colin Barr, Emily Ledder, Emma Marrington, Ian White, Jane Mackintosh, Jim Jones, John Dover, John Stratton, Katie Pollard, Megan Gimber, Rebecca Inman, Sarah Barnsley, Terence Henry and Tony Seymour.

Likewise members of the Devon Hedge Group have been a source of inspiration, information, help and companionship, among them Andrew Shadrake, Bob Harvey, Chris Townshend, Colin Clutterbuck, Corinna Woodall, Cressida Whitton, David Colman, Emily Spraggon, Fiona Hailstone, Heather Harley, Helene Jessop, Jasmine Atkinson, Lesley Strong, Martin Turner, Melanie Croll, Michelle Werrett, Mike Pearey, Paul Cottington, Peter Burgess, Peter Chamberlain, Rob Dixon, Roger Parris, Ross Gray, Sally Hope Johnson, Sarah Jennings, Simeon Day, Stephanie Knight and Valerie Darwall.

Several people have very kindly given me feedback on sections of this book: David Cann, Heather Robertson, Jo Staley, Joe Gowran, John Day, Jonathan Boaz, Keith Alexander, Lindsay Whistance, Lisa Norton, Martin Luff, Matthew Axe, Neil Foulkes, Ray Woods, Ross Dickinson, Sally Westaway, Tim Rich and Tim Sparks. I thank you all. Any mistakes remain my responsibility.

Joe Gowran, Neil Foulkes and Alan Moore provided invaluable information about hedges in Ireland. In the Netherlands, I have valued discussions and exchange of information and ideas with Jef Gielen, Lex Roeleveld, Louis Dolmans, Maja Kooistra and Robert Ceelen, while the late Georg Müller from Germany was an inspiration – I'm grateful to his wife Gisela and son Tim Müller for allowing me to reproduce a plate from his book.

Ernie Pollard helped me source the 'Thorny Questions' poem as well as generously offering encouragement. My thanks too to Colin Barr for giving me his books on hedges when he decided to move on to other things. Marcus and Sally Vergette, our good friends and farming neighbours, have shared my passion for hedges – thanks

Marcus for allowing me to include one of your artworks, and Sally for advice on bucolic cattle painters. My thanks too to Rob Hutton, who has trimmed our farm hedges for many years, teaching me by example a good deal about the practicalities of flail cutting.

My sincere gratitude to the following for providing photos and allowing me to use them: British Geological Society, Dawn Painter, Geoff Foale, Ian Andrews, Iorwerth Watkins, Jean Wynne-Jones, John Walters, Jon Stokes, Jon Valters, Kevin Cox, Lianne Joynson (National Memorial Arboretum), Lex Roeleveld, Matthew Axe, Mike Ashworth, Nigel Adams, Paula Wolton, Peter Tailby, Ray Woods, Steve Bloomfield (Worcestershire Wildlife Trust), Steven Falk and Tony Seymour.

I remain in the debt of all those who helped me identify species from Hedge 2070, including those named in Chapter 5. My thanks to Robert Aquilina for information on the numbers of flies emerging from hedge ditches, to Barbara Benfield for her article on lichens on hedgerow trees, to Becky Willson for sharing information on carbon cycles, and to Catriona Bass and Kevan Martin for help relating to their hedges in the Thames floodplain. Thanks too to Geoff Wilkinson for identifying my photo of a hoverfly larva to species, and to Augustijn De Ketelaere and William Penigot for identifying photos of Parasitica.

I have valued chatting about and exchanging information on hedges with members of the Devon Fly Group, including Andrew Cunningham, Martin Drake and Richard Lane among others mentioned elsewhere in these acknowledgements.

The team at Bloomsbury Wildlife have been unwaveringly supportive and encouraging. Massive thanks to you – David Campbell, Jenny Campbell, Heather Bradbury and Laura Browning-Brant. I have also benefited enormously from working with Hugh Brazier, copy-editor. The book is much stronger for your advice and improvements to the manuscript – thank you, Hugh. Thanks too to Carry Akroyd for her splendid cover illustration.

Others have helped and influenced me while writing this book, and over the years. I fear I may have forgotten some I should have mentioned, in which case I beg forgiveness.

Above all, I thank my wife Paula Wolton, self-professed hedge witch, for all her help, support and encouragement over the years, as well as for the use of some of her superb photos. I could not have written this book without you.

The ever-useful hedge

chapter one

Fifty years ago, in 1974, Eddie Pollard, Max Hooper and Norman Moore wrote their seminal book *Hedges*. They were scientists working at the Nature Conservancy's Monks Wood Experimental Station in Cambridgeshire, and all three had a passion for hedges, Dr Pollard specialising in mammals and invertebrates, Dr Hooper on hedge history and Dr Moore on the effects of pesticides on wildlife. One of the first things they did, in 1961, was to arrange for a mile (1.6km) of experimental Hawthorn hedges to be planted across an arable field, and a year later a further mile of different kinds of hedges to be planted in a neighbouring one. What foresight! These hedges, especially the Hawthorn ones, have been much used for research right up to the present day.

Drs Pollard, Hooper and Moore's aim in writing their book was to provide an authoritative, up-to-date account of hedges that would appeal to both scientists and non-specialists with a general interest in the subject. My hope is that this book will follow in the same line, although theirs is a hard act to follow. I remain in awe of what they achieved, especially since in those days they had none of the advantages of computers or the internet. As I write, although Max Hooper and Norman Moore have passed on, I am pleased to say Ernie Pollard is still with us and has given me his blessing to write this book, using the same title as theirs.

So much has changed in the last 50 years. Back then, hedges were still being removed without heed or hindrance and in large numbers, reflecting a complete lack of protection and little recognition of their value to wildlife. True, grant aid for hedge removal may have ceased to be available in the UK after 1970, but it was not until 1982 that public money was on offer to plant and restore them. The focus of the time was very much on the utilitarian role hedges had to play

OPPOSITE PAGE: Hedges studded with oaks border the country lane that runs through our farm in Devon.

in farming, or rather the lack of it. While recognised still to have some worth as boundary markers, living fences, shelter for livestock, windbreaks for crops, cover for game birds and jumps for fox hunters, by and large landowners and farm managers judged hedges to be of insufficient value to warrant retention. To the contrary, they considered them to be a barrier to efficient food production by keeping field sizes small and occupying land that could otherwise grow crops. The fields they enclosed may have been of a size and shape fit for working with horses, but not with tractors. Worse still, they were costly to maintain. So grubbing them out was the order of the day.

Ernie, Max and Norman expressed understanding and empathy for the removal of hedges by farmers and were resigned to their further loss in the name of agricultural efficiency. Effort, they suggested, should be expended only on keeping those of special historical or wildlife value, farmers being compensated from the public purse for doing so. Yet, reading between the lines, it's clear enough that these three authors of the first *Hedges* were deeply uneasy about the rate at which 'good British hedges' were being lost. Their book, with its focus on history and fauna and flora, is surely a plea for their retention, even if not explicitly stated.

Back then, climate change was not on the radar and there was much less concern about the loss of biodiversity. Farmland was still rich in wildlife compared to the present day. Still, the writing was on the wall. For example, although Rachel Carson's 1962 book *Silent Spring* had alerted the public to the danger of indiscriminate pesticides use, nonetheless DDT, dieldrin and other organochlorine pesticides had yet to be banned and were still being applied liberally to control crop pests. I suspect that in 50 years' time, people will look back on today with equal dismay at our current widespread use of even more powerful toxins such as neonicotinoids, even though they are applied more selectively.

It was not until 1997 that hedges anywhere in Britain and Ireland were given legal protection, with the introduction of the *Hedgerows Regulations* in England and Wales. The use of taxpayers' money to encourage and support farmers to do good things for the environment was still a pipe dream when *Hedges* was published in 1974. Even now, legal protection is far from adequate in Scotland, Northern Ireland and the Republic of Ireland.

Hedges may no longer be as useful as boundary markers or living fences as they once were, but if cared for they can still, as we shall see, be of genuine help to today's farmers – by conserving soil, increasing crop pollination and reducing crop pests. Moreover, reflecting the harsh reality of climate change, there's a resurgence of interest in the agricultural sector in their value as a source of green, renewable energy, as windbreaks for crops and as a source of shade and shelter for farm animals. They even increase land sale value. For society at large, they are hugely important assets for wildlife conservation and nature's recovery, for carbon capture and storage, for the removal of pollutants from air and water, and more.

A further new and powerful reason for hedges to be valued today is that they, along with other accessible green features, succour our health and wellbeing, especially in urban areas. All these various benefits – 'ecosystem services' or 'public goods' to use the contemporary jargon – are presented in the table at the end of this chapter. Few other habitats or landscape features boast such a long list, if any (Land Use Consultants 2009).

Overall, we now have quite a few more good reasons to keep and care for hedges than there were evident back in 1974 when the original *Hedges* was published. The breadth of subjects covered by this book has therefore had to be considerably greater. Even so, to keep within my word allocation I have given scant attention to a few, such as the role hedges have in helping to keep the air of our towns and cities clean and noise levels low, and how useful they can be educationally, as outdoor schoolrooms. Suffice to say that in urban areas hedges, particularly those that are evergreen or that have hairy leaves, can be effective at mopping up air pollutants like particulates and asthma-causing pollen, especially when positioned between pedestrians or school grounds and traffic. They can also help to keep built-up areas cool and, when wide and dense, act as a noise barrier.

It's not that everything is fine now with hedges – far from it, most are in poor condition. Still, in Britain at least, hedges are very rarely being removed any more to enlarge fields in the name of food production, although alas it seems that much loss continues across the island of Ireland. Public opinion, legislation, rules linked to farming subsidies and environmental agreements between government and farmers have collectively made a big difference. Indeed, there's much to be positive about – the gloomy forecasts expressed by those writing about hedges 40 or 50 years ago, such as the renowned author Henry Williamson in his heartfelt narrative *The Vanishing Hedgerows* (1970), have, thank goodness, proved unduly pessimistic.

Nevertheless, it remains the case that few farmers feel they can afford to spend much money on the upkeep of hedges unless paid to do so from the public purse: the benefits of farm hedges to food production and business profitability are, for the most part, intangible and hard to place an economic value on, at least in the short term. As a consequence, just as they were 50 years ago, hedges are still considered by many farmers to be a drain on resources rather than a business asset, to such an extent that neglect is now a major threat to their survival. The challenge remains to present clear and convincing arguments, based on good evidence, that it really does make sound economic, as well as environmental, sense for farmers to care for hedges. My hope is that this book will help with this.

There have been a number of good books published on hedges in the years since the first *Hedges* appeared in 1974. In 1986, Oliver Rackham, founding father of historical ecology, wrote his hugely influential *The History of the Countryside*, which covers the origins and fortunes of fields, hedges and hedgerow trees in detail. I have drawn heavily on the wealth of facts and analysis in that ground-breaking book. A few years later, in 1987, Richard and Nina Muir published *Hedgerows: Their History and Wildlife*, a readable and informative account of these two subjects. In the same year Professor Wilfred Dowdeswell published his rather more technical book *Hedgerows and Verges*, focusing on topical areas of controversy between farmers and the general public.

More recently, Tom Williamson has taken forward our knowledge of hedge history through his book *Hedges and Walls*, published by the National Trust in 2002, and again, this time together with Gerry Barnes, in *Hedgerow History*, published in 2006. The curious obsession the British have with hedges is the theme of Hugh Barker's 2012 *Hedge Britannia*, a delightful and fascinating book including accounts of extraordinary and often bizarre topiary. This was followed in 2016 by John Wright's *A Natural History of the Hedgerow, and Ditches, Dykes and Dry Stone Walls.* In this he gives an excellent overview of hedge history, including original research on hedgerow loss, and of many aspects of hedge wildlife, with a strong focus on fungi and plant galls. I must also mention Murray Maclean's practical guides to hedge management, including his 2006 book *Hedges and Hedgelaying.*

And then there is the late Georg Müller's truly epic work of 2013, *Europe's Field Boundaries.* This was his life's great work. It all started in 1977 when he revisited banked hedges in his home parish of Ganderkersee in northern Germany, where in his youth he had collected rare fungi. To his shock, many of the hedges had gone. Worried, he began a meticulous survey of those that remained in the parish.

To his great dismay this confirmed huge losses, with only 19% of those present in 1842 surviving. Worse still, only 4% were in good condition. If this was the terrible situation in his homeland, what, he wondered, was happening in the rest of the Europe? Were field boundaries being erased obliviously elsewhere too, without care or concern for the loss of cultural heritage or damage to the natural world? Unable to find the answers in any published works, he set about finding out himself, a labour of love to which he devoted the next 30 years. During this time, usually accompanied by his wife Gisela, he visited 32 European countries, travelling an incredible 238,531km in the process, including some 10,000km on foot. The result is a richly illustrated compendium of information on field boundaries across Europe, presented in two hefty volumes. That he recorded the distance travelled over the 30 years to within a kilometre is a testament to his close attention to detail. Alas, Georg found field-boundary landscapes across Europe to be in a very perilous state: his book has done much to alert people to this dire situation and to the urgent need for conservation measures. We are fortunate indeed that Great Britain and Ireland have fared better than most countries – our hedged and walled landscapes have remained relatively intact.

ABOVE: Hedges change through the seasons, a source of constant delight.

All these books I have found hugely informative and helpful, and all have influenced me. However, wisely enough, none of the authors have tried to emulate and update Pollard, Hooper and Moore's *Hedges*. This has been my ambition in writing the present book, to describe our current state of knowledge 50 years on, as far as space allows. Over the last few years, levels of research interest have been so high that just about every week there's a new publication or two on hedges in scientific journals – it is hard to keep up. As I write, an article has just arrived in my inbox on the factors affecting breeding birds in East-Fryslân in the Netherlands. It looks interesting. There's so much to be gained from cross-Channel learning and exchange of ideas.

Above all, my hope is that this book conveys just how extraordinary and fascinating hedges can be. We have, literally, millions of them in Britain and Ireland, yet no two are the same. These thin green bands that weave through our countryside and settlements, that both separate and connect, are part of our psyche. How impoverished our nations would be without them!

What hedges do for us – an overview of their main benefits

Aesthetic appeal	Hedges are a defining and attractive feature of the great majority of our lowland countryside. They even increase the sale value of farms because of this.	Chapter 1
Health and wellbeing	Nature is a very effective stress reducer. Especially in urban environments, hedges benefit us all by softening and greening the environment and providing healthy possibilities for practical and active community engagement. They also increase our enjoyment of the countryside, whether on foot or bicycle, or in a car.	Chapter 1
Cultural heritage	Many, perhaps most, are ancient. Through their rich and intricate patterns they tell the story of the countryside and farming traditions over many centuries, even millennia.	Chapter 3
Wildlife	Our most widespread semi-natural habitat, they provide critical resources for a great many plants and animals, especially in intensive farmland. They are also vital highways for movement, including dispersal.	Chapters 5 to 9
Boundary markers	Although less important in this role than they used to be, hedges are still used to announce property ownership, especially in built-up areas. They are still much used as screens, for privacy and to hide eyesores.	Chapter 10
Livestock control	Again, this role is less important than in former times, but hedges are still sometimes used as living fences, to keep cattle, sheep and other farm animals where they are meant to be.	Chapter 10
Windbreaks	They reduce wind damage to crops and help prevent their dehydration, especially in horticulture.	Chapter 10
Shelter and shade	Increasingly important with climate change, they provide protection for lambs and other farm animals from harsh weather, and shade from the summer sun.	Chapter 10
Soil conservation	They help to retain soil in fields that would otherwise be lost through water erosion or wind blow.	Chapter 11

Firewood from a coppiced Grey Willow hedge.

Water quality improvement	Through removing fertilisers, pesticides and sediment from water flowing or percolating off the land, they help to keep our streams and rivers clean.	Chapter 11
Flood risk reduction	Well-sited and well-constructed hedges can hold back storm water, releasing it slowly into watercourses, reducing the risk of land and properties below flooding.	Chapter 11
Pollination	They provide pollinators with nectar, pollen, flight lines and places to breed and overwinter, and so can increase crop yields.	Chapter 12
Pest control	They boost populations of predators of crop pests, and so can increase crop yields.	Chapter 12
Carbon capture and storage	At a landscape scale, they have a major role to play in removing carbon dioxide from the air and locking it up for long periods.	Chapter 13
Green energy	They can be cost-effectively managed to produce sustainable crops of wood fuel, including charcoal, offsetting the use of fossil fuels.	Chapter 14
Air quality improvement	In towns and cities, hedges help capture particulates and other pollutants, including pollen, that are damaging to human health. Wide thick ones can also abate traffic noise and moderate the urban heat-island effect.	Abhijith *et al.* 2017 Abhijith and Kumar 2019
Foraging	They are popular sources for some iconic British foods like blackberry jam, crab apple jelly and sloe gin.	Many hedgerow recipes available online
Education	Hedges are accessible outdoor classrooms, providing learning opportunities across a wide range of subjects, including history, maths, biology and geography.	Devon Hedge Group and Hedgelink websites

The flowers of Hogweed and other umbellifers are an excellent source of nectar.

What counts as a hedge?

chapter two

A hedge is just a line of bushes or shrubs, isn't it? Simple enough! But hold on, when does a hedge become a strip of woodland? Is it still a hedge if it turns into a line of trees? What about any underlying banks or flower-rich margins – are they part of the hedge too? Indeed, what's the difference between a hedge and a hedgerow? Or an avenue? Questions like these have led to many a long night's discussion among us hedge enthusiasts. Believe me, fuelled perhaps by a beer or two, we can talk with passion for hours on end about such mysteries, to the bewilderment of the uninitiated (or the sane).

For hedges are as variable as the British weather. There's a multitude of different types and forms, and they grade seamlessly into other linear features such as bands of scrub, shelterbelts and avenues, even stone walls. As I drive around our countryside, and especially our towns, I am constantly thinking: Is that a hedge or something else? Frequently, I am baffled. In truth, it rarely matters much. Nevertheless, we do need some sort of shared understanding if we are to talk about hedges with one another meaningfully, and for facts and figures like the length of hedge in Britain or Ireland to make any sense.

This chapter attempts to generate this shared understanding, so that you and I can agree on what this book is about. First, I delve into definitions, bravely starting with the vexed question of whether there's any difference between a hedge and a hedgerow, before attempting to answer mind teasers such as: How long is a hedge? How high? After providing a definition of a hedge fit for most, I give a flavour of their huge variety, reflecting upon basic splits such as ancient or recent, single-species or species-rich. Finally, I describe some hedge types so distinctive to particular places that they are defining features. Subsequent chapters will build on all this.

OPPOSITE PAGE: My wife, Paula Wolton, admires the Giant's Hedge, a massive pre-Norman defensive boundary bank that runs between Lerryn and Looe in Cornwall. 'Jack the Giant had nothing to do, so he built a hedge from Lerryn to Looe'.

Hedge or hedgerow?

The Oxford English Dictionary is not much help on this prickly issue. A hedge is 'a row of bushes or low trees (e.g. hawthorn, or privet) planted closely to form a boundary between pieces of land or at the sides of a road', while a hedgerow is 'a row of bushes forming a hedge, with the trees, etc. growing in it; a line of hedge'. So, effectively, one and the same thing. In practice most of us use the words interchangeably, but others are insistent that there are differences. When I first arrived in Devon I was firmly told off for referring to hedgerows – 'We only have hedges here!' The Cornish are even more vehement about this – woe betide anyone who refers to Cornish hedgerows. But in the Home Counties, those lines of bushes between fields are firmly hedgerows – while those in gardens are usually hedges. It's the National Hedgelaying Society, not the National Hedgerow Laying Society, perhaps simply because the latter is more of a mouthful. Personally, I don't think it matters at all which word one uses, provided you don't offend the locals. It's only because I live in Devon that I usually refer to hedges – but then I like the term 'hedgerow trees'. Somehow that sounds better.

The confusion about which word to employ is nothing new; back in Anglo-Saxon times both *hege* and *haga* were used for hedge, *haga* being mainly southern and *hege* western (Rackham 1986). The Saxons also used the word *hegeraewe*, presumably the origin of hedgerow, although we remain none the wiser as to the difference between hedge and hedgerow. It is possible that, long ago, the difference was that hedges were just lines of bushes while hedgerows had occasional mature trees along their length as well.

The boundary bit in the dictionary definition is important. It implies that the line of bushes or trees has been created, or at least is welcomed, and is being maintained for a purpose, namely to divide up an area of land in some way such as marking a field boundary or partitioning a garden. Lines of trees that have grown up unplanned alongside natural features, such as those that fringe so many of our rivers and streams, are excluded by the definition – they are not hedges – at least as far as anything can ever be black and white where hedges are concerned. But if they have grown up spontaneously alongside another form of boundary marker, such as a wire fence, they are hedges. In essence, a hedge is only a hedge if maintained by human agency, if not now at least in the past, to perform some useful function or another.

The Dark Hedges

These create an awe-inspiring avenue of ancient Beech trees in County Antrim, Northern Ireland. The trees, 150 of them, were planted in about 1775. The sweeping boughs of the remaining 90 or so now form a spectacular arched tunnel, made famous by featuring in fantasy drama television series *Game of Thrones*. The avenue has become such a tourist attraction that the road has had to be closed to vehicles for the sake of the trees' health. Legend has it that the hedges are haunted by a ghost called the Grey Lady, who travels the road and flits across it, from tree to tree.

The Dark Hedges, County Antrim, Northern Ireland.

How wide, how high?

When is a hedge too wide to be a hedge? When it's a wood! For most purposes that's good enough. But if you are trying to map the hedges in an area, you'll need something better. Since wide hedges grade seamlessly into narrow woodlands, a cut-off point is required. In the second edition of the *Hedgerow Survey Handbook* (Defra 2007), this is taken as 5 metres between major woody stems. Anything narrower than this is a hedge. The definition has stuck, and now become the standard in Britain, although, as we shall see, 10 metres is often used in Ireland. Note the caveat that the measurement must be made between major woody stems, so excluding small-diameter woody outgrowth such as Blackthorn suckers. This means that in practice hedges can be much wider than 5 metres.

Unless deliberately managed to keep them short, lines of bushes will, by and large, turn into lines of trees. Only those consisting of the likes of Gorse or Box, or those which are frequently blasted by coastal or upland gales, naturally remain short. Despite this, in popular parlance, hedges or hedgerows are seen as low affairs, perhaps with tall mature trees scattered along their lengths. Boundary lines of trees are considered something different – yet, in reality, they are just short hedges grown tall. I've stood many a time with people looking at a well-hedged landscape to hear them bemoan that all the hedges are being flailed to within an inch of their lives – yet many, sometimes half of them, have been allowed to grow into what I call 'line-of-trees hedges'. For some academics the workaround for this is to talk about 'woody linear features', a term that clearly includes lines of mature trees as well as lines of bushes.

BELOW: Still a hedge. These Beeches in Dartmoor have just been allowed to grow to maturity, forming a line-of-trees hedge.

The Meikleour Beech Hedge

According to *Guinness World Records*, this is the world's tallest and longest hedge. It certainly is not the longest, but it may well be the tallest *managed* hedge, averaging a spectacular 30m high over its 540m length, and peaking at 36m (118 feet). There are certainly taller line-of-trees hedges, including Beech ones, but these have been allowed to grow freely, unconstrained by trimming. Located in Perthshire on the Meikleour estate, it runs alongside the A93 Perth–Blairgowrie Road and was planted in 1745. It is said the hedge grows towards the heavens because the men who planted it were killed at the Battle of Culloden the following year, in April 1746. The hedge is usually trimmed once every 10 years, but when it was last cut in 2020 by estate foresters there had been a gap of 20 years, costing the owners, Sam and Claire Mercer Nairne, an estimated £90,000 to get the job done.

Trimming the world's tallest managed hedge, the Meikleour Beech hedge in Perthshire, in January 2000.

So, when I use the term hedge or hedgerow I am not discriminating between the short and the tall – all heights are included. I would urge others to do the same. Not recognising that short bushy hedges and line-of-trees hedges are both equally valid types of hedge/hedgerow has been, and continues to be, the root of much unnecessary confusion. The lack of consistency is especially frustrating when one attempts to compare statistics generated by different surveys about the length or state of hedges. What should be a straightforward matter becomes difficult if not downright impossible.

Where does a hedge start and stop?

This is easy to answer if the hedge is isolated. But if it is joined to others in a network, as most are, it's trickier. The *Hedgerow Survey Handbook* says simply that a hedge stops whenever it is joined by another, and in practice this works well enough. Be aware, though, that when walking along a thick bushy hedge it can be very difficult to see other hedges joining it on the far side – your hedge may in fact be several hedges.

What happens if the hedge has a gap in it? How wide does the gap have to be before the hedge becomes two hedges? The answer, according to the *Hedgerow Survey Handbook*, is over 20 metres. This figure was set to allow for gateways and to ensure all but the most gap-ridden lengths are counted as single hedges. Nonetheless, the 20m limit is based on pragmatism rather than anything more scientific linked to the natural world.

What if a gap is filled with brambles and little else? The answer: it is not a gap. One can spend happy or edgy hours debating such matters, but to no great purpose: in the end common sense usually works well enough.

BELOW: Basal and marginal flora is just as much a part of a hedge as the shrubs and trees.

More than just trees and bushes

Hedges are, or should be, much more than lines of bushes or trees. All good ones include strips of tussocky grasses or flowering herbs at their bases or in their immediate margins, and many of the best have banks or ditches too. Indeed, thinking about all these structural components – the trees, shrubs, margins, banks and ditches – is vital for effective hedge conservation and management. Far too often the focus is narrowly on the woody bits alone, overlooking the fact that the basal and marginal structural parts are essential to hedge health too. I do admit, though, that in practice taking such a holistic view of a hedge is often more easily said than done, but one should always try. I return to this matter later in this book.

A working definition

To pull all this together, a workable definition of a British hedge is:

> *Any boundary line of trees and/or shrubs less than 5m wide between major woody stems at the base, however short or tall. Any gap between the trees or shrubs more than 20m long marks the end of one hedge and the beginning of another. Any bank, wall, ditch or tree within 2m of the centre of the hedge is part of the hedge, as is any basal and marginal herbaceous vegetation.*

This is close to the definition used in the *Hedgerow Survey Handbook* and works well enough for most hedges, whether in farmland or in gardens. This definition works for Ireland too, except that here a width of 10m is more normally used.

Inevitably there are grey areas, requiring judgement calls. In a formal setting such as a park or beside the drive to a stately home, lines of mature trees with no undergrowth between them are avenues, not hedges, even if the canopies of the trees touch: in a less formal setting they may well be termed hedges.

Looking further afield, there are many other types of linear boundary feature that are, to all intents and purposes, hedges. In Morocco I have seen fearsome ones made solely from the cactus Prickly Pear. Here too, as in many places in southern Europe, one can see boundaries made from brushwood, that is cut branches, often interwoven and staked into place – we call these 'dead hedges' but abroad they are often referred to simply as hedges. Colin Barr and Sandrine Petit (2001), on behalf of the International Association for Landscape Ecology, organised a conference in Birmingham in 2001

The Great Hedge of India

The most astounding, if infamous, hedge I have ever heard of is one planted in India in the first half of the 19th century, to control the movement of people. Roy Moxham has written a fascinating account of it, *The Great Hedge of India* (2001), well worth reading. It was planted by the East India Company to control the trade in salt for the purpose of levying tax, an inhuman endeavour that caused huge amounts of suffering through depriving countless people of salt essential for their health. Incredibly, this one hedge ran for 2,400km (1,500 miles) southwards from the Himalayas through northern India, passing Delhi and Agra on route, the same distance as from London to Istanbul as the crow flies. The greater part of a customs line manned by 12,000 men, the hedge consisted of an immense impenetrable barrier of thorny trees, shrubs and creepers. By the 1990s, however, Moxham was scarcely able to find any remnant of this testament to the staggering ambition and, it must be said, greed of the British Raj.

The Great Hedge of India. The green line shows the hedge; the red lines at each end show the continuation of the Inland Customs Line.

wonderfully titled *Hedgerows of the World*. Here we heard about Vetiver hedgerows, Vetiver being a south Asian grass that has a massive deep root system. It grows well in tropical and subtropical areas and is used to form thick hedges to conserve soil and water. So, there are cactus hedges, even grass hedges. But plastic hedges are fake hedges and don't count.

Immense variation

Hedges are almost infinitely varied. No one to my knowledge has ever attempted an overall classification; it would be a daunting and likely impossible project. One hedge may be labelled as being on a farm, of ancient origin, banked, high and species-rich. Another may be in a garden, recently planted with just Cherry Laurel, kept short by regular trimming. A third may be totally different again, perhaps a line of Tamarisk trees bordering a Cornish field, exposed to the full fury of westerly gales. And so on, with just about every combination in between possible. There are though, four basic splits which are helpful, as described below.

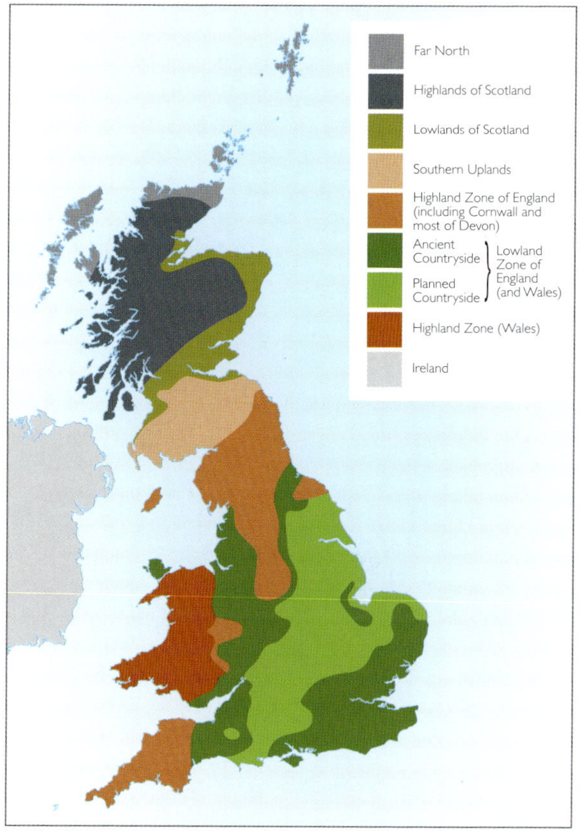

ABOVE: The distribution of ancient countryside and planned countryside according to Oliver Rackham (1986).

1. Ancient or recent?

In England, Wales and Northern Ireland, ancient woodlands are defined as those that have existed continuously since 1600 or before. This date is not used in connection with hedges: in the UK we usually refer to any hedge in existence before 1750, roughly the date of the start of the main period of Enclosure Acts, as being ancient (the same date, as it happens, as is used in Scotland for ancient woodlands). To muddy the waters, the *Hedgerows Regulations* (1997), which apply to England and Wales, give a measure of protection to 'historic' hedges, defined to include all those established prior to the first Inclosure Act – and the accompanying guidance gives the date of the first Act with Inclosure in its legal title as 1845 (DETR 1997). The words 'inclosure' and 'enclosure' are frequently used interchangeably, but there is a technical difference. I'm grateful to John Wright (2016) for explaining this. The process of enclosure is that of placing a boundary around a piece of land, while inclosure is a legal term that refers to the conversion of common land into private land, so might not involve the creation of physical boundaries at all. In view of this, it seems odd that the lawyers should have used inclosure rather than enclosure in the *Hedgerows Regulations*.

Oliver Rackham (1986) neatly and usefully divides the countryside of the English lowlands into two categories, 'ancient countryside' and 'planned countryside'. Ancient countryside is characterised by irregular small fields, typically established during medieval times, if not earlier. Here the hedges characteristically contain a mix of woody species. In contrast, planned countryside, which predominates in the Midlands, is one of larger regular fields with straight sides established mainly by the Enclosure/Inclosure Acts, most of which were passed between 1760 and 1820. These hedges tend to have fewer species of tree and shrub and are usually dominated by Hawthorn. The two landscape types occupy a land area of roughly equal size, although Rackham's English lowlands excludes Devon and Cornwall (as well as the Pennines, Cumbria and Northumberland). Devon and Cornwall have for the most part the intricate field patterns typical of ancient countryside and are among the most densely hedged counties in Britain – if they are included in the mix, then perhaps two-thirds of England have field patterns typical of ancient countryside, one-third planned countryside.

2. Garden or farmland?

Garden hedges and farmland hedges tend to be quite different, as one would expect from their different roles. Garden ones typically have an ornamental or screening purpose and so are composed either of shrubs that have attractive foliage, flowers or berries, or of evergreens that remain dense and bushy if trimmed at least once a year. Privets, Box, Yew and Cherry Laurel are common here – all, as it happens, species that are toxic to farm animals – along with the infamous Leyland Cypress. This hybrid conifer is frequently the cause of great strife between neighbours because it grows very fast and very tall, up to 40m high, if not kept under tight control.

RIGHT: Hornbeam makes an excellent hedge in a garden setting. Like Beech, it stays dense after many years of trimming and retains its dead leaves in the winter.

ABOVE: Ancient countryside: note the small irregular fields. East Devon.

BELOW: Planned countryside: note the dead straight field boundaries. Cambridgeshire.

Farm hedges serve different purposes: to mark field boundaries, to keep livestock in or out of fields or to provide firewood – or at least those are the reasons why most were established (see Chapter 10). They usually have a high proportion of thorny or prickly shrubs like Hawthorn, Blackthorn and Holly. Not that the distinction between the two hedge types is absolute: there are overlaps between garden and farm hedges. For example, both can serve as boundary markers and both can be good for wildlife, while a few native trees like Beech, Hornbeam, Field Maple and Holly make good functional hedges in both environments.

3. Banked or unbanked?

Hedges that have earth banks at their base are found throughout Britain and Ireland, but it is in south-west England and Wales that they are best developed. In these places they are a defining feature of local landscape character – you know you are in the West Country when you are driving along a narrow lane between high banked hedges that threaten to scratch your car and frustratingly obscure the view apart from the occasional glimpse through a gateway. These hedge banks may be turf-faced or stone-faced.

BELOW LEFT: High banked hedges are a feature of West Country lanes, such as this one in Luppitt, East Devon.

BELOW RIGHT: A Cornish hedge topped with Thrift. The sides of the bank that form the hedge are stone-faced.

In Cornwall, many 'Cornish hedges' are banks with at best a little bit of Gorse or Heather on top of them – shrubs or trees are an accessory. Attractively, along the county's exposed coasts many are topped just with a line of Thrift, painting an attractive pink ribbon in the spring. Likewise, the essence of a true Devon hedge is its bank, not the stuff that grows on it. The uncertain origin of the massive hedge banks found in the West Country is discussed in the next chapter.

Banked hedges are found elsewhere too. County Kildare in Ireland is renowned for its double ditches – large earthen banks with a drain and often a line of shrubs at each side, many of which are thought to date back to Norman times (Foulkes 2006a). They are a distinctive feature of the racecourse at Punchestown. In Wales, stone-faced earth banks have their own name, *clawdd* (plural *cloddiau*), and are particularly characteristic of Anglesey, the Bangor/Caernarfon coastal strip and the Llŷn peninsula.

4. Single-species or species-rich?

This split is all about the number of different trees and shrubs, collectively termed woody species, present in a hedge. The commonest type of single-species hedge by far is that dominated by Hawthorn,

BELOW: A fine Yew hedge in a formal garden setting, at Brawda in Cornwall.

typifying Enclosure Act hedges. Much less frequent, although locally common across appreciable swathes of countryside, are elm and Beech hedges; the dominance of elms reflects ancient hedges being taken over by the trees' aggressive suckering, while in places like Exmoor and Perthshire it was fashionable for estates to plant pure Beech hedges in the 19th century. In gardens, pure Yew, Box and privet hedges are commonplace, as are many others composed of single species, often exotics such as Snowberry, Firethorn and cotoneasters.

Species-rich hedges, on the other hand, have a range of different trees and shrubs along their lengths, sometimes as many as 14 species in a 30m stretch. This is explored in detail in Chapter 7.

Variation in basal and marginal plant communities

Hedges can be broadly grouped not just by their woody species but also by the herbaceous flora of their bottoms and margins. One might think these floras would be related in some way to the tree and shrub species present, but this is rarely so, and instead they are usually determined by adjacent land use. Next to farmland that has received high dosages of artificial fertiliser for many years, the flora is likely to be dominated by a few aggressive species that thrive under such conditions, notably Nettle and Goosegrass, an all-too-frequent occurrence. Next to intensively grazed pasture, a short grassy sward containing few flowering plants is usual. Alternatively, if the land has not been disturbed for many years and grazing levels are light, strips of tussocky grasses may develop. Often dominated by Cock's-foot or Yorkshire-fog, this is perhaps not a colourful community but one very useful for overwintering insects and for small mammals and their predators, such as Weasels and Barn Owls. Where land has been recently disturbed, as through cultivation, then a sparse weedy flora is likely to develop – here the open ground may be appreciated by scarce annuals, by sun-loving invertebrates like bees and wasps, and by reptiles.

Some of the most exciting herb communities are those normally found in ancient woodland. You can find these in some ancient hedges, especially those that are woodland relicts or that join onto ancient woods. Here, provided there's been no fertiliser or herbicide contamination, fine displays of the likes of Bluebell or Ramsons may be present, as well as plentiful ferns. Such floras are often to be seen along the banks of West Country hedges, especially along ancient lanes. I explore this further in Chapter 9.

ABOVE: Ramsons (Wild Garlic) lines the hedged banks of this sunken Cornish lane.

Locally distinctive hedge types

Some parts of the country have highly distinctive and unusual types of hedge associated with them, so much so that they are a defining feature of the immediate landscape. Here are some examples, based largely on work commissioned by the Countryside Agency (2000):

- Alder and poplar windbreaks – Kent, planted to shelter orchards and hop gardens.
- Box hedges – north Dorset.
- Fruit or orchard hedges dominated by Damson/Bullace and other *Prunus* species – central Nottinghamshire; parts of Herefordshire, Worcestershire, Staffordshire and Nottinghamshire; north Kent.
- Holly hedges – Ampthill and Woburn in Bedfordshire, on greensand deposits; in Staffordshire and on the Pennine moorland fringe, where short stretches in hedges are thought to have been planted for winter fodder; and within the historic Forest of Arden on the east Warwickshire plateau, including Solihull.
- Mixed hedges with Holly standards – East Sussex, said to have been planted to stop witches running along the hedge tops.
- Duke of Argyll's Teaplant hedges – scattered across Suffolk and Norfolk, mainly in western parts. This shrub is thought to have been used originally to fill in gaps, later spreading by suckering to dominate (Barnes and Williamson 2006).
- Lilac hedges – the Sandlings of eastern Suffolk and the Brecklands of Norfolk, probably rarely planted but colonising existing hedges by seed from nearby gardens and estate woodlands (Barnes and Williamson 2006).

TOP: A Scots Pine hedge or 'deal row' growing on sandy soil in the Breckland, Norfolk.

ABOVE LEFT: Fuchsia hedges line a lane on the Dingle Peninsula, County Kerry.

ABOVE RIGHT: A Tamarisk hedge growing atop a beautifully made 'Jack and Jill' stone-faced bank close to the north Cornish coast, at Porthcothnan.

- Pittosporum, Evergreen Spindle and Escallonia hedges – western Cornwall and the Isles of Scilly, planted to shelter bulb and flower fields.
- Laburnum hedges – west Wales, especially in Carmarthenshire, Ceredigion and north-east Pembrokeshire, a tree perhaps planted for its wood or simply because in the 19th century there was a ready supply, even though the seeds are poisonous to livestock.
- Fuchsia hedges – south-west Ireland; western Cornwall, around the Helford estuary.
- Scots Pine hedges – the Brecklands of south Norfolk and north Suffolk, and the eastern Suffolk Sandlings. Known as 'deal rows' in Breckland, these are now line-of-trees hedges, although they were originally planted, mainly in the late 18th and early 19th centuries, on the dry sandy soils to become thick bushy hedges and were managed by trimming and laying (Barnes and Williamson 2006).
- Tamarisk hedges – along the coasts of Cornwall, Devon, Dorset and Hampshire, planted because the tree is salt-tolerant.
- Willow hedges – Trent Valley to Newark, Staffordshire and Nottinghamshire, thought to originate from the basket-making industry.

Concluding remarks

Seeking the answer to the question of what counts as a hedge, we've peered through the intricate tangle of a multitude of different structures and variations to find a definition that serves well enough. In its simplest form, this is any row of trees and bushes that is roughly less than 5m wide at the base, always bearing in mind that the ground at the bottom and any associated bank and ditch are part of the hedge too – we must not be too woody-species-centric. One important caveat is that the hedge must have been established or have been retained to serve some boundary function, even if it is no longer managed. Thus lines of trees that have sprung up alongside rivers are excluded – although even here I can think of grey areas.

Crucially, line-of-trees hedges are still hedges. So, windbreaks and shelterbelts are types of hedge, provided they are not too wide. Inevitably there are exceptions: for example, formal lines of regularly spaced trees with no bushy growth between them are avenues, not hedges. There's no minimum length in the definition, but it can all get rather confusing in urban contexts. Here Hugh Barker, in his delightful book *Hedge Britannica* (2012), uses the term 'hedge-like objects' to describe all those short unconnected hedge fragments – 'hedges to nowhere' as he describes them – that one sees in public and municipal areas: in supermarket carparks, retail parks, roundabouts, petrol stations and industrial estates.

But to go any further with this is to run the risk of getting bogged down in semantics. What is important is that we should have sufficient mutual understanding of what a hedge is to be able to talk meaningfully with one another about them, and to generate hedge data and statistics that are consistent and comparable.

LEFT: Hedge-like objects, to use Hugh Barker's terminology. Doddington, Lincolnshire.

Origins and history

chapter three

Hedges link us back to the past as powerfully as do ancient church towers and steeples, thatched pubs, and twisted, sunken lanes. There have been hedges in Britain and Ireland for thousands of years, and they are deeply embedded in our culture. When I go to the highest point of our Devon farm it gives me huge satisfaction and a very strong sense of place to know that the hedged field patterns I look out over were much the same five or even six hundred years ago. The same can be said for much of the West Country, and indeed for the two-thirds of England and Wales that lie within Oliver Rackham's ancient countryside.

Indeed, the history of farmed landscapes is written in the patterns of their fields, described by hedges, walls and ditches. In some places, such as Dartmoor, archaeologists and historians, aided by old maps and aerial imagery, are able to deduce centuries, even millennia, of landscape history from a palimpsest of field patterns – traces of the old remaining visible beneath the new, layer upon layer of history.

Still, the myth persists in some quarters that most, if not all, hedges were a direct result of the 18th- and 19th-century Enclosure Acts. Even if these Acts had covered all of Britain and Ireland, which they did not, this is nonsense. There's an abundance of evidence to the contrary. The majority of hedges in Britain are ancient, originating before AD 1750, often much earlier. Only in Ireland may it be true that most are less than two or three hundred years old, but even here there are many that are far older. The Pale Ditch, for example, in part at least a 'double ditch', a massive earthen bank with hedges on each side, was built at the end of the 15th century to help guard the four eastern counties of Louth, Meath, Dublin and Kildare against pillaging raids. This area, known as the Pale, was the only part of Ireland under English control at that time. To this day we use the phrase 'beyond the pale'.

OPPOSITE PAGE: Hedges are often as old as our historic churches, if not older, and just as symbolic of ancient landscapes. Iddisleigh, Devon.

ABOVE: The history of farmland is written in hedges and other field boundaries. This photo captures the view over fields near Llangorse, Bannau Brycheiniog (Brecon Beacons) National Park in Wales.

This chapter explores the ways in which hedges have come into existence, that is their origins, before providing a brief account of their history from the Neolithic to the mid-twentieth century; the next chapter covers changes since the end of the Second World War in 1945. Others have written at length and authoritatively about hedge history, notably Oliver Rackham, Tom Williamson, Richard and Nina Muir and John Wright. I have drawn heavily on their books for the overview provided here, and I am indebted to them.

The origins of hedges

Hedges come about in three main ways. A fair proportion are 'spontaneous hedges', the result of fences or walls being colonised naturally by trees and shrubs. Another fraction comprises 'ghost hedges', remnants of the original woods that covered much of our landscapes – as our ancestors cleared this woodland they left strips to serve as boundaries for one purpose or another. The great majority of hedges, however, are the result of deliberate planting or, in the Celtic West, bank creation. In short, they are human-made from the start. Here I explore these three origins and address the enigma of why West Country hedges are set upon such substantial, often massive, earth banks.

Spontaneous hedges

When working on our farm hedges I have discovered deep within a few the remnants of oak posts with old-fashioned thick woven wires strung between them. These hedges are mainly composed of Grey Willow, a tree that's a rapid coloniser of open ground on our wet land. It's clear enough that they were never planted, rather that they are the result of the growth of wind- or bird-dispersed seeds: old maps and aerial photos confirm the hedges to be of a young age. Once you get your eye in for such 'spontaneous hedges' they are, in my experience, quite frequent. Initially a few oaks, thorns or willows spring up along a fence line, too scrappy to be called a hedge, but over time the gaps are infilled, and a new hedge is formed. Soon the old fence is lost, and you might never know the hedge was not planted.

While in Britain and Ireland the great majority of hedges have been deliberately planted, in other countries, by contrast, natural colonisation by trees and shrubs along fences, ditches or banks may be their main mode of origin. Oliver Rackham (1986) describes how the hedges that now cross the Texas prairies in the United States are spontaneous: naturally dispersed seeds have led to trees springing up along the barbed-wire fences erected in the 1880s, and these have been tolerated or even welcomed to become the hedges of today. Rackham describes similar processes in Michigan and Massachusetts and says Americans believe nearly all their hedges arose by default.

BELOW: Here the curved strips of an ancient open field have been effectively fossilised by enclosure with hedges. Hartland in Devon, with Lundy Island in the background.

ABOVE: A spontaneous hedge in Meeth, Devon. Oak, Guelder-rose and brambles have colonised this fence line alongside a disused railway.

Ghost hedges

Other British hedges are remnants or 'ghosts' of ancient woods, perhaps even of the original wildwood that colonised these islands after the last ice age. Strips were deliberately retained when woods were cleared to serve as boundaries for one purpose or another. Occasionally these hedges are referred to as 'assart hedges', assarting being a historical term used to describe the clearance of forest to create fields. In my reading I have come across references to such ghost hedges from many places: Worcestershire, Herefordshire, Middlesex, Essex, Dorset and the Weald of south-east England, among others. Such hedges are typically broad and sinuous, and the fields they enclose small and irregular.

In the West Country, where banked hedges prevail, ghost hedges are rare: the banks must surely have been made from scratch by humans. Assarting may, though, explain the existence of an unbanked hedge shown to me by Geoff Hearnden within Dartmoor National Park made up of a line of Wild Service-tree, an uncommon species with such poor dispersal abilities it is considered a good indicator of ancient woodland. The trees are not likely to have been planted, since it is a poor hedging plant from a stock-proofing point of view, its berries are not valued as human food, and it is not especially useful for firewood. This idea came to me after reading that Oliver

ABOVE: A 'ghost' hedge? Does this line of Wild Service-trees in Bridford, Dartmoor, reveal the hedge to be a remnant of an ancient wood?

Rackham discovered that a 600-yard-long (550m) hedge in Suffolk composed, most unusually, almost entirely of Small-leaved Lime along with a few Wild Service-trees was originally the edge of an ancient woodland. Small-leaved Lime is another ancient woodland indicator.

It is ironic (but, to my mind, pleasingly cyclic) that while once woods, through clearance, contracted to leave hedges, now, through rewilding, hedges expand to make woods. More on this later.

Planted hedges

Most British hedges have over the centuries been planted to enclose or divide parcels of land, whether fields or gardens. As such, they are human-made – but not in the same way as wooden fences or stone walls, since they are living, changing entities. As time progresses, they become more and more natural and increasingly ecologically complex, given the chance.

They may be planted with just a single woody species, most often Hawthorn but sometimes Blackthorn, Beech, privet, Yew or any number of others, or planted with a range of trees or shrubs right at the start. It would be strange indeed if people in times past did not use whatever suitable saplings they could dig up from local woods,

ABOVE: A new hedge planted with Hawthorn and a few Hazel – a greater diversity would have been better for wildlife and increased resilience to future stresses like climate change.

existing hedges or waste land, especially since hedges were often intended to serve not just as stock-proof barriers but also as sources of firewood and timber. This is discussed further later, in Chapter 7, where I write about Hooper's rule, a well-known method for estimating the age of hedges using the number of woody species present.

Nowadays multiple-species planting is the norm in rural locations, largely for wildlife reasons and to increase resilience to diseases, pests and climate change. When I planted a new hedge on our farm 20 years ago, I used 12 different shrubs and trees, and a couple of years later added Aspen (my favourite tree) and two Devon Whitebeams. The Holly and Ash I transplanted from farm woodlands, the Aspen and Whitebeam I propagated from suckers and seeds respectively, the rest were bought from a nursery. Grey Willow has colonised the ditch dug to generate the soil for the low bank on which the trees and shrubs were planted. All are doing well, except for the Ashes, which are diseased and dying back, and a few Alder Buckthorn that never recovered from being defoliated by Brimstone butterfly caterpillars.

The origin of banked hedges

Why, it is often asked, do West Country hedges have such massive banks? I wish I knew! The banks are frequently 2 metres high or more, and in pre-mechanical excavator days took a stupendous amount of labour to build. Elsewhere in Britain perfectly good stock-proof hedges exist with just low banks, if any. There must be some reason other than keeping livestock in or out of fields why West Country farmers went to so much trouble. Some say that the banks are simply the result of digging ditches for drainage, yet most contain much more material than could be provided by ordinary farm ditches. At least, that's true now – the banks may have been added to and raised over the years – but even so the question remains: why should farmers want to have massive banks?

Others say that the banks were created when clearing fields of stones, but there are stony parts of the country like Cumbria where the hedges sit more or less flat upon the land. Besides, most West Country banks are filled with earth, not stones, and banked hedges remain dominant even in areas where there are few stones. Our farm in West Devon sits on heavy clay with not a surface stone to be seen, yet all the old hedges have big banks. Some people, including William Marshall (1796), have suggested that the banks allow the hedges to remain stock-proof even after the trees and shrubs on top have been coppiced for firewood. But it takes a high and very well-maintained bank to be sure of keeping cattle and sheep where you want them to be. Surely, if you wanted firewood it would be much simpler to plant directly on the ground and protect coppiced stools with hurdles or dead hedges rather than go to all the trouble of first building and then maintaining steep-sided banks?

The most plausible reason I have heard was recounted to me by a French anthropologist, François Beaulieu, working in Brittany where, along with Normandy, they too have hedged banks. He explained that it was a Celtic belief that the souls of ancestors resided in trees growing upon banks, so banks were created for cultural reasons rather than practical ones. Whatever their origin, high banked hedges are now ubiquitous in southern Somerset, western Dorset, Devon and Cornwall, as they are in parts of Wales and in north-west France. They are as distinctive of the Celtic fringe as are pasties or clotted cream in Devon and Cornwall and much cherished – even if they do obscure many a fine view, to the frustration of tourists.

Just how massive these banks can be is exemplified by Charles Vancouver's accounts of the construction of hedges in his book on agriculture in Devon, published in 1808. He describes how, when an area of land was enclosed in the north of the county, the banks were built 9 feet (2.75m) wide at the base, the material being sourced from ditches dug 3 feet (0.91m) wide on each side – so the total land take was a strip 15 feet (4.60m) wide. The banks were faced with stone for the first 4 feet (1.22m) and with turves for the top 3 feet (0.91m), giving a total height of 7 feet (2.10m) from the ditch bottom. Examples he gives from other parts of the county are smaller, but nevertheless still impressive. The cost of the bank just described was 15 shillings for a perch (16½ feet) – that's roughly £15 per metre in today's money – including protecting the new hedge with hurdles. Stone-facing excepting, that's much the same as today: new West Country hedges, with their banks built by excavator and the saplings protected by post-and-wire fences, cost about £20 a metre.

RIGHT: West Country hedge banks can be mighty. Here the author (1.88m tall) stands next to a stone-faced Cornish hedge near Port Isaac in north Cornwall.

BELOW: A magical ancient sunken lane in Horndon, on the edge of Dartmoor, lined by moss-covered stone-faced hedge banks.

A potted history of hedges

Prehistory

The Devon Hedge Group held a meeting a few years back on Dartmoor, when we stood for a while in awe beneath the branches of a Hazel growing on a bank that was first built some 3,500 years ago. A moment that put our own short lives and troubles in perspective. We knew the bank was very ancient because it was a continuation of a reave. These are low earth and stone banks laid out in the middle Bronze Age in a distinctive rectangular grid pattern across what is now open moorland (the Bronze Age in Britain lasted from about 2,000 to 650 BC). They form long parallel lines about 100 metres apart with occasional cross-pieces, and are thought to have been about a metre tall when built. The traces of ancient field patterns they leave behind them are frequent on the southern and eastern sides of Dartmoor, covering an area of some 10,000 hectares. Archaeologists presume they had an agricultural function (what else?), perhaps to control cattle. If so, a metre high is not enough to serve as a decent fence, so surely it's likely that, even way back then, shrubs and trees were encouraged to grow on them – if not deliberately planted then naturally colonised by seed blown in by the wind or carried by birds. Or perhaps the animals were hobbled (the practice of tying two legs together to restrict movement)

BELOW: Ross Gray, Don Gaskins and Michelle Werrett standing beside a hedge bank, part of a Bronze Age reave on Dartmoor, some 3,500 years old.

ABOVE: The banks or lynchets forming diamond shapes in the middle distance reveal prehistoric field patterns on Bodmin Moor in Cornwall. A medieval hedge system, still in use, has been superimposed on them.

or under the watchful eye of a herdsman all day, the banks serving mainly to define ownership (Menneer 2007).

Although less clear, and overridden by more recent field patterns, prehistoric reave systems have been discovered in several other places in England, including some with landscapes that are very different from Dartmoor, such as in Suffolk, Norfolk, Berkshire and Nottinghamshire. In Ireland, the Céide Fields in County Mayo are marked out by a reave-like pattern of walls found buried beneath a bog – this system is believed to be even older than the Dartmoor examples, created in Neolithic (New Stone Age) times.

Very oddly, reave systems are laid out in ways that completely ignore local topography and landscape features. They are not, for instance, aligned to contours or rivers, nor do they run consistently north–south or along any other compass bearing. Where they come to an obstacle like a steep rocky gorge they temporarily stop, only to resume in exactly the same orientation on the other side. Oliver Rackham calls such field patterns 'mindless', although the planners of the time must have had some specific reasons for acting as they did, perhaps mystical ones.

The remains of small, squarish, irregular Neolithic, Bronze Age or Iron Age fields, bounded by banks, are widespread across Britain, surviving in places like moorland and chalk downland where medieval reorganisation has not obliterated prehistoric features. They are often referred to as 'Celtic fields', although strictly

speaking most are older than the Celtic period, that is Iron Age. One such classic prehistoric field pattern is found close to the tip of Cornwall, in the stunning area of granite moorland and exposed coastal farmland known as West Penwith, close to Land's End. Here, the blocky but irregular fields are tiny and defined by sizeable stone and earth banks, that is Cornish hedges. What is truly remarkable is that these hedges and field patterns have survived more or less intact since they were made in the Bronze Age. They are thought to be amongst the world's very oldest human-created features still in everyday use.

ABOVE: Bronze Age fields bounded by Cornish hedges, banks of earth and granite stones, many topped with Blackthorn and Gorse, at West Penwith.

My wife Paula and I often go down to West Penwith on special occasions like birthdays, in part because there's a fine pub at Zennor, but mainly because we savour the immense feeling of great antiquity as we walk through the fields, crossing hedges by way of granite cattle grids and styles. Many of the hedges have only grasses and perhaps some Heather growing at the top, but quite a few sport gorses and Blackthorn, providing bright streaks of colour in the spring. The granite boulders and stones are covered in rich communities of crustose and foliose lichens in many shades of grey.

Some semi-regular field systems that survive to this day are thought to have been created in the Iron Age (650 BC to AD 43). Such systems cover a large part of south-east Essex, about a quarter of the county, although in part they may be of later, early Roman, times. Here in Essex, the landscape was laid out in a planned grid of roads

and hedges on a scale even larger than that of known reave systems. Oliver Rackham expresses the view that 'nobody for at least a thousand years has been in a position to organise the landscape on such an immense scale'. These field systems differ from those of reaves because the two axes are equally strong, boundaries running east to west being as long as those running north to south.

The Roman period (AD 43 to about 420)

When the Romans succeeded in colonising Britain following their invasion in AD 43, they found well-established 'Celtic field' systems, and it seems they largely accepted them as being fit for purpose rather than trying to impose the precise 'mindless' land planning called centuriation that they introduced in southern Europe. By and large, they built their famously straight roads across field systems which were already in existence, and which it seems they did not wish to change. The Romans did, however, remove some boundaries to produce longer fields which were more efficient to plough – they needed the extra food for a growing population and for export. Doubtless too, further woodland was cleared for agriculture. Still it is surprising, given that they occupied the country for four centuries, that only a few field systems of probable Roman origin are known to exist in Britain.

ABOVE: Bronze Age Cornish hedge at West Penwith, Cornwall.

In their 1974 book *Hedges*, Pollard, Hooper and Moore recount how, in order to find out something about typical hedge length, they measured random samples of field boundaries from across Britain using early Ordnance Survey maps, made before the recent era of large-scale hedge removal. They discovered hedge lengths typically to be multiples of chains (22 yards or just over 20m), the most common being close to 10 chains (that's a furlong – 220 yards or 201m), followed by 12 chains and 8 chains. However, near the north Norfolk coast they encountered a surprising anomaly. Here, between Holme and Brancaster, the fields were of remarkably uniform size and the most frequent hedge length was 410 yards (375m), quite different to anywhere else and not a neat multiple of a chain. Since similar field grids are found in Italy and in parts of Africa colonised by Rome, and the Norfolk ones are aligned to tracks known to be of Roman origin, Pollard, Hooper and Moore suspected them too to be of Roman origin, a premise that subsequent historians have accepted. Similar field systems thought to be of Roman age are found elsewhere in England, as at Ripe and Chalvington in Sussex. Although 410 yards does not fit any known Roman measurement precisely, it is close to two stadia (405 yards), the stadium being an ancient Greek and Roman unit of length.

When they arrived in England, the Romans would have been familiar with hedge laying, since in 55 BC Julius Caesar, before he tried unsuccessfully to conquer Britain, wrote about encountering this practice in Flanders. I return to this in Chapter 15.

Early Medieval (roughly 420 to 1066)

The early medieval period in Britain, also called the Anglo-Saxon period and the Dark Ages, covered the six and a half centuries between the departure of the Romans and the Norman Conquest in 1066. Just as the Romans largely accepted the field systems they inherited, so to it seems that the Saxons and other invaders were generally happy enough with those they found, at least until the 9th century. There may have been some land abandonment in early years, reflecting a declining population, and a further enlargement of those fields that were still in use. Clearly, though, hedges remained important for farming – as testified by the frequent references to them in Saxon charters, documents under which land ownership or sole rights of use were granted or transferred to individuals.

Open-field farming

In the 9th century a new and radically different system of field organisation started to be introduced, one that led to much hedge removal over the following five centuries or so, in effect to de-enclosure. This was the open-field system, or strip farming as it is sometimes known. It may have been invented by the Anglo-Saxons in southern England, from where it spread through England into Scotland and Wales and was exported to Ireland and mainland Europe (Rackham 1994). Why exactly it was developed and became so popular is unclear, but its attraction was probably related to changes in settlement patterns, away from isolated farmsteads and hamlets towards villages, and because it delivered improved efficiency to mixed arable and livestock farming within a communal social system. Whatever the reasons, it seems to have become the norm across most of Britain. This is true not just for what we now refer to as planned countryside but for ancient countryside too. Even in Devon and Cornwall, counties where it was once thought to have been rare, it is now believed that most of the farmland was divided up into strip fields (Turner 2007). To this day large unenclosed fields divided into strips remain common in parts of eastern Europe.

The open-field system was complex. Typically, each parish or manor was divided up into three large fields, covering a hundred or more hectares each; these may have had hedges around them but had few if any internally. The fields were managed on rotation, at any one time one being fallow and the other two under crops. Each was divided up into broad strips called furlongs, which in turn were subdivided into long thin strips called selions, each selion typically being a furlong long, that's 220 yards (201m), said to be the length an ox can pull a plough before he needs to rest, and two perches (11 yards/10m) wide, so occupying half an acre (0.20ha). In any one year, the same crop was grown on all the strips in a furlong. Individuals or families held rights to plough and grow crops on strips scattered throughout the fields, perhaps as many as 70 of them, paying rent for the privilege to the lord of the manor or the Church. Land that was not fit for cultivation, often referred to as 'waste', and that which was fallow, was held 'in common', that is all the farmers had the right to graze their cattle and sheep over it, within limitations. There were many variations on this basic pattern.

The strips were ploughed by oxen, which could not be turned about tightly. Turning therefore had to start well before the end of each strip, and this resulted in strips developing a characteristic reverse 'S' shape or, if the strip ran down a slope so steep that ploughing was only feasible in a downhill direction, a reverse 'J' shape, with the curve being at the bottom of the hill. With each pass of the plough, soil was shifted towards the centre of the strip, resulting in a ridge being formed, and thus strip farming was responsible for many of the ridge-and-furrow patterns visible within fields today. In places, on steep slopes, the strips followed contours and resulted in the formation of lynchets (terraces).

The conversion of enclosed land to open-field systems continued after the Norman conquest in 1066, with the open-field system probably reaching its heyday in about 1250, although in places

ABOVE: A rare survival of an open field with strips, maintained by the National Trust. Forrabury Common, Boscastle, Cornwall.

enclosed systems continued to be converted to open-field ones for two or three centuries afterwards. Today, however, very few examples still exist, notable ones being at Laxton in Nottinghamshire, Haxey in Lincolnshire, Soham in Cambridgeshire, Portland in Dorset, Braunton in Devon, Boscastle in Cornwall and the Gower peninsula in Wales.

Late medieval enclosure

Starting in the 13th century and continuing into the Tudor period (1485–1603) , farming fashion reversed once more, and fields started to be re-enclosed. This appears to have been largely a piecemeal voluntary process, farmers agreeing to exchange or sell strips so those under their management control were next to one another and could be combined and hedged, and to forgo rights of common grazing. Sometimes, however, tenants and those holding rights of common were coerced to accept enclosure by their landlords, resulting in much dissent and hardship, even to bloody uprisings such as the 1549 Kett's Rebellion in Norfolk and the Midland Revolt of 1607.

Enclosure enabled the land to be managed more efficiently, farmers being free to make their own decisions about what to grow at any one time, and what rotations to adopt. From a livestock perspective, it meant it was much easier to improve breeding because one could be certain which bulls and rams sired the next generation: previously there was a village bull or two to serve all the cows, and village rams for the ewes. Being able to enclose your land also meant it was easier to control diseases and to avoid over-grazing – you could move your livestock from field to field without having to worry that others had taken all the grass already.

The process of enclosure seems to have begun early in the south-west of England, and is believed to be one of the main reasons Oliver Cromwell said in 1658, after being in the region with the New Model Army between 1645 and 1646, 'I have been in all the counties of England and the Devonshire husbandry is the best' (Stanes *et al.* 2008). Cromwell would have encountered a well-established landscape dominated by small fields enclosed by hedges growing on earth banks. Indeed by the end of the medieval period, that is towards the end of the 15th century, across much of what we now call ancient countryside, hedges were probably as numerous as they had been before the introduction of the open-field system (Rackham 1986).

The enclosure of open-field strips, or more likely bunches of strips, has led to many present-day fields being long and narrow and retaining the reverse 'S' and 'J' shapes created by oxen-ploughing. Examination of maps and aerial photos of many, perhaps most, patches of ancient countryside will reveal these fossilised traces of strip farming. As I drive and walk around Devon, now that I have my eye in for them, I see them practically everywhere away from the high moors.

BELOW: Late medieval field systems, East Devon.

ABOVE: These narrow fields below the village of Sheepwash in Devon are the legacy of the enclosure of strips within an open field.

Enclosure Acts (parliamentary enclosure)

By the start of the 18th century, open fields remained common only in the places we now refer to as planned countryside, its characteristically fertile farmland continuing to be largely unenclosed and unhedged. It's not that the landowners and landlords did not want land consolidation and enclosure; they did, for they believed it would make agriculture more efficient and generate greater income. The decline in the profitability of sheep ranching as opposed to controlled stock grazing, together with the introduction of new arable crops like turnips, made such change particularly desirable in their eyes. An increasing demand in these poorly wooded areas for firewood and timber, especially elm, may have been an additional driver for enclosure, since these materials could be generated from hedges.

But here the landowners and landlords encountered barriers that could not be overcome by voluntary agreement. Foremost there was, understandably, strong opposition from those who stood to lose common rights without adequate compensation, especially among the poor. After all, open-field farming had been in existence in places for close on a thousand years. And then there was the sheer difficulty of doing anything new in the face of the complex systems of land ownership, tenure and rights in existence. In short, it was proving impossible to gain agreement from all parties. So landowners resorted to petitioning Parliament to pass laws mandating enclosure – the Enclosure and Inclosure Acts (for the technical difference between enclosure and inclosure, see Chapter 2).

These Acts empowered commissioners to decide how the land should be divided up and apportioned – and they did so in a manner that was heavily biased towards those with the most land and that often dispossessed the poor. To add insult to injury, those allotted land were required to pay for the new hedges, walls or fences needed to enclose their land out of their own pocket, an expense only the wealthy could afford. It is not surprising that frequent riots and much damage to hedges ensued, perpetrated by those angry about the unjust loss of common land over which they held rights, and the consequent loss of their livelihoods.

The English rural poet John Clare (1793–1864), who started life as an agricultural labourer in the village of Helpston, near Peterborough, is famed for his powerful lines berating enclosure. He laments the loss of the wide, open spaces of common land where people and livestock were free to roam, the excessive imposition of order against the grain of nature, and the way the poor were made landless and their way of life meaningless:

Inclosure came and trampled on the grave
Of labour's rights and left the poor a slave …

Fence now meets fence in owners' little bounds
Of field and meadow large as garden grounds
In little parcels little minds to please
With men and flocks imprisoned ill at ease

John Clare, *The Mores*, c. 1820

While the first of the Enclosure Acts, which affected only England and Wales, was as early as 1603 and the last as late as 1914, the vast majority were in Georgian times, between 1760 and 1820, during the reign of George III. And there were a lot of them. In all, 5,265 separate Acts, covering some 6.8 million acres (2.8 million ha) in England, and 1.7 million acres (0.7 million ha) in Wales. That's 31% of the agricultural land currently present in England, and 40% of that in Wales. Oliver Rackham reckons that about 200,000 miles (320,000km) of hedge were planted in Britain between 1750 and 1850.

In Ireland, Rackham says that enclosure started in earnest a century earlier, in the mid-17th century, while making the point that a range of field systems, some hedged, must already have been in place here, just as in Britain. Nevertheless most hedged enclosure, at least in Northern Ireland, took place in the 18th and 19th centuries, between

1750 and 1850 (Henry 2012). However, the boundaries of townlands, the most basic unit of land division in Ireland, date back to the early Christian period (AD 400–800) if not earlier: most such boundaries were originally hedges or banks (known as ditches in Ireland), and many still are (Nicholls 2003). There are over 64,000 townlands, ranging from just a hectare or two in size to a hundred or more. Nevertheless, Neil Foulkes tells me that first-edition Ordnance Survey maps, produced in the 1830s, show that the landscapes in the west of Ireland remained largely unenclosed at that time, in contrast to the south and east which already had more complete networks in place. By the time of the second-edition OS maps, around 1910, the enclosure of much of the country appears pretty much complete, excepting uplands and bogs.

Enclosure Act fields are characterised by dead-straight hedges or walls laid out in precise rectangles. Occasionally existing curved or wavy boundaries were retained, such as those which marked the edges of open-field furlongs – these may already have been banked or hedged so it made little economic sense to realign them. Even though Hawthorn is difficult to manage for firewood, most of the new hedges were planted solely with this species, perhaps with some Blackthorn,

ABOVE: Typical Enclosure Act landscape, in Uppingham, Leicestershire, with straight-sided rectangular fields bordered by Hawthorn hedges, here in poor condition.

BELOW: Most field enclosure in Northern Ireland occurred after the mid-17th century. These fields in County Antrim have been enlarged through later hedge removal.

because these two species create a good stock-proof barrier. Also, in the fox-hunting areas of the English Midlands, Hawthorn was favoured by the gentry because it could be trimmed to a size easily jumped by horses. Trees such as elms, oaks and Ash, destined to become standards for timber, were also planted in these hedges.

The Acts were not evenly distributed across the country. Counties like Norfolk, Northamptonshire, Lincolnshire, the East and West Ridings of Yorkshire – areas of planned countryside – were heavily affected, while in others such as Devon, Dorset, Somerset, Middlesex and Essex – ancient countryside – there were virtually none. As already remarked, it is not that these latter counties were free from open fields (to the contrary, in some such as Devon they were commonplace) rather that they were enclosed, often earlier, not by statute but by voluntary agreements or heavy-handed landowner action.

Between the Enclosure Acts and the Second World War

Between 1870 and 1945, according to Oliver Rackham, there was very little change in the overall extent of hedges or field patterns in Britain. Of course, some hedges were removed, but others were planted. Using our farm in Devon as an example, the 1841 tithe map shows one field that by 1884 had been divided by a new hedge into two fields, while an open area of 'moorland' was not divided into three fields until the 1940s. Meanwhile, in the 1940s, two small fields and a larger one were amalgamated to create a single big one

BELOW: The kinks in these hedges and shadow marks on the ground reveal where hedges have been removed from this medieval landscape in Crackington Haven, Cornwall.

through hedge removal. Then in the 1990s, I planted a new hedge. Overall, there are now more hedges on the farm than there were in post-medieval times. While such net gain is not the norm, losses and gains of hedges at a farm level over time is.

Concluding remarks

Hedge history is nothing if not complicated. It varies greatly across the different parts of Britain and Ireland, even from farm to farm, reflecting the multiplicity of field patterns and farming practices that have existed through the ages. Hedges come and go, ebb and flow, over the course of millennia and centuries. It is also clear that there remain many gaps in our knowledge. Even in historical times there are periods covering many centuries where few if any contemporary records exist, leaving much to deduction and conjecture.

The next chapter picks up the story after 1945, when the next great phase of hedge removal began, this time precipitated by the change from horses to diesel. Powerful tractors and combine harvesters operate best in large fields, unwanted hedges being easily grubbed out using excavators and bulldozers. The increasing number of farms specialising in arable crops was another driver for removal, hedges no longer being needed to control cattle and sheep. Overall, the outlook for hedges was grim. They seemed largely redundant, taking up space which could be used for food production and harbouring vermin and diseases.

BELOW: Modern enclosure in East Lothian, photographed from North Berwick Law.

Post-war losses, current extent and condition

chapter four

How many hedges have we lost since they were at their peak in the 1940s, and how many do we have now? In how good a condition for wildlife are they? If we want more, better connected and healthier hedges, then having answers to these and similar questions is essential. Otherwise, how can we argue effectively for more resources to be devoted to hedge conservation and management? This chapter provides the requisite facts and figures on the current extent of hedges in Britain and Ireland, the net change in length over the last 70 years or so – losses may not be as great as you think – and on the condition of hedges today. Subsequent chapters go into more detail on such matters in relation to specific features such as standard trees, the shrub layer and herbaceous margins, and explore the impact of different management activities. Here the aim is to give a broad overview.

Extent

We have 700,000km (435,000 miles) of rural hedge in Great Britain. That's enough to go around the Earth 17 times, nearly the distance from the Earth to the Moon and back, or, if you prefer, to surround about 1.6 million football pitches. Ireland has even more hedges than Britain with 803,000km (499,000 miles), the Republic having 689,000km (428,000 miles) of rural hedge and Northern Ireland 114,000km (71,000 miles). Altogether in Britain and Ireland we have about one and a half million kilometres of countryside hedge – two return trips to the Moon. Truly staggering!

In the UK, in 2007, there were 547,000km of rural hedge in England, 114,000km in Northern Ireland, 106,000km in Wales and 46,000km in Scotland. These figures come from Countryside Survey, a rigorous and comprehensive sampling programme led in England, Wales and Scotland by the UK Centre for Ecology and Hydrology

OPPOSITE PAGE: All good hedges should have uncultivated margins and be free from the influence of fertilisers and pesticides.

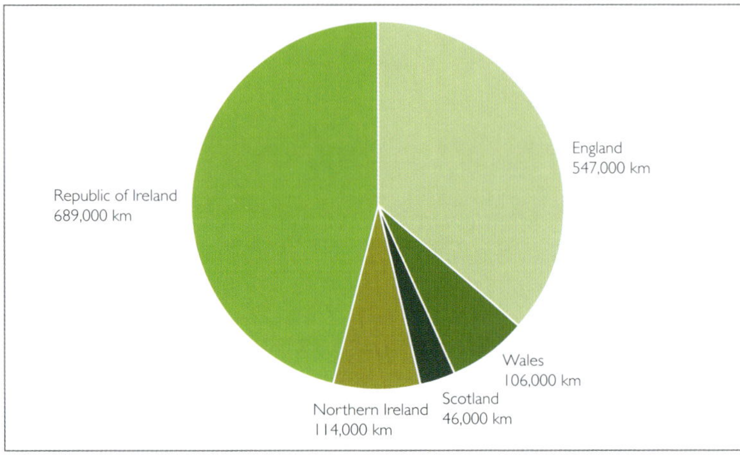

RIGHT: Length of hedges in each country, in 2007 in Great Britain and Northern Ireland (Carey *et al.* 2008, McCann *et al.* 2012) and in 2015 in the Republic of Ireland (Green *et al.* 2019).

(Carey *et al.* 2008) and in Northern Ireland by the Environmental Sciences Research Institute at the University of Ulster (McCann *et al.* 2012). The survey method relies on field workers walking the ground, recording numerous landscape features, rather than on using remote techniques like aerial photos. As shown in the map opposite, hedge density varies greatly across the three countries of Great Britain, as it does in Northern Ireland. The length of hedge today is likely to be much the same as it was in 2007 – there will have been further losses due to agriculture and built development but gains too, from new hedge planting. In Northern Ireland, however, there may have been continued significant losses in recent years, as we shall see.

Another survey, the Forestry Commission's National Forest Inventory (NFI), carried out a few years each side of 2016, gives substantially lower figures: 452,000km in Britain (compared to Countryside Survey's 700,000km), of which 336,000km are in England, 76,000km in Wales and 41,000km in Scotland (Brewer *et al.* 2019). This difference between the two surveys is not due to any dramatic decrease in hedge length between 2007 and 2016 but rather is explained by differences in sampling methods and definitions. Notably, the definition of hedges used by the NFI did not include those that have grown into lines of trees, whereas thankfully Countryside Survey covered all 'woody linear features'. The NFI also depended heavily on hand mapping information from aerial photographs, a method less accurate than a ground survey. Reflecting these points and the greater number of plots sampled, the Countryside Survey data are likely to give the truer picture. The same disquieting degree of difference between Countryside Survey

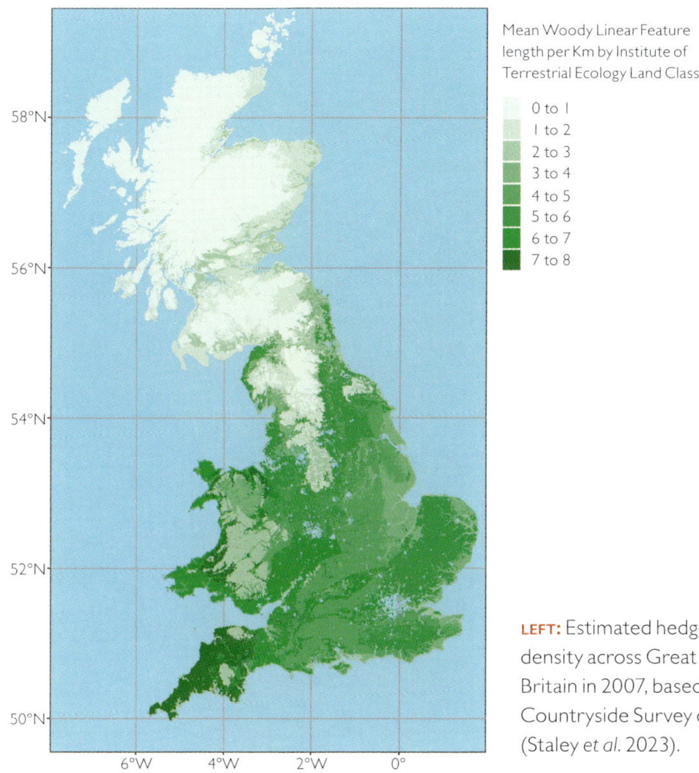

LEFT: Estimated hedge density across Great Britain in 2007, based on Countryside Survey data (Staley *et al.* 2023).

and NFI results is to be found in estimates of hedgerow tree numbers, as we shall see in Chapter 8.

The NFI does, however, give us our only estimate of the current length of urban hedge, reckoning there to be 43,000km in Britain, with 31,000km in England, 6,000km in Scotland and perhaps 6,000km in Wales too (confidence levels are low here). Thus the length of urban hedge is about a tenth of that of rural hedge. Of course, many of these urban hedges will be composed mainly of non-native species. A 2009 survey of 63 hedges in Bristol found only 51% to be composed mainly of native trees and shrubs (Forster 2010).

The 689,000km estimated length of rural hedge in the Republic of Ireland was made in 2015 and based on aerial imagery (Green *et al.* 2019). Even though a broader definition of hedge was used, the estimate can roughly be compared with the UK figures generated by Countryside Survey. The definition used in Ireland was 'woody field, forest or settlement boundaries, including remnant or outgrown hedgerows and treelines, and including species such as gorse, less than 10m in width'.

The estimated density of hedges across the entire Republic of Ireland was 9.8km per square kilometre: this compares to 4.2km/km^2 in hedged parts of Britain (Staley *et al.* 2023), the equivalent figure in England being 4.5km/km^2 (Staley *et al.* 2020). Since there are parts of Ireland where most of the field boundaries are stone walls and hedges are scarce, such as Mayo and Galway, it's clear that on average Ireland is markedly more densely hedged than the hedged parts of England. In County Leitrim one sample 1km square contained over 22km of hedge (Foulkes 2006b). Likewise the hedged landscapes of Scotland, and probably Wales too, have sparser networks of hedges than those in Ireland.

How much have we lost?

In Britain, although parts of what we now know as ancient countryside were fully hedged in the 15th century, overall peak hedge length was probably not reached until about 1870, after which there were only a few late Enclosure Acts (Rackham 1986). Between about 1870 and 1945, the end of the Second World War, there was very little change in hedge length – aerial photographs taken in 1940 show a still almost complete network of hedges, even in arable areas. Then over the next three decades farmers removed ('grubbed out') hedges in large quantities. This was in response to technological advances in agriculture, including larger machines, the conversion of much grassland to arable, which meant that hedges were no longer needed to keep livestock in or out of fields, the national desire for food security, and public money being on offer for hedge removal.

Between 1954 and 1957 the Forestry Commission measured hedge lengths in 469 sample strips 80 chains long and 16 chains wide (a chain being a length of 20.1m) across eleven counties, eight in England, two in Scotland and one in Wales (Locke 1962). These data allowed the total length of rural hedges in post-war Britain to be crudely estimated as 991,000km (616,000 miles). Later Pollard, Hooper and Moore (1974) carried out a parallel piece of research in England and Wales using aerial photographs taken in 1946 and 1947: they measured the length of hedge per square mile of farmland in photographs chosen at random. This gave them an estimate of 800,000km (500,000 miles) of hedge in the two countries, reasonably close to that derived from the Forestry Commission data, bearing in mind that Pollard, Hooper and Moore did not cover Scotland. Of

course, neither they nor the Forestry Commission will have used the same definitions of a hedge as Countryside Survey or probably each other – but nevertheless it is reasonable to compare the two estimates with 2007 Countryside Survey data. This comparison suggests that we have lost between 18% and 29% of our hedges since the time of maximum extent.

ABOVE: Ireland is more densely hedged than Britain. This photo shows the border country from Slieve Gullion, County Armagh, Northern Ireland.

Most people familiar with the scale of hedgerow loss consider this to be on the low side, but I think this is because it's not intuitive that fields can be made much bigger without removing a high proportion of hedges. A back-of-the-envelope calculation I've just done using a grid to simulate field patterns reveals that a farm's fields can be doubled in size by the removal of just one-fifth of the hedges between them. In the real world, back in 1962 two researchers, Davies and Dunsford, wrote in detail about the restructuring of a farm in a 'hilly district of south Devon'. The irregularly shaped fields occupied 40ha; before restructuring there were 30 fields averaging 1.3ha in size, afterwards 11 averaging 3.6ha. To achieve this 2.8-fold increase in field size, 2.33km of hedge had to be removed – just a fifth (19%) of the length there beforehand. So my quick calculation may even be a little on the high side – regardless, less hedge has to be removed than you might think to enlarge field sizes greatly.

Based on aerial photograph analysis, Pollard, Hooper and Moore reckoned that about 8,000km (5,000 miles) of hedge were

ABOVE: These two stag-headed oaks stand on a former hedge line in Buckinghamshire. Most in-field trees like these are remnants of former hedges.

lost each year between 1946 and 1963 in England and Wales, while recognising that this figure might be an overestimate because hedges recently cut close to the ground, as when coppiced, appeared to have been removed altogether on aerial photographs. Still, if we accept the figure of 8,000km per year and assume that it applied, on average, across the major period of hedge removal, 1946–1970 (when grant aid for removal ceased), then that would suggest that 192,000km were removed in England and Wales, a 24% loss. (In 1982, following pressure from the public and conservation bodies, government policy was reversed, and grants were offered for new hedge planting and for the renovation of existing ones.)

In Scotland, the rate of hedge removal may perhaps have been greater than in England and Wales, but doubtfully so. In 2003 the government agency Scottish Natural Heritage estimated that between about 1947 and about 1988 the total length of hedges in the country declined by 54%, an average annual loss of 560km. However, looking at the figures I suspect that much of this loss was not down to outright removal but to managed hedges turning into line-of-trees hedges or relict features. The source document suggests that in the 1980s there were 19,500km of hedge in Scotland, a figure remarkably similar to the 21,200km of managed hedges estimated by

Countryside Survey to be present in 2007. But overall, including line-of-trees hedges, Countryside Survey estimated there to be 46,500km of hedges present in Scotland, a figure which would suggest a slight rise in hedge length since about 1947. Improbable perhaps, but not impossible. I think it fair to assume that back in the 1940s few hedges were unmanaged, despite the war taking men away from the land. Once again, we are stymied by inconsistent definitions.

Across Britain, between 1984 and 2007, Countryside Survey recorded a small but significant further loss of hedges, from 710,000km to 700,000km, a 1.4% loss over the 23 years. Surprisingly (but in line with my suspicion about the longer-term trend) the length increased in Scotland, from 38,000km to 46,000km, while it remained virtually the same in Wales. It was in England that the length fell most, from 565,000km to 547,000km, a 3.2% loss.

In Northern Ireland, the length of hedges decreased over the decade between 1998 and 2007 by a worrying 4.6%, from an estimated 119,000km to 114,000km – during the same period in Britain hedge length fell by just 1.7%. This confirms the suspicion held by many familiar with land-use change in Northern Ireland that hedge loss there was continuing apace long after it had tailed off in Britain. In all likelihood this reflected the lack of legal protection in Northern Ireland, as in Scotland, that persists to this day; in England and Wales the *Hedgerows Regulations* came into law in 1997.

LEFT: Hedge removal for housing and other built development continues to be a major reason for loss.

The extent of hedge loss since the end of the Second World War in Britain, as revealed by local surveys

Area	Dates	Proportion of hedge lost	Comments	Source
Flitcham with Appleton, north and west Norfolk	1946–1970	7%	'Sheep–corn' land with light soils. Average field size in 1946 was 21.4ha.	Baird and Tarrant 1972, in Pollard *et al.* 1974
Thwaite, south Norfolk	1946–1970	20%	'Wood-pasture' land with heavy soils. Average field size in 1946 was 4.0ha.	
Rutland: arable areas	1947–1969	23.2%	Loss rate varied between 17% and 43% in different areas.	Pollard *et al.* 1974
Rutland: grassland areas	1947–1969	8.4%	Loss rate varied between 2% and 16% in different areas.	
King's Lynn, Norfolk	1903–2016	2.5%	2km lost out of 80km in an area of *c.* 10km². In 1903 this area had 90 fields, the majority of which were around 25ha in size.	Wright 2016
Stockland, Devon	1903–2016	30%	34km lost out of 114km in an area of *c.* 10km². In 1903 this area had 880 fields, giving an average field size of 1.1ha. Since then 260 fields have been lost.	
St Andrews, Fife	*c.* 1900–2016	11%	4.9km of hedges and walls combined were lost out of 45.5km, in an area of *c.* 10km2.	
Hook Norton, Oxfordshire	*c.* 1900–2016	21%	16km lost out of 76.7km, a change from 220 fields to 130 fields. Area not specified but presumably *c.* 10km².	
Near Peterborough, Cambridgeshire	1900–2014	21%	12km lost out of 58km in an area of *c.* 10km².	
Three parishes in Huntingdonshire (now Cambridgeshire): Barham, Buckworth and Leighton Bromswold	1946–1965	72%	Reduction from 114km to 32km in an area of 18.1km².	Moore *et al.* 1967
Lamberhurst Parish, Kent	1840/1860–2006	*c.* 10%	Random-sample local hedge survey covering 148 hedges within an area of 56km² (5,600ha).	Buell 2006 (see also Appendix A)
Sundon–Sharpenhoe–Smithcombe–Barton Hills, Bedfordshire	1946–2008	*c.* 0%	Full-census survey of 120 hedges within an area of 9.6km². Hedge extent and connectivity similar to post-WWII and Enclosure Acts.	Webb 2009 (see also Appendix A)
London Borough of Bromley	*c.* 1870–2008	9%	Full-census survey of 184 hedges within an area of 10km², in the Downlands landscape where Charles Darwin studied hedges in *c.* 1870. Although 2% of hedges present in 1868–1872 were missing in 2008, this was more than offset by gains in hedges, their number increasing by 7% over this 140 year period.	Hayes 2008 (see also Appendix A)
Knepp Estate, Shipley, West Sussex	1947–2010	15%	Almost complete census in 2009/10 of 116 hedges (23km) within an area of 4km². Just 11 hedges lost since 1947. Prior to rewilding, commencing in 2001, the land was mainly arable.	Anon. *c.* 2010

Rates of loss have, as might be expected, varied considerably between different parts of the UK. In the table opposite I have collated the results from local hedge surveys. They reflect a huge amount of painstaking work measuring hedge lengths on aerial photographs and sometimes maps, often by volunteers. While no doubt biased towards places that have maintained a rich hedgerow heritage, still the results support the national picture – that we have lost not half of all our hedges, as is commonly thought, but less than a third. Only in the Huntingdonshire survey carried out by Norman Moore and colleagues was a loss of greater than 30% recorded, in fact a very high 72%. However, Moore recognised that many hedges in this area may not have shown up on aerial photos because they had recently been cut close to the ground: looking at aerial images of the area now, there are certainly more hedges than shown in Moore's sketch map of the situation in 1965. Indeed, when John Wright (2016) examined maps of the same vicinity, near Peterborough, he reached the conclusion that just 21% had been lost.

Neither do the local investigations into hedge loss since the Second World War support the frequent assertion that this has been greater in areas dominated by arable farming than in those where grass fields prevail – indeed, they suggest the opposite. Thus, in Norfolk loss was considerably less in the area that used to be characterised by 'sheep–corn' husbandry than in the area that was 'wood-pasture', while the 30% loss in the almost exclusively pasture landscape of Stockland in east Devon is much greater than the 2.5% loss in the intensive arable landscape of King's Lynn in Norfolk. It seems that the loss in arable landscapes appears to be greater only because these areas tended to have had fewer hedges in the first place. Removing a small proportion of the hedges from a sparsely hedged landscape has a much greater impact visually (as well as on wildlife and landscape connectivity) than losing a larger proportion of hedges from a densely hedged area. In Devon, you can grub out 30% of the hedges and still have a well-hedged landscape with intact networks – you could not do that in much of Lincolnshire. Context is vital.

Not all loss has been down to removal by farmers. In the period 1925–1939 Pollard, Hooper and Moore estimated urban encroachment on farmland to have accounted for the loss of around 1,600km of hedge a year, with further hedges being lost to motorways, reservoirs and other developments. They suggested that in England and Wales between 1945 and 1974, when their book was published, some 32,000km (20,000 miles) of hedges were lost for non-agricultural reasons, compared with 192,000km (119,000 miles) removed by farmers. Attrition due to built development continues to this day, and the rate may even have accelerated: it is alarming, for example, to see the number of hedges being removed for the construction of HS2, the high-speed railway between London and Birmingham.

Returning to the densely hedged Republic of Ireland, Neil Foulkes, an authority on the country's hedges, tells me that peak hedge length was probably reached in the early to mid-20th century, in the window between the establishment of the Irish State (1922) and joining the European Union (1973). Across the country, between 1995 and 2015, based on 1,000 randomly selected hedges, an estimated 3.2% were lost, mostly during the first 10 years – after 2009 rules linked to EU agricultural subsidies afforded some protection to hedges (Green *et al.* 2019). Most of this loss was due to hedge removal during farm restructuring. The making of new hedges was usually associated with new housing and road developments, although such development generally still resulted in a net loss. Other hedges were subsumed

RIGHT: A banked hedge has been removed from the foreground by bulldozing it across the field to the boundary beyond. This was a frequent sight in Normandy, where this photo was taken, in 2010.

within forestry plantations, although these were not counted within the loss figures. Worryingly, hedge loss may have accelerated in recent years: in County Monaghan 0.5%, maybe 0.9%, was lost every year between 2010 and 2021, mainly because of agricultural intensification (MacElwain *et al.* 2021).

Elsewhere in northern Europe, losses have been far greater than in Britain and Ireland. In France, some 70% of hedges have been removed since 1950, a loss of 750,000km (Précigout and Robert 2022). The key driver has been the restructuring of land ownership, previously very piecemeal owing to land inheritance laws, alongside land-use change and agricultural intensification. In Germany the situation is even worse, with the length of boundary banks, most of which are hedged, decreasing from 510,000km in 1880 to 92,000km in 2012, an 82% loss (Müller 2013). In the Netherlands, Jef Gielen tells me than an estimated 280,000km of hedges were lost between 1945 and 1999, leaving just 30,000km today, an awful 90% loss (although this has been offset by significant amounts of new planting).

Assessing the condition of hedges as wildlife habitat

What does a hedge in good condition for wildlife look like? How high? How wide? What cover of non-native species is acceptable? Questions more easily asked than answered, but nevertheless important ones.

Back in the mid-1990s, in the early days of Hedgelink, I recall long and difficult conversations on such matters, out on farms poring over hedges or around the table back in Defra's headquarters in London. Hedgelink is a group that brings together organisations and individuals interested in hedges, to exchange knowledge and ideas, to engender collaboration and to take forward conservation priorities (see Chapter 16). We were tasked with helping to deliver the target in the UK Biodiversity Action Plan to 'achieve the favourable condition of 25% of species-rich and ancient hedges by the year 2000 and 50% by 2005'. Hugely ambitious, but it was a real spur to action.

Now, there are no hard and fast hedge-specific targets for those in the conservation sector to share and own, and I think we are much the weaker for it. The UK government has, though, just proposed a rather hazy non-binding target of restoring hedge length in England to 10% above that present in 1984 by 2050, which would mean a 17% increase over 2007 levels, an increase of some 90,000km (Defra 2023a). It remains to be seen whether this will gain any traction.

But to return to the matter in hand – to have a chance of delivering the UK BAP target, we had first to be able to define favourable condition. Ably led by Heather Robertson from Natural England and after a great deal of deliberation about priorities and practicalities, we reached a conclusion (Defra 2007). To be in favourable condition, a hedge must, on both sides, possess all the following qualities:

1. It should be at least 1m high and 1.5m wide and have a cross-section area of at least $3m^2$.
2. It should not have any gaps greater than 5m long, the combined length of gaps being less than 10% of the length of the hedge. Also, any gap between the ground and the base of the canopy should be less than 50cm high unless the hedge has developed into a line of trees.
3. There should be at least 2m of undisturbed ground each side of the centreline of a hedge, unless bordering a road or track.
4. Somewhere between the centreline of the hedge and the start of any cultivated ground, there should be a herbaceous margin at least 1m wide.
5. The cover of non-native woody plants or herbs should be less than 10%, although those long-established in the UK, that is archaeophytes, may be treated as native.
6. The combined cover of Nettles, docks and Goosegrass (Cleavers) should be less than 20% within a 2m-wide band beside the hedge. Any greater cover indicates that the hedge is suffering from a damaging artificial excess of nutrients.

This was as simple as we could make it: hedges are fundamentally complex. Every one is different, and the variation is huge. In truth, there's little hard science behind the thresholds we set; they are based on the views of experienced ecologists and knowledgeable hedge managers. Yet the approach has stood the test of time, and in practice it has been straightforward enough to use; it has been widely adopted. Further detail is given in the *Hedgerow Survey Handbook* (Defra 2007).

Trouble is, you could have a hedge that meets the acceptable thresholds for all these six qualities yet still be poor for wildlife. It could, for example, be dominated by just one woody species, have no hedgerow trees, never produce any flowers or berries, be contaminated with pesticides or lack dead wood. Jo Staley, Lisa Norton and I put forward measures and thresholds for such further factors in work we

LEFT: No space left for a herbaceous margin; this field has been ploughed right up to the edge – a major reason for hedges failing to be in good condition for wildlife.

undertook to define Favourable Condition Status in hedgerows in England (Staley *et al.* 2020, 2023). In our report we recommended that, in addition to the six qualities listed above, to be in favourable condition a hedge should have:

- At least three out of the following five structural components present: shrub layer, standard trees, basal flora, marginal flora, ditch.
- A minimum of 3.7 woody species per 30m sample length, on average. This is the average species-richness found by Countryside Survey in Britain in both 1998 and 2007.
- Significant quantities of flowers, berries, nuts, etc., present in at least two out of every three years.
- An average of one standard tree every 20–50m, with at least 45% of these trees being 20cm or less in diameter at breast height.
- At least one standard tree developing veteran features about every 50m. All veteran trees, stools and rotting stumps should be retained unless they pose a significant risk to safety. Likewise, substantial amounts of dead and decaying wood of all sizes should be retained *in situ* when hedges are laid or coppiced.

ABOVE: A healthy, species-rich hedge.

But much more research, thinking and consensus building needs to be done around this. There remain huge gaps in our knowledge. Is an average of at least 3.7 woody species per 30-metre stretch scientifically valid for wildlife assemblages and habitat resilience? What's a significant quantity of flowers and berries? How herb-rich should margins be? What levels of pesticides – fungicides, herbicides and insecticides – if any, are acceptable? And so forth. To complicate matters still further, there are yet more factors that should ideally be considered. Light pollution, for instance: it has recently been discovered that hedges and grassy road verges under streetlights have just half as many moth caterpillars as nearby unlit stretches (Boyes *et al.* 2021).

Even more rudimentary is our understanding of what a *network* of hedges should look like to be in favourable condition. Since a single hedge can never meet the needs of all hedge-associated wildlife in an area, to provide the best conditions for wildlife we must think at a network level as well as at that of the individual hedge. Two key questions we need answers to are: (1) how much structural diversity between hedges in a given area is optimal, and (2) how many connections between hedges should there be? New research, including modelling, is much needed here. I have a stab at answering the first question in Chapter 6, while an average density of 10km of hedge per square kilometre, as advocated in the report defining Favourable Condition status (Staley *et al.* 2020), can serve as a proxy for connectivity.

The present condition of our hedges

Sadly, most British and Irish hedges are not in as good a condition for wildlife as they could be. Far from it, many are in a very sorry state. Shockingly, 2007 Countryside Survey data reveal that across Britain only 12% of shrubby hedges, that is those that are not lines of trees, are in good condition for all six qualities in the first list above. For hedges bordering arable land, the figure drops to a miserable 6% (Barr and Norton 2008).

The good news is that in 2007 the presence of non-native trees, shrubs and herbs was very rarely a problem in rural hedges, in contrast to the situation in some other habitats, especially aquatic ones like ponds. Also, over 90% of hedges were fine in terms of their height and the absence of gaps wider than 5 metres. The big problems lay with nutrient enrichment and the width of margins: more than half of all hedges had their bases swamped with aggressive growths of Nettles and Goosegrass, and more than half had less than a metre's width of desirable perennial herbaceous vegetation at their bases or in their margins. Other worries were that over a third of hedges suffered from ploughing too close to the base, a quarter from being too narrow, while about a fifth were too gappy, either along their length or between ground and canopy.

BELOW: An all-too-common sight, more gap than hedge. The horizontal limb shows that this hedge was once laid.

I suspect the same problems persist today, other than that insufficient widths of perennial vegetation or unploughed ground are less of an issue. This reflects rules introduced in the mid-2000s linked to the main agricultural subsidy, the Basic Payment Scheme. One such rule stipulates that farmers should not cultivate the ground within 2 metres of the centreline of hedges, nor apply any fertilisers or pesticides within these 'hedgerow protection zones'. Alarmingly, such 'cross-compliance' rules may, following Brexit, be phased out entirely by 2027.

Putting the intractable problem of nutrient enrichment to one side for a moment, light at the end of the tunnel comes from the fact that a third of all hedges failed on just one quality, and well over half on just two. If we can just let hedges get wider, including both the shrub line and herbaceous margins, then much of the job would be done – the 12% would become a much more acceptable 60% or so. There would be other benefits too, such as an increase in carbon capture and storage.

Local surveys provide excellent information on hedge condition in different places. The results of some, 33 that were sponsored by Defra between 2006 and 2010, are given in Appendix A. But many others have been carried out too: Colin Barr (2009) gives information on no less than 79 in Britain. Radford (2003), for example, surveyed hedges in eastern Anglesey, and amongst other things he found that the condition of species-rich hedges was substantially better than that of species-poor ones in terms of gaps and legginess, something reported by several other surveys. More recently, Alison Hood (2014) assessed the condition of hedges in the South Devon Area of Outstanding Natural Beauty and found that two-thirds of their banks were in a poor state. There's a wealth of such useful facts and figures to be found in local surveys.

In Ireland the state of hedges appears no better than in Great Britain. In 2006, Neil Foulkes provided an in-depth fact-filled account of the hedges of County Leitrim. He found that just 25% of hedges sampled were in favourable condition for height, width, gappiness and non-native species, using the same thresholds as UK surveys. He also provides information on the percentage of hedges in favourable condition in some other midland counties: Kildare 23%, Longford 6%, Laois 20% and Offaly 5%. In Leitrim, as in Anglesey, species-rich hedges were generally in better condition than species-poor ones (40% vs 25%). Excessive gaps along the lengths of hedges

ABOVE: A hedge to nowhere …

and between canopy and ground (leading to scrawny bases) were the main issues (Foulkes 2006b).

In County Monaghan, hedge condition was assessed in 2010 and again in 2021 using the same methods that Neil employed in County Leitrim (MacElwain *et al.* 2021). Disturbingly, the proportion of hedges in favourable condition fell from a low 18% in 2010 to an even lower 12% in 2021, mainly due to hedges becoming more gappy and a large increase in the dominance of Nettles and other plants associated with high nutrient levels. Intriguingly, a new cause of concern was identified – the spread of Ivy. Whereas in 2010 Ivy dominated over a quarter of the canopy in 10% of trees, by 2021 the proportion of trees so affected had risen to 36%. The report's authors think this likely to be due to ash dieback causing crowns to thin, to the liking of light-loving Ivy. They suggest that in this way Ivy is indicative of poor tree health, noting that it can also make trees more likely to topple in strong winds.

Returning to the UK, in England and Scotland Countryside Survey has not been repeated since 2007, although Lisa Norton at the UK Centre for Ecology and Hydrology (UKCEH) tells me that a full hedge resurvey in England is due to be completed by October 2023, which is excellent news. Aligned surveys, including some squares sampled by Countryside Survey 2007, have been carried out in Wales as a rolling programme between 2013 and 2016. These found that the number of woody species per standard length of hedge fell from an average of 6.7 in 2007 to 5.4, and that a greater proportion of hedges were being managed, presumably by trimming, 88% compared to 85% in 2007, this latter change being reflected in more hedges being less than 2m high (73% compared to 64%) (Emmett *et al.* 2017). Also, fewer hedges had been newly planted, 2.3% compared to 4.6% in 2007 and 7.4% in 1998, and fewer recently laid or coppiced, 1.5% compared to 5.4% in 2007 and 7.1% in 1998. These trends paint a bleak picture for the future of hedges in Wales and are, I think, deeply alarming. In Northern Ireland funding has been received by UKCEH to complete a resurvey of Countryside Survey by autumn 2024 – we can only hope the rate of loss there will have slowed down if not halted.

Concluding remarks

In Britain all the evidence points towards fewer hedges having been lost since the mid-1800s, when numbers were at their peak, than popular belief would have it. On average it's probably between a fifth and a quarter, not half. This is reassuring – our rural landscapes are not as depauperate as feared. What is without doubt is that Ireland and Britain still have an extraordinary number and length of hedges, even if they are thin on the ground in some places. We have done much better than most other countries in north-western Europe.

It is also a myth, or so it seems, that rates of loss have been greater in arable areas than in grassland ones – it's just that arable ones had fewer hedges to begin with, so the loss of even a low proportion makes for very large field sizes and frayed networks.

Not so encouraging is the condition of our hedges. In Britain just 12% at most are in a favourable state for wildlife, and the situation is much the same in Ireland. Of particular concern, the impact of pesticides has yet to be assessed, while there's an urgent need to get to grips with defining what good wildlife hedge networks look like. Both these topics are surely priorities for research.

Despite gloomy statistics, hedges remain vital wildlife havens across most of lowland farmland. When assessing the quality of those hedges, have we at times perhaps set the bar too high? Certainly, the situation is far from irretrievable. We need to focus on two things: firstly, we must allow hedges to get wider, both the shrubby and the herbaceous parts; and secondly, we must limit the flow of nutrients into hedges, to stop the relentless march of Nettle and Goosegrass, and force these aggressive invaders into retreat.

ABOVE: All that's left! The twisted shape of the two Hawthorns on the right show that they were once part of a laid hedge.

BELOW: A Blackthorn hedge that could do with some TLC.

Reservoirs of life

chapter five

On the long train journey I used frequently to make between Exeter and Peterborough, while others were engrossed in their laptops and papers, I stared out of the window reflecting upon how, for miles on end, the countryside would be barren beyond countenance if it were not for hedges. Across vast tracts of pasture and cropland they are the last refuge for so many of our native farmland plants and animals, essential habitat for their survival.

The majority of lowland Britain is intensively farmed, these arable and grassland fields nearly being wildlife deserts. Nature is, for the most part, confined to isolated woods, to tree-lined rivers and railways and to village gardens. The saving grace is the existence of all those many miles of hedges that still border fields, and that connect the remaining fragments of woodland, scrub and flower-rich grassland, one to the other. They are the corridors that afford safe passage through a hostile landscape: the green veins, the highways, that unite the countryside.

Edge habitats, or 'ecotones', where two or more habitats grade into one another, are recognised by ecologists as among the most biologically diverse on Earth. Hedges are a prime example of this, or at least should be if given a little room to prosper, being an intricate mix of woodland, woodland edge, scrub and grassland. The central line of shrubs and trees grades down to grassland, in a blurred and gentle transition. If a ditch is present, then wetland is thrown into the mix too; if bare earth from cultivation or erosion, that as well. It is this fluidity, the soft edges, that enables hedges to support staggering amounts of wildlife.

This chapter offers an overview of hedges as a wildlife habitat and reviews the ways in which the various parts of a hedge complement one another, synergistically. Following chapters provide more detail

OPPOSITE PAGE: Paula and our dogs enjoy walking between Cornish hedges in full flower in May.

The best wildlife hedges

The very best hedges for wildlife are those that:

1. Possess multiple structural parts – standard trees, shrubs, banks, ditches, margins.
2. Grade softly down from trees and shrubs, through scrubby outgrowths like bramble patches and coarse tussocky grasses, to margins rich in flowering herbs.
3. Lie within structurally diverse networks, with short hedges, tall ones and all stages in between.
4. Have a dense shrub layer, unless just laid or coppiced or purposefully left to grow up, perhaps even into a line of trees.
5. Contain a wide range of different trees and shrubs.
6. Feature standard trees scattered along their length, open grown, with spreading canopies.
7. Produce plentiful blossom and fruits.
8. Hold plenty of dead and decaying wood.
9. Have thick basal vegetation such as tussocky grasses or similar dense perennial growth.
10. Sport flower-rich margins.
11. Encompass a ditch with at least a seasonal, if intermittent, flow of water.
12. And, perhaps most important of all, connect at both ends with other hedges or wildlife habitats such as woodlands and ponds.

Hedges like this, with soft edges, grading gradually from shrubs and tree to the field edge, are best of all for wildlife.

on the wildlife of hedgerow trees, of the shrub layer, and of hedge bottoms, margins, banks and ditches; another looks at the value of hedges as wildlife corridors. I end this one with a eulogy for dead and decaying wood. But I start with an account of the wildlife of a single hedge on our farm in Devon.

ABOVE: Hedges are vital wildlife refuges in intensively farmed arable landscapes

Extraordinary diversity: the life of a single hedge

On a car journey somewhere in Suffolk, in the autumn of 2010, Jon Stokes from the Tree Council and I were talking, as we often do, about hedges. We got to wondering just how many species live in hedges and came to the realisation that no one knew. Jon challenged to me to find out. With little idea what I was letting myself in for, and thinking that it might broaden my natural history knowledge, I accepted.

I chose a hedge on our farm, which lies in the heart of Devon, solely on the basis that I could visit it at the drop of a hat – it runs alongside the access lane to the farmhouse and starts just 40 metres from our front door. In the context of Devon, the hedge is nothing special, just an ordinary lane-side hedge with a good range of shrub and tree species and an earth bank, slumped with age. A shallow ditch runs along one side. It has three or four oaks selected some 25 years beforehand to become standard trees, but no mature or veteran specimens. The margins are flower-rich on both sides – that's the only part that is perhaps a little unusual, so many hedges in the county are cultivated or tightly grazed right up to the edge of the shrub line, giving herbaceous flowers little chance. The hedge is 85 metres long.

On 1 January 2011 I started recording. That day I recorded 14 different trees, shrubs, ramblers and creepers, four mosses, three birds (Blackbird, Chaffinch and Blue Tit) and Winter Moth. The next three days I added House Sparrow, five more mosses and four liverworts. On 5 January I identified six ferns, three winter gnats and the fly *Dryope flaveola*, and found a Mole hill. And so on until 31 December 2012, a full two years later. I searched the hedge thoroughly for plants, lichens and fungi, looked for birds (discounting anything that probably had nothing to do with the hedge like Swifts hawking high overhead) and inspected flowerheads for insects. I ran a low-powered moth trap tucked in close to the base of the hedge, used catch-live small mammal traps and Hazel Dormouse nest tubes, placed bits of corrugated tin on the ground to find reptiles, did a little bit of pitfall trapping for beetles, and worked the vegetation as best as the brambles and thorns would allow with sweep net and beating tray. A few emergence traps, simple home-made constructions of wire and fine netting, were pegged over the ground to sample insects emerging from the soil, rotting vegetation and the like.

BELOW: I spent two years recording the wildlife of this hedge on our farm in Devon. I call it Hedge 2070 after the number of species I found. The tent-like structure is a Malaise trap, used to sample flying insects.

By far the most effective means of finding insects was to use a Malaise trap. René Malaise was a remarkable Swedish entomologist and intrepid explorer. In 1934, while in Burma (now Myanmar), he invented the insect trap named after him. It consists of a tent-like structure and relies on the tendency of most flying insects to move upwards towards the light when encountering an obstacle, in this case

a vertical sheet of fine netting. The top of this vertical central sheet has downward flaps on each side and itself slopes upwards, leading to a narrow opening through which the insects pass into a collecting bottle. These traps are remarkably efficient at catching day-flying flies (Diptera) and wasps (Hymenoptera), the two most numerous orders of insects occurring in Britain and Ireland, with over 7,000 species each. Each day I would empty the trap set alongside my hedge and examine and sort the catch under a microscope, a process that would take an hour at least most summer evenings.

The final tally? 2,070! This was the number of plant, fungus and animal species, all big enough to see with a naked eye, identified from the hedge. Nearly everyone I tell this to is astonished. The reason is simple: most of the life in a hedge is small, secretive or well disguised. Among animals, we are amongst the very largest surviving in Britain: only cattle, Red Deer and horses are larger, at least in the landscape within which our farm lies. Everything else is smaller, usually much smaller. When walking along a hedge we may see the odd Rabbit and a few birds, but it's so easy to overlook everything else. We are just too big and too fast. As Gordon Corbet (2011) observed in an article about the life to be found in a dune system, if we were ant-sized we would see so much more. And even then we would overlook myriad microorganisms. There's a Devon dialect word 'palsh' meaning to 'walk slowly whilst looking'. We all need to palsh more, if we are to appreciate the full richness of life in hedges.

By far the majority of the organisms I recorded were insects. This is as expected, since there are after all over 24,000 known British species, more than within any other class of animal or plant, although fungi come close with some 15,000 species known. So while I found 125 plants (including mosses and liverworts), 80 fungi and lichens and 50 vertebrates (amphibians, reptiles, birds and mammals), the number of insects stands at 1,718. To give this number some context, no fewer than 17% each of all the true flies, of the butterflies and moths, and of the caddisflies so far recorded in Britain and Ireland were found in the hedge over the two years of sampling.

BELOW: How the 2,070 species recorded in Hedge 2070 were apportioned between major taxonomic groups.

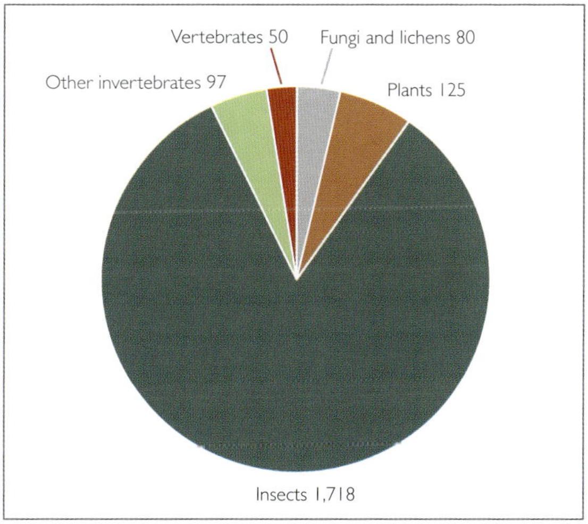

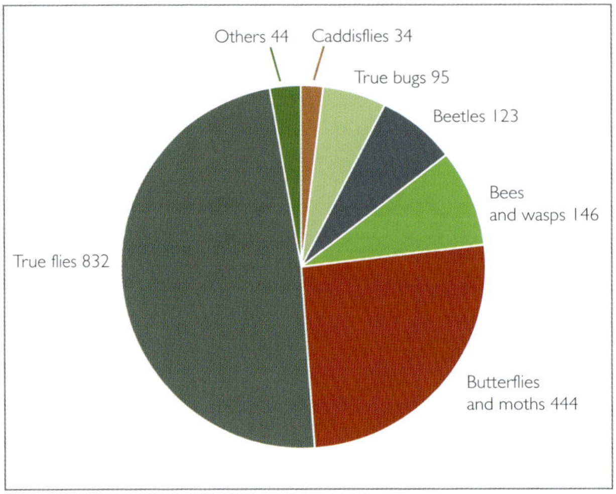

ABOVE: The number of species identified within different insect orders in Hedge 2070.

I was very reliant on a huge amount of help freely given by the many experts I sent samples to – without their willing assistance the task of identification of the myriad species would have been far beyond my abilities. Among others – and a huge thank you to you all – Peter Chandler sorted through literally thousands of fungus gnats to find 139 species, Martin Drake and Adrian Plant examined tubes of alcohol-sodden small dark flies, David Gibbs determined big-headed flies (appropriately named although their heads have a nasty habit of dropping off when under the microscope), and Alan Stubbs and John Kramer picked through piles of dried and often legless craneflies. Chris Spilling bravely tackled the spiders, Matthew Shepperd had a look at some springtails and Keith Alexander and Martin Luff keyed out a motley assortment of dried beetles, while Bob Heckford dissected small drab micromoths and Paddy Saunders helped me with the bees and wasps. Andrew Halstead and Ian Wallace gave names to pinned sawflies and caddisflies respectively. Lichen-covered twigs were carefully examined by Maxine Putnam and Tony Halwill, and Mark Pool covered the mosses and liverworts. The taxonomic expertise on offer was simply astounding, the identification skills astonishing.

Even the list of 2,070 species is far from a complete inventory. One very species-rich grouping of insects remains to be tackled, and that is the Parasitica, the parasitic wasps, of which there is a vast diversity in Britain and Ireland (some 6,500 species). Each time I sorted through a Malaise trap sample I would marvel at the number and variety present. Being unable to even make a start on identifying them, I stored them in jars of alcohol which are now on some shelf in the Natural History Museum in London, hopefully not being buried beneath too deep a layer of dust. The technology exists such that it should be possible to run them through a DNA barcoding machine to say how many species there are, if not what they are – I must wait patiently for the resources to become available to do this for my hedge samples.

OPPOSITE: Nearly all the 2,070 species found in the hedge I studied in detail (Hedge 2070) are shown on this 1,500 × 1,800mm 'map' made by artist Marcus Vergette. Each species is represented by a coloured dot, with each colour corresponding to a major taxonomic group. Around the edges are hand-written all the names.

Also, truth be told, I was not very good at catching beetles: I found only 132 beetles, just 3% of the British list. Here, I think, use of a vacuum sampler would have made a lot of difference. Surely there must have been more. I think it reasonable to suggest that, with further sampling and full identification, some 10–12% of the 24,000 insects on the British list might be found in this one hedge.

With fungi, it's a hit and miss affair as to whether the fruiting bodies required for non-molecular identification appear in any one year; 2011 was quite a good year, 2012 very poor. I am sure the number of fungi I recorded was a considerable underestimate. Georg Müller (2013) reckoned that a series of six banked hedges in Lower Saxony, Germany, supported at least 600 species of fungi, identifying 280 from a single hedge; I succeeded in finding only 56 in my hedge – there must surely have been many more. Neither of us covered rusts or smuts.

All in all, I reckon the true tally of species in the hedge to be round about the 3,000 mark – and that's just those big enough to see with the naked eye.

Do hedges always help wildlife?

Were all the species I found in my hedge benefiting from it? Or were some there by chance or mishap? Using an analogy coined by landscape ecologist John Dover, it is conceivable that hedges act as giant drift nets stretched across the landscape, capturing and waylaying some unfortunate individuals, preventing them from reproducing. Once I saw a Large White butterfly flying across the adjacent field towards the hedge, rise up over it and continue on its way – presumably it had to expend a little extra energy to overcome the obstacle, reducing its chances of surviving to reproduce by a tiny fraction. Otherwise, everything I saw or found seemed likely to be benefiting from the hedge in some way – as a source of food, for shelter from rain and wind, for protection from predators, as somewhere to nest or lay eggs or perhaps to find a mate, or as a safe highway. My reading around the subject has only served to confirm this view.

It's possible that a very few species of open ground found Hedge 2070 to be an obstruction in their efforts to disperse to find new places in which to breed: the Large White I saw may have fallen into this category. Needless to say, I never found any open-field species like Skylarks or Brown Hares ensnared by the hedge (neither is on the 2,070 list, although I could hear Skylarks singing above nearby fields). Elsewhere, waders such as Lapwing and Curlew strongly prefer wide open spaces free of the trees and hedges where Carrion Crows and Magpies nest and from which Stoats and other mustelids can sally forth. The Great Bustard is another species that requires large open, plain-like, expanses – it became extinct in Britain in 1832, possibly partly as a result of enclosure, but has

LEFT: Skylarks are among a small minority of species in farmland that avoid hedges, preferring wide open spaces.

now been reintroduced on Salisbury Plain. But these are exceptions. Most species, even if they may not live in or move along hedges, can tolerate their close proximity.

Few of the 2,070 species identified from the hedge will be totally reliant on it, or even on the farm's network of hedges as a whole, for survival in the landscape. But it is true to say that without the existence of hedges the populations of most would be lower and, doubtless, some would die out: the area of habitat remaining would not be big enough to support successive generations through time, or the landscape would be too broken for individuals to move easily enough, safely enough, and often enough between habitat patches. In support of this contention, researchers from Bristol University have found that most of the biodiversity on a Somerset farm could be conserved by retaining and appropriately managing uncultivated habitats such as hedges, scruffy areas around the back of farm buildings and so forth (Evans *et al.* 2011, 2013). They found the importance of hedges and 'waste ground' for wildlife to be out of all proportion to the 4.5% of the land area which these features occupied. The fact that the farm was organic makes this discovery even more forceful.

Some of the 2,070 will live out their entire lives in the hedge; the majority probably won't do this but need other habitats nearby to complete their life cycles. Migrant birds like Chiffchaffs or Redwings spend most of their lives not just in other habitats but in other countries – but when they are here in my Devon hedge it provides resources they need, insect food and nesting places in the case of the

ABOVE: A dance fly belonging to the genus *Hilara*. The inflated forefeet of males contain silk glands used to wrap nuptial gifts for females.

Chiffchaff, winter berries for the Redwings. Even Robins, highly territorial birds, leave the hedge to visit our bird table during the winter months. The Barn Owl I saw hunting along the hedge edge was nesting in one of our sheds. Butterflies like the Brimstone spend time in the hedge sipping nectar from flowers but have to go elsewhere to lay their eggs – the caterpillars eat the leaves of Alder Buckthorn, a shrub not found in Hedge 2070 but present in others on the farm.

The larvae of some dance flies of the genus *Hilara* are known to use open grassland, but the adults form mating swarms around bushes (as within hedges) and hunt at the edges of ponds: all three habitats need to be present near to one another for these flies to thrive (Delettre *et al.* 1998). Such examples show that hedges provide important resources for animals of all sorts, even if for many they are not enough on their own. For wildlife to thrive in farmland landscapes, a range of different habitats and features need to be present and close to one another.

Still, many of the 2,070 species I found could complete their life cycles and reproduce in a single hedge successfully. Amongst these, naturally, are the plants, fungi and lichens. The emergence traps gave an indication of the number of insect species breeding in the soil, leaf litter or living plants of the hedge bottom, ditch and margins. Although these traps covered just 0.05% of the area of the hedge they caught no fewer than 162 true fly species from 38 families, together with a few moths and other flying insects. If the 19 flies I detected through the presence of their leaf mines or galls are added to this list, then at the very least 22% of the flies recorded at the hedge reproduce within it. I suspect the true proportion is likely to be considerably higher. For instance, 42 of the 48 hoverfly species recorded as adults more than five times from the hedge could potentially have bred within it since the necessary microhabitats were present for their larvae, yet only five species appeared in the emergence traps – but then they covered only a tiny percentage of the hedge area.

Many species I recorded just once or twice. This does not necessarily mean they were strays from other habitats, just that they were present at low densities or difficult to find. Indeed, in most data sets collected to assay species-richness in particular places, a high proportion of species are recorded only a very few times. As an example of this, of

ABOVE: Green lanes like this one in Buck's Mills, Devon, provide sheltered conditions favoured by butterflies and many other insects, and afford protection from pesticide drift.

the 161 fly species identified from emergence traps I placed in a wet woodland about 1km from the hedge, as many as 53 were recorded just once and 20 just twice (from a total of 1,945 individuals). Research shows, somewhat counterintuitively, that for most sites there exists a strong relationship between total species-richness and the number of singletons and doubletons found. Using a simple equation, if one knows the total number of individuals in a sample and the number of species recorded once or twice, one can estimate the full species-richness of a site with remarkable accuracy – a process known as Chao estimation after the person who developed it, Anne Chao. The biological reasons underlying the statistical relationship are, as yet, not well understood. Readers wishing to find out more about it can do no better than look at Anne Magurran's (2004) excellent book on measuring biological diversity.

Local landscape quality is important

Almost certainly one of the main reasons why so many species are to be found in Hedge 2070 is that it has a tree-lined stream at one end, a small reed-fringed farm pond nearby, and just 130m away across a pasture field, a small mature native woodland. There may be no nature reserves or protected sites in the vicinity, but the landscape remains full of good places for wildlife.

ABOVE: The hedge where I recorded 2,070 species (marked in red) on our farm is part of a landscape of well-connected hedges and other wildlife habitats such as small woodlands and farm ponds.

Many of the animals to be found in hedges require other habitats nearby to prosper. If you are, say, a Yellowhammer, it's no good finding a safe nesting place in a hedge if the adjacent fields are devoid of the seeds and insects you and your chicks need as food. We should expect hedges lying within landscapes, whether rural or urban, rich in good wildlife habitat to support more species than those that lie within impoverished landscapes such as those with a high density of buildings or those covered by intensive livestock or arable farms.

Hedge 2070 also has a parallel hedge running beside it, on the opposite side of the farm lane. It could thus be said to be part of a 'green lane', although technically the track surface should not be tarmacked to qualify as such. John Dover and fellow researchers from Staffordshire University have demonstrated that green lanes are a high-quality habitat for invertebrates like butterflies, providing far better shelter than single hedges as well as protection from pesticide drift – unless they form a wind tunnel (Dover *et al.* 2000).

All that said, since many species can complete their life cycles in a single hedge and others have quite remarkable and unexpected abilities to move across hostile landscapes to colonise suitable habitat patches, I would expect even disconnected hedges in landscapes with little other wildlife habitat left to turn up large numbers of species. My hope is that someone soon will attempt to find out. There's a challenge.

A good hedge is far more than a line of shrubs

In 2013, I was part of a small team commissioned to explore exactly which bits of hedges are important to different species. Our conclusion was unexpected: the majority depend on not just one part, but on two or more (Wolton *et al.* 2013). By parts, I mean structural components: the shrub layer, standard trees, hedge bottom and any bank, margins or ditches. It's no good just having one of these present and in good condition, you need several for a hedge to be rich in wildlife.

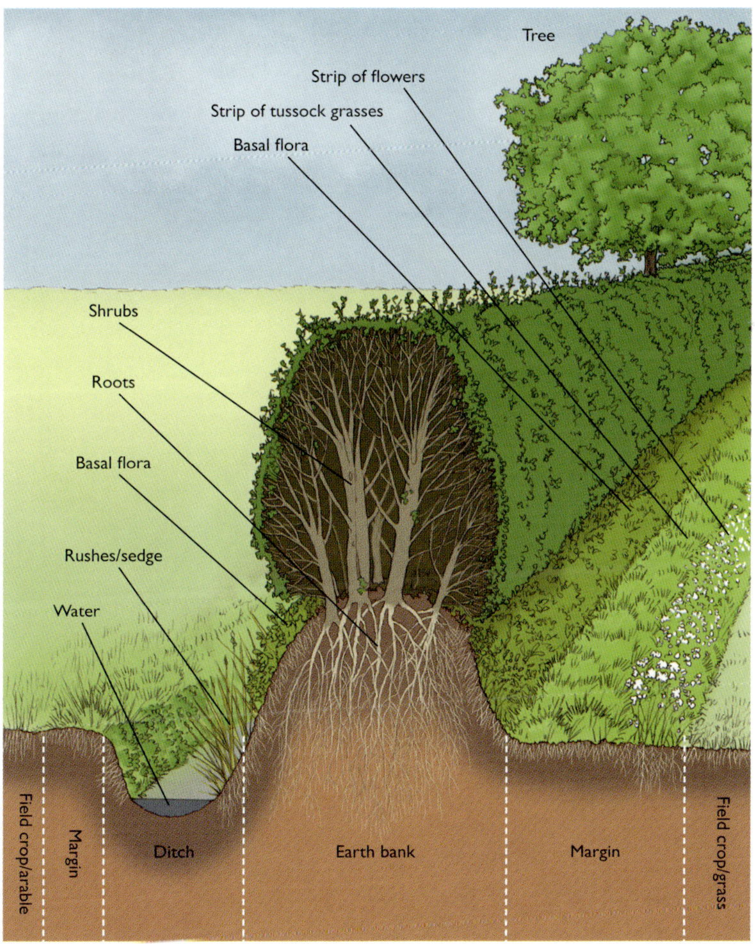

LEFT: The complete hedge. Hedges are more than just lines of shrubs – the ground beneath them is part of the hedge too, as are any standard trees, banks, ditches and flower-rich margins.

Specifically, we looked at the way 107 species use hedges, 84 of these being among the Priority Species for conservation as listed in Appendix C, and 34 of them used by the UK government as indicators of the state of farmland biodiversity (11 species fell into both categories). The farmland indicators all have a close association with hedges and comprise an additional 13 widespread species of butterfly, six birds (Kestrel, Rook, Woodpigeon, Whitethroat, Goldfinch and Greenfinch) and two bats (Common Pipistrelle and Serotine). We found the majority (65%) of the 107 species to require two or more structural components to be present, and over a third (35%) depended on three or more.

Drilling down into these results a little further, a difference becomes apparent between species that are scarce and those that remain widespread. While 57% of those that are Nationally Rare or Nationally Scarce are reliant on just one component, 81% of those that are widespread (including all the farmland indicators) rely on two or more. So if we don't want to see the populations of those species that remain widespread fall further, it is essential that hedges should have most of their bits in good health. (Nationally Rare species are those which are known to occur in 15 or fewer 10 × 10km squares in Britain, Nationally Scarce ones those known to occur in between 16 and 100 such squares.)

Standard trees come out as the single most important component for the Priority and indicator species, with 60% relying on them, either alone or in some combination with other components. Close behind, at 56%, is the shrub layer. Hedge bottoms and banks, and margins, are next at 42% and 40% respectively. Few, just 9%, of the 107 species are associated with ditches – while these support abundant wildlife, the species involved don't tend to be ones that are either conservation priorities or used as indicators of environmental health.

These figures drive home the point that a good hedge is more than a just a line of shrubs: it should have at least some standard trees sprinkled along its length and good basal and marginal floras. Without this structural complexity, any hedge will fall short of its potential. A ditch is a bonus rather than a must-have in the context of species of conservation concern – but, as we shall see later, a ditch in good condition can be a powerhouse for insect biomass.

The Yellowhammer is one of many birds that benefit from hedges that are more than just lines of shrubs hard up against cultivated land or tarmac. They advertise their territories by singing from hedgerow

Bumblebees

These important insects exemplify well the need for structural complexity within hedges. Common ones like Buff-tailed, White-tailed, Garden and Common Carder nest in hedge bases, often in old rodent nests, and hibernate there too. In the spring they gather the pollen and nectar they need for themselves and their broods from the flowering shrubs and trees in the hedge, visiting willows, Blackthorn, Hawthorn, Field Maple and limes amongst others. As these sources diminish, they switch to the flowers of the hedge bottom, margins and ditch sides.

If hedges are just lines of shrubs and nothing else, the bees are unlikely to be able to complete their life cycle and survive, especially when living in intensively farmed landscapes with few alternative habitats. And if there are no bumblebees to pollinate the flowers, then there will be fewer seeds or berries for finches like Goldfinches and Bullfinches, or wintering thrushes like Redwings and Fieldfares, to feed on; the web of life will start to fail.

A queen White-tailed Bumblebee searches a hedge bank in the spring for a place to make a nest and start a colony.

trees, feed in insect and seed-rich margins and typically raise their first brood in nests built in dense cover alongside a ditch and their second brood in nests in the shrub layer once in full leaf. Another bird, the Tree Sparrow, nests in holes in mature hedgerow trees, feeds in the margins, and flies for cover into the shrubs when birds of prey threaten. Song Thrushes sing from the trees, nest in dense cover among the shrubs, and spend much time foraging for worms, snails and fallen berries in the hedge base. And so on.

Some mammals, too, use several different parts of hedges. Greater Horseshoe Bats, for example, use hedges as highways and navigational aids, and forage along them for their insect prey of larger moths, dung beetles and craneflies. When they have caught something, they fly to a convenient branch in a nearby hedgerow tree to settle and devour it. Hazel Dormice also need multiple components, as explained in Chapter 7.

Network diversity

Hedge structure, such as whether short or tall, thin or wide, see-through or dense, has a profound influence on biodiversity. It follows that the greater the range of heights, widths and so forth present across a farm's hedges, the greater the amount of wildlife collectively these hedges are likely to support. The best illustration of this comes from the bird world. Field studies have looked in considerable depth into the hedge structures favoured by various species (Green *et al.* 1994, Macdonald and Johnson 1995, Stoate 2019). Here are some of their findings:

- Whitethroats, Linnets and Yellowhammers prefer short hedges with few trees, optimally 3.1m high for Whitethroats and less than 1.2m high for the other two.
- Blackcaps, on the other hand, like tall hedges; the more trees the better.
- Greenfinches are most abundant in wide (10m+) hedges, while Lesser Whitethroats and Yellowhammers prefer narrower ones, optimally 5m and 3m wide respectively.
- Dunnocks have a particular aversion to hedges with many gaps.
- Whitethroats and Yellowhammers like wide (2–3m) herbaceous margins.
- Goldfinches favour not just wide hedges, but also those that contain dry ditches. Yellowhammers like ditches too, but apparently Chaffinches do not.

BELOW LEFT: Whitethroats prefer short hedges with wide herbaceous margins for nesting.

BELOW RIGHT: Dunnocks, with their beautiful sky-blue eggs, dislike nesting in hedges with gaps.

To add to this, in my experience, Redstarts and Spotted Flycatchers favour hedges that have developed into mature lines of trees – here too one will find true woodland birds like Nuthatch, Great Spotted Woodpecker and Tawny Owl nesting. The relationship between hedge structure and bird communities is clearly complex and difficult to disentangle; there are all sorts of variables and unknowns. For example, the reason why Linnets and Yellowhammers prefer short hedges may simply be that these are the ones with the widest herbaceous margins where they can forage for insects and seeds – taller hedges tend to have thinner margins because they are squeezed between dense shade and the crop.

Hedge specialists

There are few true hedge specialists: the habitat is after all largely human-made and a mash-up of woodland, scrub and grassland. Still, there are quite a number of species that as so strongly associated with hedges that they have been named after them: three species of Hedge-parsley (Knotted, Spreading and Upright), Hedge Bindweed, Hedge Bedstraw, Hedgerow Crane's-bill, Hedge Mustard, Hedge Garlic (Garlic Mustard), Hedge Woundwort, Hedge Snail (*Hygromia limbata*, an introduced species), Hedge Rustic (a moth), Hedge Brown (a butterfly also known as the Gatekeeper), Hedge Sparrow (an alternative name for the Dunnock) and Hedgehog. All these are names in common use today and make good sense, with the possible exception of the Hedge Rustic, which is often found in more open, grassy places.

BELOW: Gatekeepers, also known as Hedge Browns, are well named since congregations can often be seen in gateways through hedges.

Analysis of data collected over many years by the British Trust for Ornithology has shown that seven common bird species usually prefer to nest in hedges rather than in woodlands: Dunnock, Whitethroat, Lesser Whitethroat, Linnet, Goldfinch, Greenfinch and Yellowhammer, although hedge quality and structure have a big influence on precisely where the birds choose to breed (Fuller *et al.* 2001). Some scarce birds such as Grey Partridge and Cirl Bunting may also be hedgerow specialists, the former nesting in hedge bottoms and grassy margins, the

latter breeding close in to the base of hedges and singing from hedgerow trees.

There are many local names linking plants to hedges (Grigson 1975), some descriptively apt (although some of these have in recent times faded out of use). In Yorkshire, Traveller's-joy or Old Man's Beard is Hedge Feathers. In Somerset, a local name for Foxglove is Hedge Poppy. In Worcestershire, Black Bryony is Hedge Grape, while in Hampshire it is Jack-in-the-hedge. In East Anglia, Ground-ivy is Hedge-maids, while in several other counties it is Robin-run-in-the-hedge, the same name being used for Bittersweet or Woody Nightshade in Lancashire. Blackthorn has a range of evocative names: Hedge-picks and Pig-in-the-hedge in Hampshire, and Hedge-speaks in Wiltshire, a name given to Dog-rose in Gloucestershire (where Blackthorn is called Heg-pegs). Dog-rose is called Hedgy-pedgies in Wiltshire. Hedge Bindweed has various local names including Hedge Bells and Hedge Lily after the shape of its flowers, as well as Jack-run-in-the-hedge. A Devon name for Herb-Robert is Hedge Lovers (my favourite, given my own name, where I live and my love of hedges), while Common Dog-Violet is Hedge Violet. Soapwort, not a native plant but long naturalised, has been called Hedge Pink in Hampshire. Goosegrass or Cleavers, the bane of so many hedges today, has been variously Jack-at-the-hedge, Lizzie-run-the-hedge, Rabbie-rinnie-hedge, Robin-run-the-hedge, Robin-round-the-hedge and Willy-run-

LEFT FROM TOP: Garlic Mustard, also known as Hedge Garlic and Jack-by-the-hedge.

Hedgy-pedgies and Hedge-speaks are among local dialect names for wild roses such as Dog-Rose shown here.

Hedge Bindweed has many local names including Hedge Bells and Hedge Lily after the shape of its flowers, as well as Jack-run-in-the-hedge. The hoverfly on this flower, *Melanostoma scalare*, is very common in hedges.

the-hedge, although the Devon name Cling Rascal seems most fitting to me.

Most appropriately, the word hawthorn is derived from the Old English *hagaþorn*, meaning 'hedge thorn'. We have two native species of hawthorn in this country, the common Hawthorn, which is ubiquitous, and the Midland Hawthorn, which is much less widespread, its natural range being restricted to central, eastern and south-eastern England. Strongly associated with ancient woodlands and ancient hedges, Midland Hawthorn is the less thorny of the two species, so would have been selected less for hedge planting. It can be distinguished from Hawthorn by the shape of its leaves, with the gaps between the lobes extending less than halfway to the midrib, and the lower margin of the lower lobe toothed almost to the base while in the Hawthorn this margin is smooth. Also, the fruits have two stones, not one, and the twigs are less spiny. Intermediates occur, since the two species hybridise. Given the origin of its name, it is not surprising that Hawthorn has various local names such as Hag-bush in Yorkshire or Hag-tree in Devon and Somerset, along with more evocative ones such as May, May-tree or Bread-and-cheese Tree, the last of these to encourage children to nibble the tasty young leaves and flowers. It is a tree once believed to have many strong magical properties, both good and bad (Grigson 1975) – at home, our family still thinks it is unlucky to bring the blossom indoors.

RIGHT FROM TOP: A local name in Yorkshire for Traveller's-joy, also commonly known as Old Man's Beard, is Hedge Feathers.

Herb-Robert – a Devon dialect name for this flower is Hedge Lovers.

Hawthorn flowers, known as May blossom. A local name is the Bread-and-cheese Tree.

The very rare Plymouth Pear may be thought of as a hedgerow specialist, since its only known sites are a couple of remnant hedges in the city of Plymouth (Devon) and a few very old hedges at five sites, some now roadside thickets, in farmland near Truro in Cornwall. There are thought to be just 600–700 individuals in total. The Plymouth hedges used to lie within Buckland Common but that has been entirely lost to airport and other development as the city expanded – it is remarkable that the hedges have survived at all. Another tree closely associated both with hedges and Devon is the Devon Whitebeam.

A third woody species closely associated with hedges is Barberry. This was probably introduced to Britain in medieval times as an ornamental shrub and for its edible berries and medicinal properties. It forms dense, spiny bushes and is therefore a good hedging plant – or was believed to be so until the 19th century, when it was discovered to be a host for a black stem rust, *Puccinia graminis*, which also infects wheat and barley, damaging crops badly. To complete its life cycle and survive British winters, the rust needs both hosts, shrub and cereal. Thus an eradication campaign was waged against Barberry, and it is now scarce. Not being a true native, this would not be much of a conservation worry were it not that the shrub is the sole foodplant of the caterpillars of a rather attractive moth, the Barberry Carpet. This moth now occurs in only about 12 places spread across Dorset, Wiltshire, Gloucestershire and Oxfordshire, and is rated Endangered. In response, led by Butterfly Conservation, conversation organisations and community groups have propagated, planted and looked after thousands of Barberry bushes to help expand and link up these moth populations (Fox *et al.* 2021). The rust is currently absent from Britain and Ireland but there have been recent outbreaks of new virulent strains in southern Europe, and it has the potential to re-emerge here. If this happens, we may see another war waged against Barberry, placing the Barberry Carpet at grave risk of extinction (Barnes *et al.* 2020).

BELOW: Barberry Carpet moth, an Endangered species closely associated with hedges.

Devon Whitebeam

This special tree, like the Plymouth Pear, is closely associated with hedges. An attractive small tree with white flowers and orange-brown berries, it is endemic to Britain and Ireland, found nowhere else in the world. The species' origins lie in the hybridisation of a complex of whitebeam species, probably on the north Devon coast, and its heartland now is the Culm Measures of northern Devon, an extensive area of heavy clay land between Exmoor and Dartmoor. Otherwise there are a few outlying populations in southern Devon, Cornwall and south-eastern Ireland: botanist John Day tells me that these may be the result of long-distance seed dispersal by migrating thrushes such as Redwings moving south-westwards in the autumn. There are also a couple of records from western Somerset.

Across northern Devon, most Devon Whitebeams now occur in hedges, although some grow in open woodland, favouring woodland edges and internal banks, and others in heathy areas. Its fruit is edible. Indeed, there are newspaper and other reports of the berries beings sold in Barnstaple pannier market in the 1870s through to the late 1920s, as researched and recounted by species specialist David Cann (2016). It must have been quite sought-after, for in 1925 a bunch was being sold for one penny while a pound of blackberries, a much larger quantity, was making just three to four pennies. Various local names were used for the tree: French Hales, French Hails and French Eagles. I have been told that 'French' related to anything strange and 'hales' or 'hails' to the resemblance of the leaves to those of Hazel.

Devon Whitebeam berries. They are edible and quite tasty if picked at just the right time.

Turtle Doves

I have nostalgic boyhood memories of the soft purring calls of Turtle Doves coming from tall thorny hedges and thickets in Suffolk on hot summer days. Alas, that's now a sound of the past nearly everywhere. Critically Endangered, the dove has the awful status of being England's most rapidly declining breeding bird, undergoing a shocking long-term population decline of 98% between 1970 and 2018, with an 81% loss just between 2008 and 2018.

A summer migrant, the Turtle Dove has specific nest-site requirements, namely tall thorny bushes, so overgrown hedges are often ideal. Along with scrub and woodland edge, these are the nesting habitat of choice. Indeed, so important are hedges to Turtle Doves that their nesting density has been shown to be linked to the 'hedginess' of farmland landscapes (Browne *et al.* 2004). Suitable hedges should preferably have a minimum height of 4.5m and width of 3m; short, flailed hedges are of no use. Along with good nesting sites, to raise young successfully the doves require seed-rich foraging habitat nearby; wide grass and herb-rich hedge margins help to meet this need (Dunn and Morris 2012). Increasing dangers on their long migration to and from sub-Saharan Africa, such as drought, desertification, habitat loss and hunting, don't help.

On a recent visit to the renowned rewilding project at Knepp in Sussex, I was delighted once more to hear Turtle Doves calling from hedges. Here the hedges were so overgrown as to hardly be recognisable as such, having for many years been free to spread out into the fields on either side to form wide bands of scrub – altogether wonderful habitat for much wildlife, including Nightingales. It shows what can be achieved when nature is freed from oppressive management, although in the long term the habitat may cease to be Turtle-Dove-friendly – time will tell…

The Turtle Dove, a much-loved but endangered bird, is closely associated with tall, thick, bushy hedges.

Rare and threatened species

If naturalists want to find rare species they go to rare habitats like sand-dune systems, base-rich fens or mountain tops. Hedges, being common and ubiquitous across lowland Britain and Ireland, support relatively few real rarities, and those they do nearly always occur in other habitats as well. That's not to say that hedges are not important for threatened species – to the contrary, they are, in particular for ones that while still widespread have nevertheless experienced great falls in population size or range in recent decades. Farmland, sadly, is rife with such declining species. If it were not for hedges, the majority would be in an even worse state.

I reckon there are 98 UK Priority Species significantly associated with hedges, as listed in Appendix C, although one could argue that a few more should be included and some dropped. By law such species are priorities for conservation action, being either very rare or having rapidly declining populations. Of the 98 species, 72 are considered to be under threat of extinction in the UK, or close to it, according to strict criteria laid down by the International Union for the Conservation of Nature (IUCN). Further threatened species are associated with hedges but are not Priority Species, as are a great many species that are rated Nationally Rare or Nationally Scarce in Britain because they are currently known from 100 or fewer 10km squares (out of 2,913).

LEFT: The Willow Tit is now a rare bird that forages along hedges and sometimes nests in holes in their trees.

ABOVE: The very rare Scarlet Malachite Beetle is one of the 98 plant, animal and fungus species associated with hedges that are considered to be priorities for conservation action in the UK (see Appendix C).

In Hedge 2070, I recorded 25 of the 98 Priority Species: 11 species of moths along with Common Toad, Grass Snake, Common Lizard, Bullfinch, Dunnock, House Sparrow, Lesser Redpoll, Marsh Tit, Song Thrush, Spotted Flycatcher, Starling, Willow Tit, Hedgehog and Hazel Dormouse. None of these except the Willow Tit can be regarded as a real rarity since all are still widespread. Nonetheless, all are undergoing severe population declines at a UK level. Altogether at the hedge I recorded 60 Nationally Rare or Nationally Scarce species – a good tally bearing in mind that many protected sites have been designated on the strength of far fewer.

These figures reinforce the point that although hedges may be the sole habitat for very few species and even fewer real rarities, they do provide important resources such as food or breeding sites for many that are of conservation concern, rare or scarce. It is for this reason, and because of the way in which they tie the countryside together, that all UK hedges with at least 80% cover of native woody species are recognised under law as a Priority Habitat for conservation action (Joint Nature Conservation Committee). Some 84% of hedges in the British countryside have sufficient cover of native woody species to count as Priority Habitat. Thus hedges have equivalent conservation status to habitats such as upland hay meadows, lowland heathlands, lowland fens, traditional orchards and ponds, while beating them all in terms of frequency and geographic spread. They are truly our largest nature reserve.

Why dead and decaying wood matters

It's said that two-thirds of the organisms in forests are associated with decaying tree matter. I am sure the same must be true for hedges. Yet this is often totally overlooked by hedge managers – witness, for example, the fact that the removal of all dead and dying wood is still widely encouraged in hedge laying circles. It is viewed as an untidy waste product, best disposed of in a bonfire. The pejorative meaning

LEFT: A rare but pleasing sight: a dead hedgerow tree left standing.

of 'dead wood' as applied in business circles has carried over, it seems, to hedge management. Instead we should think positively about the death and decay of wood, whether stem, branch, twig or root, as an essential part of nutrient recycling and healthy ecosystems. Remove decaying wood, and a host of life dependent on it will disappear.

The main primary agents of decay are fungi. Each piece of wood will probably be decomposed by a succession of different species, often competing with one another. Hundreds, probably thousands, of fungi will occur in every decent hedge, but very few of those responsible for the decomposition of wood will be pathogens, that is attack living plant tissue. So leaving dead wood in a hedge is unlikely to decrease the survival chances of living plants unless they are already under severe stress – to the contrary, the plants will need the released nutrients to thrive. Even such prominent bracket fungi as Beefsteak Fungus (*Fistulina hepatica*) or Southern Bracket (*Ganoderma australe*), commonly but mistakenly thought to be a sign that a tree is diseased and doomed to die soon, will be feeding on the dead heartwood at the core of the tree, recycling the nutrients to the benefit of the tree and arguably making the tree stronger in the process. As the great storm of 1987 revealed, hollow trees are often better able to withstand gales than solid ones.

Earlier in this chapter I mentioned that the great researcher of European field boundaries Georg Müller (2013) had found a considerable diversity of fungi growing on hedged banks in his home municipality of Ganderkesee in Lower Saxony, Germany. Between 1985 and 1988 he detected 240 species of fungi growing on six banks; in subsequent years he visited the site sporadically and concluded that, together, the hedges supported at least 600 species, perhaps even double this. Of course, many of these will not have been decomposers of wood but reliant on other dead plant material such as leaf litter or on living plant tissue, or had a mutually beneficial relationship with the trees, shrubs and other plants through mycorrhizal associations – part of the Wood-Wide Web. Still, these figures give an idea of the high fungal diversity

RIGHT: Cobalt Blue Fungus on a dead branch lying at the base of a hedge.

LEFT: Scarlet Elfcups attached to moss-covered decaying branches in a hedge bottom.

associated with hedges. Georg concluded that hedges play an important role as refuges for fungi, including rare species.

When mature, hedgerow trees are often rich in veteran features such as rot holes, hollows, splits, dead branches and sap runs. To explore just how important for invertebrates these features are in hedges, David Clements and Keith Alexander (2009) surveyed hedges of different age in the parish of Shapwick in Somerset. They were looking specifically for species indicative either of long historical continuity of large old trees – 'Old Growth Indicators' – or of long-term woodland continuity – 'Ancient Woodland Indicators'. The hedges were searched using a combination of sweeping vegetation with a net, beating foliage with a stick and tray, and leaf-litter sampling with a modified leaf blower, as well as by carefully examining bits of dead wood, fungal bodies and the like. Even though none of the hedges contained ancient trees, to their delight David and Keith found a remarkably good range of saproxylic invertebrates, that is those dependent on dead or decaying wood, including 10 species of fly and beetle that were Old Growth Indicators or Ancient Woodland Indicators. Thirteen Nationally Scarce species were discovered and a further 10 species considered rare in Somerset. Older hedges, of medieval or earlier origin as assessed from historical documents, had more saproxylic species, including indicator species, than recent ones.

ABOVE: Rot holes like those in this ancient, short, hedgerow Ash pollard provide vital habitat for insects that need decaying wood, including many scarce species.

David and Keith concluded that old hedges can be an important habitat for ancient woodland invertebrates and those dependent on old growth, as well as for saproxylic invertebrates in general. Indeed, they noted that networks of old hedges can match substantial areas of ancient semi-natural woodland or wood-pasture in the range of species they support, including good numbers of Nationally Scarce or regionally local species. This reflects the sheer quantity of dead and decaying wood present in hedges – always provided they have not been obsessively cleaned – not just in mature trees but also in old stumps and stools and as fallen branches and twigs lying on the ground. Keith tells me that those trying to re-find the very rare and Critically Endangered Small-spotted Spineleg (*Myolepta potens*) in Somerset should probably have been searching rot holes in veteran trees in hedges rather than in woodland.

Apart from the availability of suitable larval habitat, another reason why hedges are so good for saproxylic insects like beetles and flies is that the flowers on hedgerow shrubs and in herbaceous margins provide a ready supply of essential nectar and pollen for the adults. Such flowers are often lacking in parkland and wood-pasture, and in closed-canopy woods.

In Hedge 2070, out of the 830 species of true fly (Diptera) that were identified, 163 are known to be associated as larvae with decaying wood and associated fungi (Wolton *et al.* 2014a). That's a fifth at least – the true proportion, though, is likely to be considerably higher, since the feeding preferences of many flies are unknown. By comparison, only 85 of the species I found are known to feed on living plants as larvae, that's just a tenth or so. Admittedly, had I done the same analysis for them, the figures for moths and butterflies (Lepidoptera) and for true bugs (Hemiptera) would have been very different, since the larvae of these feed mainly on plants, but far fewer species of these insect orders were found in the hedge than of true flies.

ABOVE: The larvae of this Black-spotted Longhorn Beetle and other longhorn beetles (Cerambycidae) mostly feed within pieces of dead wood, but the adults need nearby sources of nectar like this hedgerow Rowan.

One matter that we know little about at present is the impact on wildlife of the rain of debris that descends to the ground when a hedge is mechanically trimmed by flail. This can form a layer several centimetres thick each time the hedge is cut. On the one hand we might expect it to smother any low-growing plants present, including mosses; on the other hand it will benefit organisms that require decaying organic material – of which there are many, ranging from worms through springtails (Collembola) and a host of

LEFT: Research is needed into the impact of the rain of flail debris on the biodiversity of hedge bottoms.

insects to fungi. It may also serve as a mulch, helping the shrubs and trees to survive periods of drought, as well as recycling nutrients. We need new research to elucidate where the balance of interests lies. My guess is that overall the fall of chippings is beneficial for wildlife. In addition, I suspect that the decay of chippings increases the rate at which soil organic carbon accumulates, and thus is good for the climate.

Concluding remarks

Just two or three decades ago, the wildlife of hedges was largely ignored by all but a handful of keen ecologists and scientists. The focus was all on the likes of ancient woodlands, heaths, reedbeds and upland bogs – the critical role of hedges as the last refuge for a great many species across huge tracts of farmland was overlooked. True, they have been recognised as a priority habitat for nature conservation since the mid-1990s, but somehow few people in positions of influence really seemed to believe this should be so. Thankfully that has now changed, and their high importance for biodiversity is widely recognised by politicians, land managers and the public, at least in Britain if not yet in Ireland. The talk is all about nature's recovery, and the well-placed belief is that hedges have a key role in play in achieving this.

RIGHT: Scorpionflies like this *Panorpa* species are common in hedges. Although they look fearsome, especially the males, they are harmless, being mainly scavengers feeding on dead insects and other carrion.

ABOVE: Bastard Balm is a Nationally Scarce and Vulnerable plant of woodland edge and hedge banks. It is largely restricted to east Cornwall and Devon – elsewhere it is declining. It dislikes dense shade, so favours managed hedges.

The next big attitude change we need to see is people thinking about hedges as not just lines of trees and shrubs, but as wider structures, with margins, banks and ditches, and potentially rich basal floras. Together it's all these parts that make up a hedge, and from a wildlife perspective they do so synergistically. Many species, perhaps most, require two or more of them to prosper. Managing just one like the shrub line while ignoring the others is wasteful of time and resources. It's like trying to keep a building safe and sound by repairing only the walls, ignoring the state of the foundations or roof.

Likewise, to reinforce the point made at the end of the last chapter, we need to think less about the state of individual hedges and more about the state of networks of hedges and their relationship with other habitats and landscape features. Even if there was a single perfect state for a hedge to be in, you could not keep it that way. But a network of hedges can be managed to meet the needs of all species – and finding out how best to do this is a research priority. Altogether, more holistic thinking is needed.

A final thought to close this chapter: dead and decaying wood are now widely valued in woodland and parkland settings, but the same is seldom true for hedges. This we must change.

Wildlife highways

chapter six

Even if your local landscape is well endowed with patches of good wildlife habitat, if these are not physically connected, one to the other, then many species you may expect to find will probably be missing. This is where networks of hedges really come into their own – they provide vital safe passage through otherwise hostile farmland.

Animals exhibit three distinct types of movement: those made during everyday life; those made for dispersal purposes, to find new places in which to settle and reproduce; and seasonal migration. The first type includes commuting back and forth between resting or breeding places and sources of food or water. The second applies not just to animals but to plants and fungi as well, dispersal here being through spores, pollen or seeds. The third often involves seasonal movements over long distances, sometimes between continents. While it is true that the distinction between these three types of movement is often blurred in practice, particularly with invertebrates, treating them as if they were one and the same in the context of landscape connectivity has at times led to misleading conclusions.

A good example of this confusion is provided by the Hazel Dormouse. These nocturnal, arboreal mammals dislike descending to the ground to cross gaps in hedges during their day-to-day movements, presumably for fear of predators. This has led many naturalists to assume that dormice must find it very difficult to colonise isolated habitat patches. Yet research reveals that youngsters dispersing away from the place of birth can voluntarily cross 500m or more of open ground. They even traverse dual carriageways to breed in scrubby central reservations (Chanin and Gubert 2012). So, while a high level of hedge connectivity may be important for day-to-day survival, it is not strictly necessary for dispersal.

OPPOSITE PAGE:
Hedges both connect and divide. Marshwood Vale, Dorset.

ABOVE: Even though some hedges have been removed, this landscape remains well connected by those that remain. The two long narrow strips formed part of a leper colony between the 14th and 17th centuries. Taddiport, near Great Torrington in Devon.

In this chapter I consider the strength of evidence for hedges being used for everyday movement as well as for dispersal. I also discuss the downsides of connectivity – one species' highway is another's roadblock – and think about what an optimal hedge network might look like.

Day-to-day movements

When going about their normal business within farmland, most vertebrates, whether it be Common Toads, Grass Snakes, Common Lizards, Weasels or bats, choose to stick closely to hedges and similar uncropped features rather than venture into open fields. On winter walks it's a frequent delight to see mixed parties of small birds working their way along a hedge while foraging, taking care to avoid crossing large open spaces wherever possible: woodland birds are three times more likely to move along hedges than cross open fields (Bellamy and Hinsley 2005). Just the same behaviour has been found for bats: radio-tracking and echolocation studies show that many preferentially follow hedges and lines of trees during nocturnal farmland forays from their roosts (Feber *et al.* 2019). Common Pipistrelles, Brown Long-eared Bats, Barbastelles and Greater Horseshoe Bats, all species with short-range echolocation abilities, are among those reluctant to fly across open ground. Only those with very strong echolocation powers like Noctules dare to venture into open fields and skies. For others there is a real danger they will get disorientated and lost without strong linear navigational aids like hedges to follow.

Greater Horseshoe Bats, which can live for 30 years and have colonies that remain faithful to their nursery sites for decades, apparently develop lasting mental maps of their surroundings, as they can continue to fly along the lines of hedges that have been removed years ago. Presumably the knowledge is passed from one generation to the next as the young follow their mothers. These bats also obtain most of their insect food within 10m of hedges, so are reliant on them in more ways than one (e.g. Billington 2005).

ABOVE: Greater Horseshoe Bats use hedges as navigation aids as well as hunting grounds for their moth, beetle and fly prey.

The same predilection for moving along hedges holds true for invertebrates: butterflies, moths and hoverflies are well known to follow them, usually for shelter or food, but at times as navigational aids. Louise Cranmer, Duncan McCollin and Jeff Ollerton (2012) at the University of Northamptonshire have confirmed that bumblebees follow hedges closely when moving between flower-rich patches in farmland. To cast light on why the bees do this, they erected lines of black weed-control netting between flower patches. The bees followed these artificial lines just as often as hedges, suggesting that they use hedges mainly as navigational aids rather than for shelter or safety. Bumblebees are strong fliers and capable of taking bee-lines across open fields – but it seems they risk getting lost if they do so.

In another neat experiment, Cranmer, McCollin and Ollerton planted patches of Meadow Clary next to hedges with varying numbers of connections to other hedges. They found that the greater the number of connections, the greater the number of bumblebee and hoverfly visits to the plants, leading in turn to increased pollination and seed set: the degree to which hedges were connected in the landscape had a profound effect on plant reproductive success. Perhaps, the researchers suggest, one straightforward way to conserve isolated populations of plants at risk from inbreeding depression or local reproductive failure is to connect them with hedges or other linear features.

RIGHT: Bumblebees like this queen White-tailed Bumblebee follow hedges closely when moving between patches of flowers in farmland.

Pulled together, all this evidence confirms the widespread belief that hedges are important for day-to-day movements. Counterintuitively, the same cannot be said for dispersal – the evidence here is much weaker.

Dispersal movements

It is truly remarkable how species can colonise remote unconnected fragments of habitats. One of our farm ponds, excavated in the early 1990s and into which I introduced a range of native plants, is now home to some nationally uncommon insects reliant on these plants – even though, as far I know, there are no other populations of either the plants or the insects for many miles around. One is a sawfly, *Tenthredo moniliata*, dependent on Bogbean, among others are two really tiny flies of the Dolichopodidae family, *Thrypticus intercedens* and *T. pollinosus*, both dependent on spike-rushes. That such insects can find or chance upon isolated patches of suitable habitat far away from their place of origin seems incredible, yet they do so. It's highly unlikely they came in with the plants. Easier to grasp is the ability of many fungi, including lichens, and plants like orchids, to colonise remote areas, since they produce vast numbers of tiny spores or seeds which are wind-dispersed.

Perhaps this explains why careful reviews, trawling through hundreds of published papers, have found rather little evidence of hedges facilitating dispersal in either animals or plants (Kirby 1995,

Davies and Pullin 2007). I know of just a few good examples. In Sweden, trees with animal-dispersed seeds, such as Hazel, Hawthorn, Spindle and oaks, are more frequent in landscapes with well-connected hedge networks than in those without (Sarlöv Herlin and Fry 2000): one can imagine Jays, for example, flying along hedges to cache acorns. In the bocage landscape of Brittany, the carabid beetle *Abax parallelepipedus* has been found by Jacques Baudry and Françoise Burel (2019) to use hedges as movement corridors between patches of fragmented woodland, its preferred habitat. In a similar vein, a study looking into the role hedges play in boosting populations of crop pollinators and crop-pest predators found that fields next to hedges have more bumblebees and more money spiders, provided the hedges are continuous without large gaps (Garratt *et al.* 2017).

The Glow-worm is a sedentary beetle, females being unable to fly and the larvae not moving far, even though capable of crawling at a surprising 5 metres an hour. A species of edge habitats, it favours the transition between grassland and scrub or woodland where the larval food, slugs and snails, may be most abundant. Following a study into the species' distribution in Essex, Tim Gardiner (2008) believes hedges and green lanes provide corridors along which females and larvae disperse, and that they are therefore of high importance in conserving genetic diversity in Glow-worm populations.

A fine example of the importance of hedgerow trees for the dispersal of flying insects is provided by some research carried out by Tom Merckx, Eleanor Slade and colleagues in Oxfordshire farmland (Merckx *et al.* 2009, 2012). They investigated the impact of hedgerow

LEFT: Glow-worm: hedges are likely to be important for the dispersal of these beetles.

Pale Shining Brown

This is a rare and highly endangered moth of farmland. Since the 1970s its distribution has contracted to such an extent that it is now found only in small areas of Wiltshire, Oxfordshire, Hertfordshire and Norfolk – there is some concern that it may even be extinct in Britain (Fox *et al.* 2021). The caterpillars have only been found once in the wild, on Broad-leaved Dock, but it probably feeds on various herbaceous plants and possibly on woody species too.

In 2006 an unusually large population was discovered in an area of arable land in west Oxfordshire, and the opportunity was taken to investigate whether the presence of hedgerow trees had any influence on the moth's numbers (Merckx *et al.* 2010b). Through a combination of light trapping and mark and recapture, the presence of hedgerow trees, mainly oaks, was shown to lead to a marked increase in adult numbers. The researchers considered it highly likely that the moths follow hedges when moving through farmland – none were caught in the centre of fields – and suggested that they favour hedges with trees because of the increased shelter they provide. Other explanations are possible, though, such as the trees acting as navigational aids, being good places to find mates, or having warmer air beneath their canopies.

Pale Shining Brown, a rare moth more numerous in farmed landscapes that have hedgerow trees.

trees on larger moths by placing light traps beside hedges: those traps placed near trees caught 38% more species than those next to stretches of hedge lacking trees. The number of individuals caught was also considerably greater, by up to 60%. Delving deeper, they were able to show that this effect was largely due to the physical presence of the trees rather than to them offering food for adults or caterpillars (Merckx *et al.* 2010a). By marking moths and attempting to recapture them they discovered that many species, regardless of how mobile or how reliant on woodland they are, use trees outside woodlands as stepping-stones to cross large expanses of open farmland (Slade *et al.* 2013). Trees lying within hedges are best of all for this.

In the plant world, hedges provide corridors along which herbs that are poor at colonising new ground, such as those that are indicators of ancient woodland, can move. However, they may do so only very slowly. Short Wood, near Oundle in Northamptonshire, is an ancient wood with abundant Dog's Mercury and Bluebell, both indicators of ancient woodland in that county. A few hedges dating from the Enclosure Act period radiate from the wood. When new, these enclosed fields rather than woodland, so it is likely they did not contain any woodland plants at the start. Dog's Mercury is a herb that spreads either by rhizome growth or by ant-dispersed seeds, each seed having a fatty lump attached to it that is so tempting a food source that the ants carry the seed back to their nest, where it germinates. In the early 1970s Pollard, Hooper and Moore (1974) measured how far this plant had spread out from Short Wood along the hedges. They discovered that along one hedge it had taken 200 years to cover about 25 yards (23m) and in another hedge about 300 years to spread 100 yards (91m): a glacially slow rate of just 11cm to, at most, 30cm a year.

BELOW: Dog's Mercury, a plant that can spread out from ancient woodlands along hedges, but only very slowly.

Returning to the same two hedges over 40 years later, in 2017, Keith Kirby and Heather Robertson discovered that the Dog's Mercury was still there and appeared to have spread a little further (Kirby 2020). However, the shrubs, especially Blackthorn, were now tall and wide, casting the hedge bases into deep shade, seemingly to the detriment of the Dog's Mercury, which was reduced to just scattered clumps. Intriguingly, Keith and Heather found an isolated clump of Bluebell some distance along one hedge from the woodland edge. They conjectured that this was the result of Muntjac deer carrying the seed in their coats as they travelled down the distinct animal path in the centre of the hedge.

Marsh Tit

This small, rapidly declining bird offers some convincing evidence for the importance of hedges for animal dispersal. Rigorous research by Richard Broughton and Shelley Hinsley (2015) in Cambridgeshire found that young Marsh Tits have great difficulty dispersing between woods where these are further than 200m apart unless connected by hedges. One colour-ringed young bird was seen attempting three times to cross a gap of 460m between two woods, aborting each time after about 100m, returning to the wood it came from. Broughton and Hinsley concluded that insufficient hedge connectivity between woodlands where Marsh Tits nest is probably the single most important factor responsible for the recent sharp fall in population size.

Marsh Tit – the young birds depend on hedges as safe dispersal routes away from their natal woods.

Trees outside woods, such as those in hedges, may facilitate gene flow across fragmented landscapes, helping trees in small populations overcome problems associated with isolation (Breed *et al.* 2011). Thus scattered individuals have the potential to help some tree species, and the fauna and flora that depend on them, to adapt to climate change. Not so positively, these trees may also facilitate disease transmission from one woodland patch to another.

To summarise, despite the evidence being thin and despite the extraordinary ability of many organisms to cross inhospitable ground, it still seems highly likely that many species do follow hedges when they are dispersing, given the option. We know for sure that at least a few do. If you were a young Hedgehog or Robin looking for a place to establish your own territory, would you follow hedges given the chance, or risk crossing open ground?

Some downsides of connectivity

A note of caution: connectivity is not always a good thing. Hedges can be barriers to the dispersal of species of open ground. For example, they have been found to be a significant obstacle to the dispersal between fields of some carabid beetles such as *Nebria brevicollis*, a species which overwinters in hedges but reproduces in fields (García *et al.* 2000). Even strong fliers such as the Marmalade Hoverfly (*Episyrphus balteatus*) and another hoverfly, the Migrant Aphideater (*Eupeodes corollae*), both migrants capable of crossing the North Sea, have been found to have their movements between fields restricted by high (7–8m) hedge lines of poplar trees in Lincolnshire (Wratten *et al.* 2003). More obviously, Roe Deer and Red Deer clearly find hedges an obstacle, especially when fleeing a dog, creating many a small gap to ease their movement through the landscape.

LEFT: Small gaps, like this deer passage, help to prevent hedges acting as barriers to species of open fields.

For plants too, hedges may be a barrier to dispersal, of pollen, whether carried by the wind or insects, and of wind-blown seeds. A study in the foothills of the French Alps reported that the genetic diversity of Primroses was greater in densely hedged landscapes than in those with few hedges, suggesting that pollen dispersal is limited by hedges (in Burel and Baudry 2012). Another study, this time in Brittany and of aquatic non-biting midges (Chironomidae), exemplifies particularly well the complexities associated with hedge continuity. While densely hedged landscapes limit the dispersal of adults, hedges keep them close to their larval habitats and provide shelter and markers for mating swarms: on balance, the researchers considered them to be beneficial (Delettre and Morvan 2000).

Worse still, if well connected, hedges may facilitate the movement of diseases, pests and predators, to the detriment of their hosts or prey. A study in Brittany illustrates this well. The breeding success of Linnets was assessed before and after more than half the hedges in a 350ha area of farmland were removed (Eybert *et al.* 1995). The predation of eggs and nestlings *decreased* significantly, that is breeding success improved, *after* the hedges were removed. The most likely explanation was that predators – Jays, Carrion Crows and Weasels – were no longer able to search hedges so effectively once the network was fragmented, although this was not proved conclusively.

BELOW: Great Crested Newts move in straight lines between breeding ponds and places where they aestivate or hibernate; they find hedges to be a hindrance in this seasonal migration.

A further example of the downside of hedge continuity comes, surprisingly, from a study of newt movement (Joly *et al.* 2001). Contrary to expectations, populations of Palmate, Alpine and Great Crested Newts in south-eastern France were negatively correlated with the length of hedge in the area. It appears that hedges formed a barrier to annual migration between breeding ponds and woods used for aestivation and hibernation. Newts like to move in straight lines when migrating, following either chemical cues or using a magnetic compass. Putting more hedges into a landscape is likely to hinder their movements, not help.

What does a good hedge network look like?

No hedge should be an island, nor is there such a thing as the perfect hedge for wildlife. Different organisms prefer different sizes of hedge, as we've seen already for nesting birds. But, as I remarked at the end of the last chapter, we could, theoretically, have a perfect *network* of hedges. Such a network would have hedges of all different sizes, from those that have just been planted, laid or coppiced, to those that are lines of mature trees. As a starter for ten, I have suggested elsewhere that a perfect hedge network might have 50–60% of its hedges thick and bushy with occasional standards, 15–20% taller and more open because they are being allowed to grow up prior to laying or coppicing, 5–10% in the early stages of regrowth or just planted, and 5–10% as lines of trees (Staley *et al.* 2023). Thick, bushy hedges are especially good for wildlife, which is why I suggest more hedges should be in this state than in any other. If the intention is that the network should be particularly hospitable for species like Turtle Doves or Greater Horseshoe Bats, or if the aim is to deliver other objectives beyond biodiversity, such as carbon capture, then these proportions will need adjusting.

The perfect hedge network would also be one that is optimal in terms of number of connections and length per unit area, that is density. Here again there's very little research or modelling available for us to say what this optimal state might be. One might think that it's simply a matter of having as many connections and as great a density as possible, but the answer is more complex than that – some farmland species require expanses of open field and others a combination of hedge and open field with plenty of edge. Besides, as we've seen, it's possible that too many connections or too many hedges will increase the incidence of diseases, pathogens, parasites and predators, to the detriment of their hosts or prey.

The Hedgehog is one species that may not gain any further benefit from an increase in hedge density beyond a certain limit. Based on an area of intensive farmland south of Oxford, Moorhouse and colleagues (2014) modelled the impact on this much-loved mammal of field boundaries being either hedged or fenced. Assuming they move only along hedges, a reasonable presumption from what we know about Hedgehogs, when the proportion of hedged field boundaries was doubled from 25% to 50% population connectivity increased substantially – yet tripling it to 75% resulted in nothing better.

A broadly similar conclusion was reached by a team led by Rob Fuller of the British Trust for Ornithology after looking at the

ABOVE: Densely hedged landscapes like this one in Parracombe, Exmoor, with about 10km of hedge per square kilometre, may be optimal for biodiversity.

relationship between hedge density and the number of bird species present in an area. They analysed woodland and hedge bird census data from 33 farmland plots averaging 71ha in size scattered across England and Wales and found that the number of birds increased strongly up to a hedge density of around 8km per square kilometre but declined thereafter (Fuller *et al.* 2001). Another study, this time looking at birds of open fields, suggested that there should be no more than about 9.5km of hedge per square kilometre for farmland to support at least one grassland species (Besnard and Secondi 2014). Switching focus to pollinators, it has been estimated that 13.8km of flowering hedge per square kilometre is sufficient to meet the April larval pollen needs of six common bees, three bumblebees and three solitary bees (Dicks *et al.* 2015). Thus more hedges does not necessarily mean more individuals or species – there's a limit to this relationship.

From this scant evidence base, it seems that roughly 10km of hedge per square kilometre is optimal for biodiversity. To put this in context, there are currently about 4.2km of hedge per square kilometre across Britain, excluding those landscapes where hedges do not feature such as open uplands (Staley *et al.* 2023). This suggests the current length of rural hedge would have to be increased massively, by 138%, to 966,000km, to achieve an optimal density for biodiversity. Far

ABOVE: An ideal place to plant new hedges to improve landscape connectivity and reduce wind-blown soil loss? Lincolnshire Fens.

beyond what we can ever hope to achieve. However, in the Republic of Ireland, the estimated density of hedges in 2015 was 9.8km/km^2 (Green *et al.* 2019), which already represents an optimal density.

Concluding remarks

Connectivity has its upsides and its downsides. Generally, for most species, it appears to be a good thing, especially for the day-to-day movement of species that are true hedge inhabitants. But for dispersal, hedges do not seem so important for many, while for species of the open field or heath they can seriously get in the way. A compromise is called for, where small gaps are tolerated or created, frequent enough to make the landscape permeable to species of open ground while retaining sufficient continuity for those that rely on hedges. Such an approach will run against the grain for many actively involved in hedge conservation today, where gaps, however small, are viewed with antipathy, harking back no doubt to the time when the main role of hedges was to keep sheep or cattle in or out of fields.

The shrub layer

chapter seven

To most, the heart of a hedge is the line of bushes down its centre. It is these that turn a hedge white with the blossom of Blackthorn or Hawthorn, an evocative promise of spring eagerly looked forward to each year. It is here that we find the blackberries to pick on country rambles. And it is upon these bushes that much hedge life depends, in part if not in whole. The utterly charismatic Hazel Dormouse finds them very much to its liking, along with many of our most loved songbirds.

Here I explore the rich variety of wildlife associated with this line of bushes, extol the virtue of having a range of different shrubs and trees, and look into why some hedges have more woody species than others. I describe the ways in which shrub communities can be classified, pay special attention to Hooper's rule for dating hedges, unfairly much maligned, and explain why thick dense hedges are best of all – a divisive matter, since in farmland they are nearly always the product of the flail.

OPPOSITE PAGE: Autumnal plenitude: rich berry crops grace many hedges, as long as they have not been recently cut.

LEFT: Blackberries are great food not only for humans but for much wildlife too, from flies to Hazel Dormice.

The importance of tree and shrub diversity

Structure is an important determinant of biodiversity, but so too is the number of different plants present. Many insects specialise on just one or perhaps a few species, sometimes on just one bit of a particular plant, its leaves or fruits for example. So, the greater the variety of plants in a hedge, the greater the numbers of animals there are likely to be, fungi too. Another reason why plant and animal diversity go hand in hand is that generalists, those with a broader diet, are likely always to be able to find something to eat in a hedge rich in different plants. Pollinators in particular benefit from the longer period over which flowers are available.

Willows are among the very first of our native trees to flower in the year. In March, the male catkins, much loved as pussy willow, appear on the leafless twigs, usually a short time before the less impressive female catkins. They are a hugely important source of pollen and nectar for queen bumblebees freshly emerged from hibernation and for early flying hoverflies like the Tapered Drone Fly (*Eristalis pertinax*), an excellent honeybee mimic, and Common Spotted Bristleside (*Parasyrphus punctulatus*). Even earlier, the introduced Cherry Plum comes into flower, often mistaken for Blackthorn, which usually waits until late March. These two *Prunus* species can usually be told apart easily enough on close inspection, Cherry Plum leaves starting to appear at the same time as the flowers while in Blackthorn they usually wait until after the petals have fallen. As the Blackthorn goes over and

RIGHT: A Blackthorn hedge in full flower.

OPPOSITE PAGE FROM TOP: The male catkins of willows, such as this Grey Willow, are a very important source of pollen and nectar in the early spring.

May blossom, the flowers of the Hawthorn tree.

Crab Apple blossom – although this tree may, like so many others, be a hybrid with the domestic apple.

the month of May approaches, other common hedgerow shrubs and trees start to blossom – Crab Apple, Rowan and Wayfaring-tree among them. Later, in June, it is the turn of Elder, Wild Privet, Guelder-rose, Dogwood and roses. Brambles and Honeysuckle have a longer season, starting in May and not finishing until September. Holly too has a long season, May to August, its inconspicuous flowers attracting many insects including the adults of the eponymous Holly Blue butterfly, the spring generation of caterpillars feeding mainly on Holly, the summer one on Ivy.

May blossom, the flowers of Hawthorn, is intriguing. Alongside roads and other places where it has been planted in recent decades, the tree comes into leaf in March and starts to flower in mid-April, whereas in long-established hedges the leaves only appear in mid-April and it rarely flowers before mid-May, a month later; one might well think there are two separate species involved, so marked is the difference. The reason lies in genetics – the April-flowering trees have been propagated from seeds originating from trees growing in continental Europe, in more southerly, warmer regions. They have evolved to coincide with pollinator emergence there and are out of sync with emergence in Britain – or at least they were before climate change. Even now, in some years, the May blossom on our farm does not appear until the end of May: we wait for this before shearing our sheep, just as in the past countryfolk were warned 'Ne'er cast a clout till May is out' – that is, do not discard warm winter clothing before the Hawthorn is in flower. In earlier calendars, May Day coincided with the flowering of Hawthorn, the celebrations taking place during mid-May as we now know it, not the first day of May.

ABOVE: Ivy flowers are very important for hoverflies and other insects in the autumn when there are no other sources of pollen or nectar around.

Ivy berries (inset) are often not ripe until March or April and are an important food source for many birds, including returning migrants.

Just as willows are so important early in the spring, so Ivy is a vital source of nectar and pollen at another time of the year when supply is very short, the autumn. Such periods of food shortage are referred to as 'hungry gaps' by ecologists. From September to November, Ivy flowerheads swarm with hoverflies, muscid flies, blowflies and wasps. Andrew Cunningham, passionate dipterist, has recorded 69 species of fly feeding on Ivy flowers along a 400m stretch of hedge near his home town of Tiverton in Devon. Likewise its berries, ripening in the late winter or even early spring long after all others have been eaten, are of great importance to birds. John Walters, superb naturalist, tells me that returning migrants like Blackcaps enjoy them, even feeding them to their first brood. The widespread negative attitude towards this creeper is most unfortunate, and I am always sorry to see stems needlessly and harmfully cut or pulled away from hedgerow trees and bushes. The only good reason to remove Ivy is to prevent it smothering special lichen communities (Sanderson and Lamacraft 2022). I think it can rightfully be called a keystone species, so important is it to so much wildlife.

So, a hedge rich in woody species including Ivy can produce blossom from March through to November, and berries from

August to March, supporting a far greater range of wildlife than a single-species hedge ever could. But the advantages for wildlife of a diverse range of shrubs extend beyond blossom and berries. Tim Gardiner (2010), for instance, discovered that the number of species and abundance of earwigs and bush-crickets along hedges bordering 25 green lanes in Essex were related to the number of woody species present – the more trees and shrubs, the more of these insects. Earwigs and bush-crickets typically feed on a range of plants, insects or both, so it is not surprising that they favour hedges with a wide range of woody species. Such hedges are usually old, and indeed the Oak Bush-cricket is probably a good indicator of ancient hedges, just as it can be of ancient woodland. This species, together with the Speckled Bush-cricket and Common Earwig, shows a strong preference for Blackthorn and brambles, reflecting not just the abundance of these plants but also, Gardiner suggests, the way they form dense stands when in full sunlight – conditions favoured by these insects, and many others too.

Further evidence, if any is needed, of the positive link between woody species-richness and the ability of hedges to support a wide range of wildlife comes from farmland bird research. A study conducted on five farms in Oxfordshire found the number of breeding bird species to be correlated with the number of woody species, rising from on average of three birds where there were five or fewer woody species per hedge to an average of six birds where there were more than 10 woody species (MacDonald and Johnson 1995). A later study

LEFT: Oak Bush-crickets may be a good indicator of ancient hedges.

confirmed this relationship: across 22 sites, the number of breeding pairs and the number of breeding species increased broadly in line with the number of woody species per unit length of hedge (Hinsley *et al.* 1999). Nonetheless, while this relationship was statistically significant it was far from tight, indicating that many other factors were at play too.

There's another very good reason to encourage tree and shrub diversity, and that is to increase resilience in the face of new pests, diseases and environmental stresses such as climate change. Ecological theory contends that complex ecosystems should be able to cope better with shocks, and recover more quickly from them, and this has been found to hold true in practice. Pure Ash hedges are experiencing a very tough time at present, since the arrival of ash dieback (see Chapter 8 for more information on this devastating disease). Just imagine if an aggressive form of fireblight (caused by the bacterial pathogen *Erwinia amylovora*) were to get hold of Hawthorn … And will pure Beech hedges survive more heatwaves and droughts?

ABOVE: Hedgerow Ash felled because it had ash dieback. Mendip Hills in Somerset, 2022.

Why do some hedges have more woody species than others?

Precisely which and how many shrubs and trees are present in any given hedge will depend on a range of factors: its geographic location, the local geology and soil, proximity to seed sources like woodland, and what it was originally planted with – if indeed it was planted and is not a woodland relict or self-seeded from the start. Age is among the most important factors determining species-richness: the number of species in a hedge will normally increase over time as new ones colonise by natural means, the basis of Hooper's rule (see page 135). There are, it is true, occasional exceptions where the number of species declines over time, such as when hedges are overtaken by elms, their vigorous suckers outcompeting other trees and shrubs. I have seen many an old hedge dominated by suckering elms to the near-exclusion of everything else.

Occasionally some species are deliberately targeted for removal, such as Elder, a hopeless tree when you want a dense stock-proof hedge, or Barberry, which as described in Chapter 5 is a host for a rust that has been a serious disease of cereals. Elder not only lacks thorns or the dense structure needed for a stock-proof hedge but the tree is near impossible to lay and, worst of all, kills off other trees and shrubs next to it, leading to gaps: the roots are suspected to produce toxins, a process known as allelopathy. It is a plant that likes a lot of nitrogen, much appreciating the rain of droppings beneath a rookery for example, but also growing well in hedges subjected to high levels of fertiliser input. Consequently, many a hedge in intensive arable areas has now become little more than a gappy row of Elder bushes growing above a dense bed of Nettle and Goosegrass. From a wildlife point of view this is not a total disaster, as elderflowers and elderberries are much appreciated by insects and birds respectively. Surprisingly for such an open-grown tree, coming into leaf early as it does, Elder can be good for early-season nesting: mossy crooks in its branches are often used by Chaffinches for their first brood.

ABOVE: Although Elder creates gaps in hedges, its flowers and berries are excellent for wildlife. The insect visiting this flowerhead is a Common Batman Fly (*Myathropa florea*).

To illustrate the impact different soils and exposure can have on hedge tree and shrub communities I will describe the situation in Devon, a county blessed with a remarkably varied geology, mainly lowland but with two uplands and two coasts. To set the scene, the majority of the county's hedges are species-rich, with 66% having five or more woody species in a 30m stretch,

BELOW: Hedges with a lot of Gorse are frequent in exposed coastal and upland situations in western Britain and Ireland.

compared to the British average of 26%. This is not surprising given that three-quarters of the hedges present in Devon today are of medieval origin. The 66% figure rises to 76% if one excludes the Beech hedges of Exmoor and the Blackdowns, planted by landed estates during the 18th and 19th centuries to mark out the boundaries of newly enclosed fields. Otherwise, those single-species hedges that occur in Devon are mainly elm-dominated ones growing in the neutral, fertile, red soils on either side of the River Exe. Pure Hawthorn hedges are very localised in the county. On the wet, heavy, acidic clay lands of the Culm Measures that occupy much of northern Devon and where our farm lies, Grey Willow is a major component of the species mix and Guelder-rose is frequent; it is only in these hedges that Devon Whitebeam and Aspen can be encountered. Calcareous soils derived from underlying limestone and chalk are found along the south-eastern coast leading up to Lyme Regis and around Torbay, Buckfastleigh and Plymouth: these favour quite different mixes that include Wayfaring-tree, Dogwood and Wild Privet, together with Traveller's-joy. Finally, on the high windswept areas of Dartmoor and along exposed stretches of coast where the wind carries with it salt from the sea, the hedges are characterised by plentiful Blackthorn, Gorse and Sycamore.

The same story of hedge diversity can be told for many other counties. Gerry Barnes and Tom Williamson (2006), for instance, describe a wide range of different hedge types in their detailed study of the hedges of Norfolk.

LEFT: Spindle (top) tolerates a wide range of soil types; Guelder-rose (centre) favours acid soils, whereas the closely related Wayfaring-tree (bottom) prefers base-rich ones, as found on limestone and chalk.

Hooper's rule

In a classic study, Max Hooper and colleagues examined the relationship between the number of woody species in a hedge and the age of the hedge (Pollard *et al.* 1974). For their research they selected 227 hedges across Devon, Lincolnshire, Cambridgeshire and Northamptonshire, all of documented age, covering the period between Saxon times and Parliamentary Enclosure, and counted the number of woody species in 30-yard (27m) stretches. They found a clear relationship between age and species-richness. So much so that they felt confident in proposing, as a rule of thumb, that for a 30-yard (effectively 30m) stretch every woody species present represents 100 years of age. Thus a hedge with 10 species may well be a thousand years old or, more precisely, 95% of hedges with this many species are likely to be between 800 and 1,150 years old.

This rule has been widely used as an indicator of hedge age. Hooper was well aware of its limitations and urged caution in its application. He acknowledged that the rule assumes that hedges are planted, so it cannot be used safely with woodland-relict (ghost) hedges. Also, the rule assumes the initial planting was with just one species, typically Hawthorn, while we know from historical records that some hedges were planted with several species and now it's usual practice to do so. As a consequence, Hooper advised that the shrubs be counted in at least a dozen hedges of known age before drawing any conclusions about the likely age of hedges in an area. To do otherwise would, he said, be highly speculative.

Despite Max Hooper urging that the rule should be used with caution and not regarded as an immutable universal law, there have been many who have chosen to ignore his entreaties and to pour unfair criticism on it. Despite these attempts to discredit it, the rule has proved useful in many studies, has become a standard in the teaching of hedge ecology, and has helped to raise people's interest in their local hedges, encouraging them to become involved with their care.

Hooper's rule is based on native woody species and those that are archaeophytes – that is, recorded as naturalised in the wild before AD 1500. A full list of these species can be found in Appendix B. Bramble is not amongst them, nor are any climbers other than roses. Some complex species groups are lumped together – roses, whitebeams, elms and willows. Thus a hedge containing Hawthorn, Dog-rose, Field-rose, Goat Willow, White Willow and Elder would be assessed as containing four species, not six.

In 2007 the average number of woody species per 30m length across Britain was 3.7 (Carey *et al.* 2008). Sadly Max Hooper died not long ago, otherwise I would have been tempted to ask him whether this indicates that the average age of hedges across the country is in the order of 300–400 years. This might just about be right.

Hedges rich in species of shrubs and trees are likely to be hundreds of years old.

What is a species-rich hedge, and how many are there?

When assessing the conservation value of a hedge, it's often helpful to classify it as either species-rich or species-poor in woody plants. But how are these terms defined? The UK Habitat Action Plan for 'Ancient and/or Species-rich Hedgerows' first gave a definition of species-rich hedges, saying they should contain five or more native woody species on average in a 30m length, reduced to four or more in northern England, upland Wales and Scotland (Anon. 1995). Then in 1997, for England and Wales, the *Hedgerows Regulations* came along and these raised the bar, specifying that hedges with at least seven woody species or six for the north of England, again in a 30m stretch, qualified as 'important', and so were worthy of protection. It is, however, the earlier five/four definition that is used within the standard hedge survey methodology developed by Hedgelink (Defra 2007). The tree and shrub species that may be used to determine whether a hedge is species-rich or not are the much the same as those used when applying Hooper's rule to date a hedge (Appendix B).

So how many of our hedges are species-rich? Countryside Survey data suggest that in the lowlands of England and Wales 28% of hedges can be described as such (based on five or more species per 30m), rising to 55% in the uplands (four or more species); in sharp contrast, in the lowlands of Scotland only 11% of hedges are species-rich (four or more). At a British level, just over a quarter (26%) of hedges have five or more woody species per 30m length, with just 5% of those in Scotland reaching this threshold (Barr *et al.* 1999). In Northern Ireland, where species-rich hedges are concentrated in County Fermanagh, about 35% of hedges have six or more species per 30m length (McCann *et al.* 2012). Of course, there's huge variation between places in species-richness, as demonstrated by the wide spread of values reported by local hedge surveys. Appendix A presents such values for surveys in England and Wales sponsored by Defra between 2006 and 2010.

Hedge tree and shrub communities and their extent

Given the huge variety of hedges and the way different types merge into one another, sorting them into useful groups is a challenge. The gold-standard classification of British plant communities is the National Vegetation Classification (NVC), published in five volumes between 1991 and 2000. It represents the culmination of 15 years of research by eminent plant ecologists led by John Rodwell from Lancaster

University, and the first volume covers woodlands and scrub (Rodwell 1991). Unfortunately, although it works very well for most habitats, it is difficult to apply to hedges because they are edge habitats (ecotones), grading from woodland or scrub communities through to grassland and arable ones. Nevertheless it is of some help in an academic context. Tim Rich (personal communication) lists the principal woodland and scrub NVC communities associated with hedges in Britain as:

- W8 Ash–Field Maple–Dog's Mercury woodland
- W9 Ash–Rowan–Dog's Mercury woodland
- W10 Pedunculate Oak–Bracken–Bramble woodland
- W14/W15 Beech–Bramble, and Beech–Wavy Hair-grass, woodlands
- W21 Hawthorn–Ivy scrub, with the Ivy–Nettle subcommunity W21a being by far the commonest hedge type
- W22 Blackthorn–Bramble scrub
- W23 Common Gorse–Bramble scrub
- W24 Bramble–Yorkshire-fog underscrub
- W25 Bracken–Bramble underscrub

A more straightforward and practical classification was developed by French and Cummins (2001) using 1978 and 1990 Countryside

ABOVE: Wonderful Beech hedges in Powler's Piece, Devon, now in the form of lines of trees, almost luminescent with moss on a damp winter's day.

Survey data. They sampled 1,213 plots spread randomly across rural Britain and classified their tree and shrub floras into 11 different classes. The table opposite gives a brief description of each of these, and their relative frequencies are shown in the pie chart. Nearly half of all hedge samples were Hawthorn-dominated, and nearly a quarter had more Blackthorn than any other tree or shrub. The next most frequent class of hedge was mixed-Hazel, about one in every seven hedges falling into this category: such mixed-Hazel hedges are often species-rich, as indeed are those in several other classes including the Blackthorn example. The other eight hedge classes all had frequencies of about 5% or less.

The particular merits of managed, dense hedges for wildlife

On farm walks in the summer, and especially in the autumn after many of the leaves have fallen, I keep a careful eye on our hedges for bird and dormouse nests. In hedges that are kept dense by trimming every few years, 2–3 metres high with bramble or rose thickets in their margins, I find the nests of many species: Song Thrush, Blackbird, Chaffinch, Goldfinch, Bullfinch, Blackcap, Garden Warbler, Long-tailed Tit, Wren and Dunnock. In sharp contrast, in those hedges that have tall leggy stems I see only the nests of Woodpigeon, Magpie and Carrion Crow. Birds that typically nest on the ground or close to it, like Robins, Chiffchaffs, Willow Warblers and Coal Tits, probably breed in both hedge types, although it's difficult to be sure because their nests are much more difficult to spot. Hole-nesting birds like Blue Tits and Great Tits are associated with hedgerow trees or large stumps with suitable cavities. Sadly, we no longer have any Greenfinches and have never during our 30 years on the farm had Linnets, Whitethroats or Yellowhammers, the land being too wet and without arable fields, but all these species too are associated with short dense hedges. As for Hazel Dormice, I find their nests almost exclusively in trimmed, thick, impenetrable hedges; taller ones lack the cover they need.

BELOW: Chaffinch nest in a thorny hedge.

The 11 different types of rural hedge tree and shrub communities identified by French and Cummins (2001)

Class	Name	Description
H1	Planted exotics	Mostly exotic conifers or Garden Privet: unplanted species are very rare in these hedges.
H2	Wild Privet	At least 20% cover of Wild Privet. Hawthorn and Sycamore can be frequent.
H3	Beech	Beech dominant, Hawthorn up to 40% cover, and Box may be present. Other species only incidental.
H4a	Hawthorn-dominant	Hawthorn dominant. Many other species occur incidentally, but only roses, Elder and Blackthorn occur in more than 10% of samples.
H4b	Rich Hawthorn	Similar to H4a but not always dominated by Hawthorn and usually with more woody species present, especially Elder, Field Maple and Ash.
H4c	Elder–Hawthorn	Elder dominant or co-dominant. Hawthorn or other thorny species frequently present but never dominant.
H5a	Willow	Willows generally dominant, roses often subdominant.
H5b	Mixed–Hazel	No clear dominants, although Hazel is usually frequent and some Hawthorn usually present. There tends to be a relatively even mix of the main species present. Only H4a has more incidental species.
H6	Blackthorn	Blackthorn is the most frequent shrub but is not always dominant, while Hawthorn is the main secondary species. Moderate numbers of other species occasionally present at low frequencies.
H7	Elm	Elms dominant or co-dominant. Few other species present.
H8	Gorse	Gorse dominant, although often providing only patchy cover.

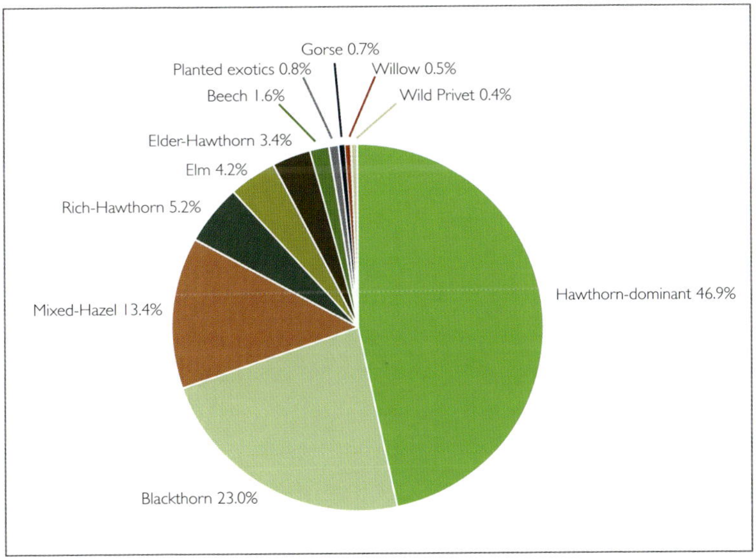

RIGHT: The relative frequencies of the 11 different types of woody plant communities found in hedges, as classified and recorded by French and Cummins (2001).

Hazel Dormouse

To many, this charismatic and charming small mammal is closely associated with large blocks of ancient woodland, but it is really a creature of woodland edge and scrub, both habitats for which hedges are an excellent proxy. To my constant delight, the hedges on our family farm in Devon support a healthy population. Every autumn, after years of practice, I now easily find a dozen or so nests as the leaves drop and their camouflage fails, and when hedge laying I've occasionally come across hibernating individuals. Fortunately, some hibernating dormice at least have a habit of squealing when disturbed yet still in deep torpor, drawing attention to their presence in time to stop you treading on them.

Research carried out by Paul Bright and Don MacPherson (2002) has shown that hedges can provide high-quality habitat for dormice, supporting populations as strong as within any wood, if not better. While individuals living in high-quality woodland typically have ranges of 1 hectare or less, perhaps in the case of females as little as 0.2ha, in hedges ranges may be an order of magnitude smaller, as little as 0.05ha, equivalent to a stretch of hedge 3m wide and 170m long. The resources dormice need are more concentrated in hedges than in woods, reflecting the greater density of woody growth and far richer crops of flowers and berries.

Dormice like hedges with a wide range of different trees, shrubs, climbers and ramblers, catering for their varied diet of insects, flowers, fruits, berries and nuts right the way through the active season that stretches from March/April to October/November. But above all, dormice need safe places to rest – this is, after all, a mammal that is famously sleepy, spending most of its life in slumber, torpor or hibernation. They particularly like hedges that are thick and dense to conceal and protect their breeding and shelter nests, woven balls of grasses, Honeysuckle bark, Bracken, tree leaves and other material, built in tangles of stems. Favourite nesting places are in bramble clumps and outgrowths of suckering Blackthorn, and in the flail line where growth is especially close and intertwined, impenetrable to predators like Grey Squirrels and corvids (Wolton 2009). They will use tree holes for nesting, but these are infrequent or absent in most hedges and by no means necessary. Hibernation nests, the size of tennis balls, are built on the ground, in cold damp depressions in unremarkable places at the hedge base, on bank sides or in margins.

'Hazel Dormouse' is something of a misnomer: Hazel is by no means a prerequisite of good dormouse habitat. Within the range of the dormouse, most Hazel nuts these days are taken by Grey Squirrels, and in any case dormice like them when the shells are still green and soft, which is in August, long before dormice need to

A young Hazel Dormouse. Hedges can be prime habitat for these increasingly rare rodents.

start to accumulate fat reserves for hibernation. Hazel bushes themselves are too open structured to be attractive as nesting places – unless, that is, one attaches an artificial tree hole, in the form of a nest box, to one of the stems. 'Bramble Dormouse' would be a much better name, since brambles not only provide excellent opportunities for safe nesting, but are also rich in insects and produce abundant nutritious blackberries, a favoured food.

Being arboreal mammals, dormice appear reluctant to come down to the ground except for hibernation, presumably out of predator fear, although the strength of this reluctance remains unclear. We do know that they can cross gaps if need be, dispersing young traversing up to 500m of open ground. Even adults are willing to cross 50m of open ground to occupy isolated nest boxes. Early research radio-tracking dormice suggested they are unwilling to cross even narrow gaps, those of 6m or more never being crossed (Bright 1998). However, the dormice involved had been caught elsewhere and released into the hedges shortly before radio-tracking, so found themselves in strange territory; they were unlikely to have been behaving naturally.

Within Britain and Ireland, Hazel Dormice are found only in England and Wales. Sadly, over the last century they have declined in both distribution and number. Although they are thought never to have been present in Scotland, their range once extended far into northern England; now they are restricted to the southern half of the country and Wales. It is of grave concern that this decline continues. Analysis of data derived from the UK's National Dormouse Monitoring Programme strongly suggests that dormice suffered a 72% population crash between 1993 and 2014, equivalent to a mean annual rate of decline of 5.8% (Goodwin *et al.* 2017). Consequently the species is now classified as Vulnerable to extinction within Britain. The reasons for this decline are not well understood; certainly loss and declining quality of habitat, whether hedge, wood or scrub, is one factor, but others must be at play too, such as warmer winters interfering with hibernation and increasing deer numbers removing undergrowth.

Bramble patches are a favoured place for dormice to nest. This is a breeding nest.

ABOVE: The Stripe-legged Robberfly (*Dioctria baumhaueri*) is typically a hedge predator, resting on shrub and bramble leaves in sunny spots while waiting to pounce on its quarry of smaller insects.

While it is often said that more birds favour tall hedges than short ones, this is an oversimplification: many hedge and farmland species don't much care for the open leggy structure characteristic of tall hedges. In the round, the evidence suggests that the optimal shrub height for nesting birds is 2–3 metres: higher than this, nest density levels off and the species count starts to fall. This is fortuitous, since in many farmed landscapes hedges of just this height are the norm, intensive arable units excepting – here the hedges are often kept just a metre high.

The preference that songbirds have for nesting in managed, thick, dense hedges of the type created by regular mechanical trimming reflects the protection such hedges offer against predators. When the fate of 399 songbird nests was followed on 11 farms in eastern England, failure rates were much higher (82.4%) in open-structured hedges than in dense hedges (32.7%) (Dunn *et al.* 2016). If they had the choice, birds nested in places with dense cover above them yet from where they could see out reasonably well, so predators like Carrion Crows and Magpies could be spotted early and prompt evasive action taken if needed. Bramble was used most frequently to support nests, presumably because it is dense and thorny, while higher up Ivy provided important cover, especially early in the season. Roses were avoided, perhaps because some common species like Dog-rose come into leaf late and provide relatively sparse cover.

Hedges with outgrowths of brambles or tangles of roses, or of suckering Blackthorn or elm, are especially favoured by birds. In a study of single-species Hawthorn or elm hedges, Norman Moore, Max Hooper and Brian Davis (1967) found far more birds in such hedges than in other types. Where outgrowths were present, in 1,000 yards (914m) of hedge they recorded an average of 33.9 pairs of 19 species in Hawthorn hedges, and 14.4 pairs of 15 species in elm hedges. In comparison, in dense but closely trimmed hedges with no outgrowths, there were, in Hawthorn and elm hedges respectively, just 9.3 pairs of seven species, and 5.7 pairs of five species. So the presence of outgrowths increased both the number of birds and the number of species by about three times. Moore, Hooper and Davis attributed the preference shown by birds for Hawthorn over elms to the denser cover it provides, as well as to it coming into leaf earlier in the year, generally two weeks earlier, and supporting a richer and more abundant insect fauna.

Concluding remarks

Hedges sporting a range of shrubs and trees clearly have the capacity to support much more wildlife than those with few such species. They are also more likely to be able to cope with the storm of new pests and diseases arriving in Britain and Ireland, the seemingly unstoppable juggernaut of ever-increasing agricultural intensification, and the forces of climate change. The days of planting single-species hedges must surely be over. No longer can the maintenance of local landscape character hold sway – we must look to the future, not the past.

It is also clear that while a diversity of different hedge structures at a farm level is highly desirable, those with a dense shrub layer are of particular importance for wildlife. This means regular cutting or trimming – and, in an agricultural context, this entails, like it or not, the use of flail cutters. I shall return to this, along with hedge planting, in Chapter 15.

LEFT: A Blackthorn branch laden with sloes.

Hedgerow trees

chapter eight

Mature trees scattered here and there along the warp and weft of hedges generate landscapes that are a joy to the eye. They make even the most monotonous countryside, vast tracts of intensive arable land, fields separated by no more than thin lines of Hawthorn, so much more appealing. The trees, free from the intense competition for light typically experienced by those in woodlands, develop into fine specimens with large spreading crowns that flower, seed and fruit prolifically. Some become ancients, many develop veteran features like hollow trunks and rot holes, others are pollards. If neither too many nor too few, they create a parkland effect which is quintessentially British or Irish and of great value for wildlife.

Although most trees in Britain are in woods or forests, it is surprising how many are not. 'Trees outside woods', as they are termed, are abundant. According to the National Forestry Inventory, a sample survey carried out by the Forestry Commission over a period spanning a few years either side of 2016, these trees make up 30% of all native broadleaved tree cover in Britain (Brewer *et al.* 2019). This surprising fact bears repeating – nearly a third of all our native broadleaved tree cover is down to trees that don't grow in woods or forests. Even when conifer plantations are included in the figures, 20% of total tree cover is provided by trees outside woods. In rural areas, as we shall see, about half of these non-woodland trees lie within hedges, but a much lower proportion in urban situations.

Some say there are more hedgerow trees now than there have ever been – yet, paradoxically, others say there are far fewer than in the past. This contradiction is the result of the first set of people talking about all hedgerow trees, including those forming uninterrupted lines, while the second set are referring only to standard trees, those that are largely free-standing with well-developed canopies. You can

OPPOSITE PAGE:
Trees outside woodland, like this magnificent hedgerow Ash, still healthy in 2022, provide an astonishing 30% of broadleaved cover in Britain.

ABOVE: Idyllic lowland Britain, with mature standard trees scattered along hedges. Shute, East Devon.

see why it is necessary to make a clear distinction between the two types of hedgerow tree.

My focus here is on standard trees, variously called stand-alone trees, lone trees or emergents. I prefer the term 'standard', mainly because the crowns of the trees are often not entirely free-standing but overlap with those of neighbours. The key point is that they have wide spreading canopies with far-reaching branches. Nearly always these trees have been deliberately selected to be allowed to grow to maturity. On the other hand, trees in line-of-trees hedges are usually the result of neglect, reflecting a lack of hedge management over a couple of decades or more; having to compete fiercely for light, such trees tend to be leggy, with narrow trunks and small crowns. In practice, I appreciate it's not always easy to tell whether a tree is a standard or one of a line of trees: hedgerow trees often grow in twos or threes, sometimes tall ones abut short ones, and so on. One year a tree's crown may be free, the next, after a little bit of growth, it may touch its neighbour's – in years to come they will distort each other's form. Still, I find it is normally feasible to decide whether a tree is a standard or one of a line. Given the difference in their origins, and in their landscape and wildlife values, it's an important distinction to try and make.

In this chapter I cover the rise and fall of hedgerow trees, the reasons why standard trees are in trouble, and the numbers that remain. I follow this by contemplating whether there is an optimal number of standards and what this might be, and by reviewing the best means of recruiting new ones. I finish by explaining why standard trees are

of such great importance for wildlife. The roles hedgerow trees play in providing shelter and shade, in facilitating the movement of insects and other animals through farmed landscapes, and in helping to reduce climate change, are covered elsewhere in the book.

ABOVE: A fine standard hedgerow Pedunculate Oak.

The rise of line-of-trees hedges

Old postcards, photographs and paintings of rural landscapes rarely in my experience show lines of trees with touching canopies other than along river and stream banks: the same holds true for those wonderful old maps where every mature tree is meticulously shown. In those utilitarian days hedges were far too valuable to be allowed to become unkempt lines of trees. They were kept short, thick and dense as stock-proof barriers, and carefully managed to provide small-diameter firewood. Only well-spaced oaks, elms, Crab Apples and other trees were allowed to grow old, for timber, fruit or fodder, or sometimes to serve as important boundary markers. How different matters are today!

Oliver Rackham (1986) suggests that the process of change in hedges from a short, managed state into lines of trees may have started as early as 1870, noting that the period between then and 1951 was, on the whole, one of agricultural depression where there was less money to spend on maintaining (or destroying) hedges. Neglect, Rackham observes, gave innumerable saplings the opportunity to grow into trees.

ABOVE: About a third of all the hedges in Britain are now line-of-trees hedges like these near Hatherleigh, West Devon.

Certainly the conversion of managed hedges into line-of-trees hedges has continued apace in recent decades. Countryside Survey, the comprehensive countryside monitoring programme described in Chapter 4, tells us what happened in Britain between 1984 and 2007 (Carey *et al.* 2008). During this period, while the total length of hedge scarcely changed, the length of *managed hedge* in Great Britain fell from 624,000km to 477,000km, a 24% decrease, while the length of *lines of trees* rose from 90,000km to 228,000km, a 153% increase. Despite the greater prosperity of post-war agriculture, it seems that the rate at which hedges have been allowed to develop into lines of trees has only accelerated. By 2007 line-of-trees hedges comprised no less than a third of all hedges.

How many standard trees do we have?

While the number of trees in line-of-trees hedges has increased greatly, the number of standard trees has fallen, and substantially so. But it's very difficult to say how many standards we still have because different surveys give very different results.

In 2007, Countryside Survey estimated there to be 1.6 million standard trees in hedges across rural Britain, a figure that included even small ones down to 3cm in diameter, with England having

1,427,000, Scotland 30,000 and Wales 144,000. Only young trees clearly selected to become standards were counted, not saplings or stems hidden within the main hedge mass.

On the other hand, the National Forest Inventory (NFI), from data collected around 2016, estimated the number of *all* 'lone trees' taller than 3m in rural Britain to be 18.8 million. No breakdown of the numbers of trees according to whether they are in hedges or elsewhere is given in the published report, although the figures for canopy areas are split in this way. These suggest that about half the lone trees in rural land are in hedges, the other half being associated with fences and other open land. Thus, in rural Britain, the NFI data suggest there are about 9.4 million standard trees in hedges, a figure six times higher than that given by Countryside Survey. This considerable disparity presumably reflects differences in definitions, in sampling frequency and in methods; whatever the reasons, it is unhelpful.

The NFI found urban areas typically to have far higher densities of lone trees than rural ones – indeed, the number of lone trees estimated to be in urban areas, 11.3 million, was not far short of that found in rural ones, 18.8 million, a result that held true for all three British countries. However, the proportion of trees in urban areas that are in hedges is low in comparison to rural areas.

A further source of information on recent numbers of hedgerow trees comes from 33 local hedge surveys in England between 2006 and 2010 – see Appendix A. All bar one of these surveys were either full censuses or based on random samples so may be considered to present a more or less true picture, although coverage across the country was uneven, with a bias towards the south-west. Together these 32 surveys covered 6% of England and provided data on 5,767 hedges with a total length of 1,091km. A total of 10,483 standard trees were recorded – that's one for every 104m of hedge. Extrapolating this to the whole country and using the Countryside Survey 2007 estimated total hedge length of 547,000km, this suggests a total of about 5 million hedgerow trees in England, not far off the National Forest Inventory (6 million) estimate.

I hope we will not have to wait too much longer for the results of a new rigorous survey to settle the question 'Just how many standard trees do we have in Great Britain in the early 21st century?' Is the figure closer to 2 million or 9 million? Likewise, how many standards do we have in Ireland? Such surveys may either be based on people

ABOVE: Usk Valley, Powys – a landscape rich in hedgerow trees.

walking the ground, as with Countryside Survey, or on improved remote sensing technology with ample ground truthing: either way, they will hopefully first seek and gain consensus on what constitutes a hedgerow tree.

How many standards have we lost?

Oliver Rackham (1986) considered the period 1500–1750 to have been the heyday of hedgerow trees in Britain, reflecting the high value placed on them at that time as a source of timber and firewood. Subsequently, he believed that standard trees probably rose in number with the Enclosure Acts (mainly 1750–1850) as more hedges were planted. However, since then all the evidence points to numbers having fallen sharply.

Rackham counted the hedgerow and field trees shown on first-edition OS maps within 550 plots of 50 acres (20ha) in England, no mean task. From this he calculated that between 1845 and 1888, the period over which the maps were made, there were 23 million individual (i.e. standard) trees in the country, just under one per acre (2.4 per ha) of farmland. Assuming that the great majority of these trees were in hedges, this is a very much greater figure than suggested either by Countryside Survey (1.4 million) or the National Forest Inventory (6 million) to be present in the early 21st century.

Have we four times fewer trees now than 160 years ago, or perhaps even fifteen times fewer?

In recent decades it is very difficult to say just how many standard trees have been lost, because different surveys record different sizes of tree, some record just standard trees, others all trees outside woodlands, and so on. The most reliable data come from Countryside Survey, which reveals a statistically significant 3.9% loss of 'lone' hedgerow trees across Britain between 1998 and 2007 – that's a loss of 65,000 trees in just nine years. In England, the loss was 4.5% (62,000 trees); the changes for Scotland and Wales were not statistically significant (–3.1% and +2.3% respectively). As many as 10% of larger trees, those with a diameter at breast height of 20cm or more, disappeared over this short period.

The Woodland Trust has looked in detail at changes in the numbers of trees outside woodland in the Eastern Claylands of Essex and Suffolk (Reid *et al.* 2021). This is 5,000km^2 of farmland within which trees outside woods are particularly important to the area's character – it is the romantic landscape captured by John Constable in his paintings. In 2021 volunteers digitised the position of over 100,000 trees marked on first-edition (*c.* 1850) Ordnance Survey maps. Recent aerial photos were then examined to see how many of these trees survive to this day or have been replaced by a tree in the same location. Over half, 54%, of trees on boundaries, the majority of which must surely have been in hedges, were lost, the number falling from 0.86 million in 1850 to 0.40 million in 2021. Unfortunately, standard trees were not distinguished from those in rows, nor were the Trust's volunteers asked to count the number of trees established since the 1850s in new locations. So, while the study tells us that over half the boundary trees present in the Eastern Claylands in the 1850s are no longer with us today, or at least have not been succeeded by new trees on the same spot, it does not unfortunately tell us anything about overall changes in numbers, whether standards or otherwise. Given that probably a quarter to a fifth of all the hedges will have been grubbed out since about 1850, it is perhaps surprising that only half the trees growing in the area 170 years ago have been lost and not replaced, not many more.

Of course, when a hedge is grubbed out not all its mature trees may be felled; a few are sometimes retained, perhaps because they are too big to remove easily, perhaps because they are fine specimens, or maybe because they are valued for the shade they cast on a hot

ABOVE: This fine Ash in Perthshire is, like most lone trees in the middle of fields, almost certainly a relic of a former hedge, now grubbed out.

summer's day. The result is that most lone in-field trees one sees outside parkland are remnants of former hedges. How else did they get there? Farmers have long been loath to plant trees in the middle of fields, for practical reasons, although the current rise of interest in agroforestry may be changing that mindset.

The reasons for standard tree loss

After the Second World War, hedgerow tree numbers started to fall back sharply as hedges were removed in large numbers, but there were other major factors at work too. In the late 1960s and 1970s a fresh outbreak of Dutch elm disease had a devastating effect, resulting in the loss of millions of mature elms, such beautiful trees. It is difficult to comprehend now just how many elms we used to have. This is brought home by a sentence in Pollard, Hooper and Moore's 1974 book, where they write 'The trees most associated with hedges are elms. It is possible that there are as many or even more oaks in hedges …' A 1971 Forestry Commission survey across much of southern England estimated the population of rural elms (all three main species) outside woods to number 8.7 million. Presumably, most of these were in hedges. Nowadays it is a rare sight indeed to see a mature hedgerow elm, whereas oaks remain common.

But above all, the demise of standards reflects the loss of value, real or perceived, of hedgerow trees to farm businesses – they have no worth any more for timber, firewood, fruit or fodder. To the contrary, they are expensive to maintain, they reduce crop yields by casting shade and drawing away nutrients and water, and their limbs and branches have an annoying and sometimes dangerous habit of dropping onto productive land, roads or public footpaths. Many have therefore been felled for agricultural or safety reasons, and there's been no reason to replace those that die. Even if not felled, the demise of many has been hastened by agricultural intensification – through ploughing their roots, through land drainage causing drought stress, and doubtless through fertilisers harming mycorrhizal fungi. To make matters worse, the advent of the flail trimmer means that saplings are destroyed annually *en masse*, while Grey Squirrels stunt and deform the growth of many young trees by stripping their bark – although less so than in plantations. Urban expansion has not helped either.

Currently, we are experiencing a dearth of young standard trees – there are just not enough around to replace those that die. Based on tree mortality and size data available through Countryside Survey, scientists at Forest Research (2009) have been able to model the population dynamics of standard hedgerow trees. For a population of hedgerow trees to be stable through time, 45% need to be young, that is to have a diameter at breast height (DBH) of 20cm or less. However,

LEFT: What a loss! Mature elms in hedges, once a common sight across southern England, are now very rare. The brush-like growth on the trunks suggests that the side branches have been repeatedly lopped, probably for tree hay. Hedgerow elms in Devon photographed by Charles E. Brown, taken before 1939.

ABOVE: Fine Ashes, as yet, in 2022, unaffected by ash dieback disease, over a Holly shrub line kept dense by regular trimming.

in 2007, across Britain only 19% of trees were in this size range, less than half the number required: population decline is inevitable. The Forest Research model predicted that 30,000 new trees need to be recruited annually to maintain the 1.6-million-strong British population estimated by Countryside Survey, whereas just 15,000 were actually gained each year between 1998 and 2007.

And now, to add to all these pressures, we have ash dieback, a lethal disease caused by the fungus *Hymenoscyphus fraxineus*, first

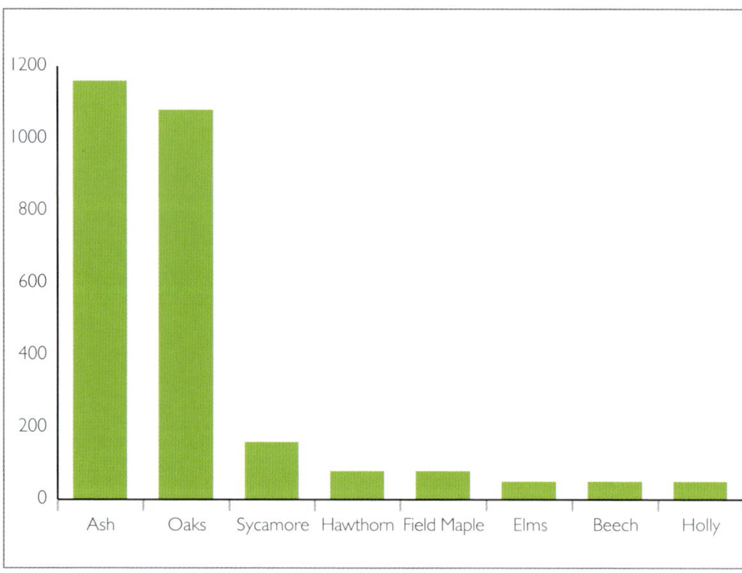

RIGHT: The numbers of mature standard trees of each common species in hedges across Great Britain, as found in 2007 Countryside Survey sample plots. Adapted from Maskell *et al.* 2013.

detected in both Great Britain and Ireland in 2012. At a British level, Ash is our commonest standard hedgerow tree, followed closely by oaks: now that the elms have gone, no other trees come anywhere near these two in number – except in Scotland, where Beech is the second commonest tree after Ash (Barr *et al.* 1999, Maskell *et al.* 2013). A Forestry Commission inventory suggests that in 1998 there were 5.4 million Ash trees in narrow linear features such as hedges across England (Smith and Gilbert 2001). Of course, many of these will not have been standards. By comparison, the same inventory estimated there to be 4.1 million willows (a surprisingly high number), 3.9 million oaks, 3.5 million elms (presumably very nearly all of small stature), 2.4 million Sycamores, 1.8 million Beeches and 1.5 million birches.

Ash dieback is anticipated to kill 90% or more of Ashes: it remains to be seen how many are sufficiently resistant to survive and produce seed. As I write, in the early 2020s, trees throughout Britain and Ireland are infected and a great many are dying if not already dead. Large numbers are being felled alongside roads and in other places where they are hazardous. In Devon, my home county, by 2021 it was unusual to see a healthy mature tree with a full crown. In places such as parts of the East Midlands, where Ash trees are just about the only hedgerow standards, the disease is having a devastating impact on the landscape and no doubt on much wildlife. At least Dutch elm disease rarely kills its host outright, the roots usually surviving to send up numerous suckers, and individuals can survive multiple bouts of the disease. Ash trees, if the main stem succumbs, die outright. To maintain the overall population of standard trees, we now need to recruit far more than the 30,000 new trees a year that Forest Research estimated to be necessary back in 2009.

BELOW: Hedgerow Beech badly affected by drought in 2022. The following spring the tree flushed green with new leaves, but it may have been fatally weakened.

Other tree pests and diseases are waiting in the wings, such as acute oak decline. If our two oaks, as well as elms and Ash, succumb to some pandemic, the impact will be terrible. Our three most numerous hedgerow trees will have largely gone, at least as mature specimens. Climate change is creating further stress, particularly through heatwaves and droughts. Beech, a prominent hedgerow tree in many places, is particularly susceptible to drought, and some land managers are already avoiding planting it in new hedges for this reason. Following the record-breaking heatwave and associated drought in 2022, the leaves of many hedgerow Beeches turned crisp and brown, the trees appearing dead by the end of the summer – as I write, it remains to be seen how many survive.

How many standards do we want?

A mature oak tree with a full canopy could easily cover 15m of hedge, and the same is true for Beech, Sycamore and Ash. That means that a 100-metre stretch of hedge could only host six or seven such trees if their branches were not to touch. Of course, you could fit in quite a few smaller trees like Rowan, Crab Apple, Hawthorn or Holly, and in any event it would be sensible to start with many young trees to allow for losses. The advice usually offered by ecologists is that one should aim for one mature large tree every 20–50m, or perhaps one small one every 10m. Any more than this and you might end up with a continuous line of trees that casts such dense shade on the shrubs beneath that they dwindle away and the precious undergrowth is lost. Only a few shade-tolerant species like Holly and Beech would survive.

That there may an optimal number of hedgerow trees per given length is suggested by a detailed study of the way songbirds use hedges, carried out on 46 farms in lowland England (Green *et al.* 1994). The abundance of Wren, Robin, Blue Tit and Chaffinch tended to level off, and of Great Tit, Greenfinch and Tree Sparrow to decline, when there were more than one or two trees per 30m stretch. The more trees the better for Blackcap, while the fewer the better for Whitethroat and Linnet. As ever with hedges, there's no one perfect state – variety across the farm and local landscape is the key.

Assuming one standard tree per 40m is optimal, then given that there are some 477,000km of managed hedge in Britain, that means we should ideally have about 12 million such trees. If the Countryside Survey data are right and we currently have just 1.6 million, we need a further 10.4 million, a six- to sevenfold increase.

How to get more hedgerow trees

ABOVE: For large trees like these oaks, one standard every 20–50 metres is about right. Any more, and the shrub line beneath risks being shaded out.

The answer to this question may seem obvious: plant them! But there's a better, simpler way. Back in 1986, in his *History of the Countryside*, Oliver Rackham wrote, in the context of hedgerow trees:

> *By 1970 … the planting mentality had become established and prevailed over the conservation of existing trees. Official trumpets were blown for the planting of tens of thousands of trees, but no voice was raised against the destruction of hundreds of thousands of natural saplings.*

This remains true to this day: the same mentality persists in many quarters. Remarkably little recognition is given to the scope for recruiting millions of new hedgerow trees, with all the benefits they bring to society and wildlife, without planting a single one.

To increase the number of hedgerow trees, we need only to protect the growth that's already there. Hidden in most hedges one can find young oaks, Beeches, Crab Apples, Rowans, Hollies and so forth. They just need to be given a chance – to be liberated from the flail. In practice this means finding and marking promising saplings and stems so that those operating hedge trimmers can avoid them, allowing them to grow towards maturity. Otherwise, they can be selected when a hedge is laid or coppiced, or perhaps when a line-of-trees hedge is being thinned. Bob Harvey, a passionate advocate of this approach of finding and leaving existing stems and saplings to

ABOVE: I deliberately left this Rowan uncut when the hedge was laid a few years before the photo was taken, so it could develop into a tree.

grow, estimates that a substantial part of our national tree-planting targets could be met in this way. Even if you only have a single-species Hawthorn hedge, then letting one thorn grow to maturity every 40 metres or so will do the job.

As I have found out, trying to plant a tree in an already established hedge is difficult. You have to find or create a suitable gap to stand a reasonable chance of success. Even then, competition for nutrients, water and light from existing shrubs and trees means you may well have to feed and water the tree, and prune back encroaching growth for several years before it can survive unaided. I know of one Devon farmer, Glen Bealey, a hedgerow-tree enthusiast, who uses an excavator to make gaps in the top of hedge banks, each a metre of two wide, importing soil from the adjacent field in which to plant. The new tree is then protected within a triangle or square of stakes, each brightly tagged, to ensure whoever is trimming his hedges does not inadvertently chop the tree's top off. It works, but it's a lot of effort and probably only worth it if you want to enrich your hedge with a species not already present in the immediate landscape.

Part of the mindset behind planting is the desire to achieve the perfect specimen tree, for the chances are that at any promising sapling or stem you find in a hedge will have some imperfection. Perhaps it has been cut before by the flail and has a kink. But does this really matter? The tree will still grow, and the odds are that it will become a decent tree with the potential to reach the species' expected lifespan. Many of the young oaks in our farm hedges, selected when the hedges were laid 30 or so years ago, have developed from damaged stems but you would not know it now: they appear strong and straight. We are not after fine timber trees with long straight trunks – if we were, then we would be growing them densely in plantations. Instead our aspiration is that those who follow us should be able to enjoy trees with wide spreading canopies, full of character and wildlife.

ABOVE: Holly makes an excellent hedgerow standard, as do many other small native trees.

BELOW: Windswept Beech trees near the north Devon coast.

The wildlife of hedgerow trees

Mature, open-grown trees enhance any hedge immensely, providing a wealth of additional ecological niches and attracting an abundance of wildlife. They are important for fungi and insects of dead and decaying wood, as covered in Chapter 5, while in Chapter 6 I described how the rare Pale Shining Brown uses hedgerows with trees as corridors when crossing agricultural landscapes. Other moths that do the same are Bright-line Brown-eye, Heart and Dart, Flame Shoulder and Flame: these are all widespread species that in open farmland are more abundant at hedgerow trees, oak in particular, than elsewhere (Merckx *et al.* 2010a).

In the dark, often cramped conditions of closed-canopy woodland, most of our native trees flower and fruit poorly, and some can barely survive, for example elms, Large-leaved Lime, Buckthorn and many fruit trees. When did you last see a Crab Apple or even a Hawthorn in full blossom beneath a canopy of mature trees? Hedges, on the other hand, provided excellent conditions for flowering and the

BELOW: Hedgerow trees are keystone structures for wildlife in intensive agricultural landscapes.

production of large crops of seeds and berries. Hedgerow trees and shrubs have the added attraction of coming into flower earlier than those in woodland, to the benefit of flower-visiting insects and birds like Bullfinches. As I write this, in mid-winter, our hedgerow oaks are attracting great flocks of Woodpigeons from up country to feed on acorns, while Redwings, Fieldfares and Blackbirds, having polished off the haws and sloes, are relishing fallen crab apples.

Lichens

Trees outside woodlands, including those of wayside and hedge, are often more important for lichens than trees inside woodlands, because more light reaches their trunks and lower branches (Coppins 2001). Indeed, hedges can be important refugia for lichens of ancient trees, as confirmed by expert lichenologist Barbara Benfield (2021). Prompted by COVID-19 lockdowns in 2020/21, she searched hedgerow oaks along lanes near her home in east Devon, focusing on *Cresponea premnea*, a lichen that grows on the bark of old trees. Although it is usually thought of as a species of ancient woodland, Barbara found it on the trunks of many hedgerow trees judged to be 200 years old or more. Another lichen that can be found on hedgerow trees is the spectacular and well-named String-of-sausage Lichen (*Usnea articulata*). This beard-lichen hangs from twigs, often high in the canopy. It is a nationally declining species graded Near Threatened with extinction.

LEFT: *Usnea articulata*, the aptly named String-of-sausage Lichen.

Orange-fruited Elm-lichen

This lichen was formerly widespread across lowland Britain, but is now found only occasionally, mostly in eastern England and Scotland. Its favoured habitat, as its name suggests, is the bark of mature elms, in particular trees growing in hedges and other well-lit places outside woodlands. Its population declined rapidly after the outbreak of Dutch elm disease in the late 1960s and 1970s, and the species is now rated Endangered. Fortunately, the lichen can also live on Sycamore, Field Maple, Ash and Horse Chestnut so, with a fair wind, it should hang on. However, faced as it is with ash dieback, and with ammonia and nitrogen oxide pollution generated by traffic and agriculture, this is by no means certain. Ray Woods tells me that sometimes it can be found at the base of trees where the bark has been knocked off by passing heavy machinery, livestock or the like.

Orange-fruited Elm-lichen (*Caloplaca luteoalba*).

Hairstreaks

BELOW: Female Purple Hairstreak. This butterfly can often be seen flying high in the canopy of hedgerow oaks.

Among butterflies, hairstreaks are well known to frequent hedgerow trees. I have often seen Purple Hairstreaks flying high around the crowns of large oaks, where they feed on honeydew and the males compete for females, and I once watched a female egg-laying on the twigs of a large hedgerow oak whose boughs swept conveniently down to head height. Hedgerow trees can also support colonies of the much more local White-letter Hairstreak. This butterfly is dependent on elms as the sole foodplant for its caterpillar and has declined greatly because of Dutch elm disease. It shows a strong preference for mature, flowering elms; fortunately it favours Wych Elm, less affected by the disease than other species, and colonies can persist on small trees and suckers. It is, however, a third species, the elusive Brown Hairstreak, that is most closely linked to hedges.

Brown Hairstreak

Brown Hairstreak nectaring on Common Fleabane.

This attractive butterfly has a close association with hedges. The caterpillars eat the leaves of low-growing Blackthorn twigs, while the adults feed and mate around the tops of tall trees. The same trees, sometimes called master trees, are used year after year. Ash is the species most often selected, although the butterflies can use others – fortunately so given the arrival of ash dieback disease. The males congregate around the tops of the master trees, spending almost their entire adult lives there waiting for virgin females to arrive, feeding on honeydew and only descending to the ground occasionally, to collect nectar from flowers when honeydew is scarce. On the other hand, once mated, the females descend to sup from low-growing flowers like Common Fleabane, brambles and thistles, in between bouts of egg-laying. Eggs are usually laid on low-growing Blackthorn at the point where one-year-old wood branches from a two-year-old stem. Consequently, the eggs are very vulnerable to hedge trimming: cutting hedges hard back every year is believed to be the main driver behind the steep decline experienced by the butterfly in recent decades.

Trees as insect beacons

Further evidence that standard hedgerow trees play a very important part in the insect life of farmland comes from a study into their influence on the distribution of true flies (Diptera) (Peng *et al.* 1992). A tower was built next to an 11m-tall lime tree and suction traps suspended from it for two weeks. An impressive 80,000 flies were caught and identified to family level. Members of all but one family, the Scatopsidae, were significantly more abundant on the side of the tower next to the tree than on the opposite side, 4.5m away. Most flies were caught at mid-level, 5m high, the exception being window-gnats (Anisopodidae), which preferred the top of the tree, and fungus gnats (Mycetophilidae), which preferred the bottom. For the vast majority, there was no direct link to the tree itself as a source of food for either larvae or adults. Rather, the abundance of flies seemed to be a response, above all, to the physical presence of the tree – it's as if hedgerow trees are beacons to which insects are inexorably drawn.

RIGHT: Hedgerow trees appear to act as beacons for flying insects, in turn becoming a focus of activity for bats and insectivorous birds.

This attraction may be explained in part by the great value trees have as stepping-stones, enabling insects to move across inhospitable farmland that would otherwise be unpassable, as has been demonstrated for moths. Trees probably also act as swarm markers, attracting crowds of insects intent on finding partners with which to mate. Perhaps too the higher parts of trees are less contaminated than the shrub layer by pesticides applied to nearby fields.

Another plausible idea is that the trees are used by some insects for tree-topping. This is behaviour analogous to hill-topping, where insects with small, thinly dispersed populations pursue a strategy of going to the highest point in the landscape to find mates. It is a well-known phenomenon with tropical butterflies. In pursuit of interesting insects I have climbed up to Dartmoor tors on still summer days and been rewarded by seeing Wall butterflies, queen wasps in balls of eager males, the Yellow-barred Pond Fly (*Sericomyia silentis*) and, best of all from my point of view, Horse Bot Flies (*Gasterophilus intestinalis*), large furry bumblebee-like insects that are now scarce. You would not normally expect to see any of these species in such high exposed places; they are driven there only by the need to find a mate. Tree-topping is less well known than hill-topping because it is difficult to observe insects at the top of a tall tree, but it is suspected to occur, for example in some butterflies, flesh flies (Sarcophagidae) and parasitic flies (Tachinidae). Tall, isolated hedgerow trees would seem well suited to this purpose. We need further research into exactly why hedgerow trees are so attractive to insects.

Birds and bats

The presence of trees is one of the most important factors influencing hedgerow birds (Hinsley and Bellamy 2019). As well as being a rich source of seeds, fruits and invertebrates, trees provide song posts for birds like Yellowhammers and Chaffinches, cavities for hole-nesting birds like Kestrel, Jackdaw, Stock Dove, owls, woodpeckers and tits, and nest sites for Buzzards, Magpies and Carrion Crows. On the downside, if the trees shade out the shrub layer beneath, birds that favour nesting in thick undergrowth, of which there are many, will be lost. Their presence may also render nearby land unsuitable for birds of wide, open spaces like Skylarks, and waders such as Lapwing, Snipe, Redshank and Curlew.

Bats too make extensive use of hedgerow trees. Soprano Pipistrelles, but not Common Pipistrelles, forage more often along hedges with trees than those without, and it is likely that the same holds true for other species associated with structurally complex habitats such as Brown Long-eared Bat and Natterer's Bat (Boughey *et al.* 2011).

BELOW LEFT: Little Owls nest in cavities in hedgerow trees.

BELOW: The rare Bechstein's Bat is among the many species of British and Irish bats that use holes, splits and cavities in hedgerow trees for roosting.

Old woodpecker holes, splits and cavities can be used as roosts or even nursery sites. The rare Bechstein's Bat, although usually occupying old woodpecker holes in veteran trees in woodland, is among the species recorded as using hedgerow trees for roosting. One evening while at university, by chance when setting out moth traps, I found a Noctule roost in a hole high in a hedgerow oak. Greater Horseshoe Bats perch on the branches of trees to devour their insect prey caught whilst in flight along hedges.

Concluding remarks

In the past, just a century or so ago, there were many more standard trees in hedges. Our landscape and biodiversity are much the poorer for their loss. They have gone with the hedges in which they once stood, through diseases like Dutch elm disease and ash dieback and through the stresses that accompany agricultural intensification; and they have not been replaced. They have lost their use for timber, firewood or fodder just as they have lost their importance to local culture and sense of place. In short, they have become unloved and redundant.

For all that, we still have a few million left. In fact, they constitute a significant part of the nation's tree cover: as we have seen, together with other trees outside woodland, hedgerow trees make up no less than 30% of all broadleaved cover in Britain. In Wales trees outside woodland match ancient woodland for cover, while in Ireland half or more of the broadleaved cover is outside woodlands.

We now know, and appreciate more and more, that a tree in a hedge is worth many in a wood. Hedgerow standards are often magnificent specimens, among the most distinctive features of our lowland landscapes. Unlike trees in line-of-trees hedges, they don't tend to shade out the shrubs beneath them, so wildlife has the best of both worlds – thick undergrowth and mature trees. What's more, as we shall see later, they provide much-needed shelter and shade for livestock (and us!) and remove large quantities of carbon dioxide from the air.

The open-grown structure, bounteous flowering and fruiting, and abundant veteran features of standard trees are all supremely good for wildlife. They serve as stepping-stones for safe passage across intensive farmland and, even if we may not fully understand why, they act as beacons around which insects congregate, in turn attracting birds

ABOVE: A hedgerow Sycamore sculpted by Atlantic gales in Kilkhampton, Cornwall. This tree is salt-tolerant and thrives in exposed coastal situations.

and bats. As Tom Merckx and colleagues (2010a) commented after noting the strength of the association between moths and hedgerow trees, they are, ecologically speaking, keystone structures in intensive agricultural landscapes. The benefits they bring to farmland ecosystems far outweigh the small area they occupy.

But hedgerow standards are facing a storm of threats. Ash dieback is the latest of many. If we lose still more of them, then the consequent loss of wildlife will make it that much more difficult to achieve nature recovery goals. Landscapes will be impoverished, and we'll be more vulnerable to climate change. Urgent action is needed now if the decline in hedgerow trees is to be halted, if they are not to become just a memory – as they have over large tracts of continental Europe.

But the outlook is by no means all gloom and doom. Attitudes are changing fast, led by concern over biodiversity and climate crises. Hedgerow trees are surely on the verge of making a comeback. Fortunately, there are millions of saplings already growing in our hedges, just waiting to be given a chance. All we need to do is to find them and save them from the flail.

Bottoms, banks, margins and ditches

chapter nine

Colourful displays of flowers gracing hedge bottoms, banks and their margins are one of the great joys of country walks and cycle rides in spring and summer. At times they can match the very best of meadows, so profuse and varied are their flowers and accompanying insects. Ditches add further variety and excellent wildlife habitat, real powerhouses when it comes to insect biomass. Yet these non-woody hedge parts are often wholly overlooked by those concerned with hedge management and conservation.

It's challenging to describe the plant communities involved. Not only do they vary greatly with geographic location, soil and above all management, but many hedges contain several different communities, grading into one other – we are, after all, dealing with ecotones. At the base of the hedge, if the shrubs are thick and dense above, the ground is often so heavily shaded that few plants grow – apart, that is, from Ivy, Ground-ivy and the occasional Lords-and-ladies (Cuckoo-pint), ferns like Broad Buckler-fern or, perhaps in damper areas, mats of mosses. Instead there's a thick litter layer of fallen leaves, decaying sticks, twigs and flail chippings that has its own distinctive wildlife value.

Once beyond this zone of dense shade, hedges may have scrambling plants like Hedge Bedstraw, Tufted Vetch and Greater Stitchwort growing up their flanks, together with brambles and various wild roses. Next, many a good hedge has a band of tussock-forming grasses like Cock's-foot or, on wetter soils, clumps of Soft-rush, perhaps with some tall perennials like Hogweed and thistles growing through them. Beyond that, the very best hedges have a strip rich in native flowers such as Common Knapweed, Yarrow or Common Bird's-foot-trefoil before crops or tarmac begin. Banks and ditches add both

OPPOSITE PAGE:
When allowed to develop, broad bramble margins like this are, for wildlife, a very valuable part of any hedge.

ABOVE: Uncultivated flower-rich hedge margins like this next to crops or pasture are rare but of great value to wildlife, especially pollinators.

physical area and many more niches, allowing a greater diversity of plants to grow and so more wildlife to thrive.

Arguably, these floras reach their zenith on the steep sides of high hedge banks in the West Country. Here the season starts in March with Primroses and Lesser Celandines, soon to be followed by further beauties like Common Dog-violet and Early-purple Orchid. A little later, a stunning patriotic display appears, the red, white and blue of Red Campion, Greater Stitchwort and Bluebell, perhaps with patches of white-flowered Ramsons in damper places or interspersed with the vivid yellow of buttercups amid the lush greens of unfurling Hart's-tongue and other fern fronds. Cow Parsley forms a white ribbon on the margins, to be followed in the summer by Rough Chervil, Hogweed, Wild Angelica, Hemp-agrimony or, in places, the deep pink of Rosebay Willowherb.

Increasingly these days, in arable farmland, one sees wide and often colourful field headlands where various flowers and grasses, often non-natives, have been purposely cultivated for the benefit of wildlife, usually birds or insect pollinators. Such carefully nurtured but often short-lived headlands are outside the scope of this book, but where present clearly complement the habitats provided by hedges. They need frequent management and regular renewal if they are to persist and retain their original purpose, just as cropped land does to yield well over many years, necessitating farmers to acquire and apply new knowledge and skill sets.

LEFT: The patriotic red, white and blue of Red Campion, Greater Stitchwort and Bluebell on a West Country hedge bank.

This chapter explores the non-woody, that is herbaceous, plant communities associated with hedge bases and edges, including banks, ditches and margins, together with their associated animal life. I explain the reasons why so many of these communities are in a poor state of health, and the difficulties faced by anyone attempting to restore them. Finally, I make a plea for far more attention to be paid to these hedge parts – they are not mere accessories, they are vital components for good health.

LEFT: Spring at last! Primroses and violets.

ABOVE: Bluebells atop a Devon hedge bank in Dartmoor National Park. Displays of spring flowers like this are dependent on flailing tight to the bank.

The diversity and origin of basal and marginal plant communities

Many hundreds of native flowering plant and fern species have been recorded in British hedges, 500–600 according to Pollard, Hooper and Moore (1974). That's over a third of our vascular plant flora. This is true too in parts of continental Europe: in northern Belgium, even though hedges, lines of trees and small forest patches (<0.1ha) occupy a mere 0.7% of the land, they support about 45% of all plants in the surrounding countryside (Van Den Berge *et al.* 2018).

Even a single farm can boast a remarkably species-rich hedge flora. Michelmore and Proctor (1994) counted the number of plants associated with the banked hedges of Farley Farm, near Chudleigh in Devon. Despite this being, in their own words, 'a very ordinary small farm' (12 fields covering 28ha), they found no fewer than 293 species, a figure that did not include mosses and liverworts.

In contrast to a hedge's woody flora, its herbaceous make-up is likely to depend far more on recent management practice and adjacent land use than on hedge age, soil type or location. There's little relationship between the woody species above and the herbs below: to all intents and purposes their compositions are independent. It is herbicides, fertilisers, shade, cultivation, mowing and grazing that are the main factors determining which plants you will find at the bottom of a hedge and in its margins, and in turn which invertebrates and other animals may be present.

That's not to say that origin and age have no impact. Many old hedges have plants that are indicators of ancient woodland, noting that these vary from district to district: for example Dog's Mercury may be a good indicator of ancient woodland in Northamptonshire but is not in Norfolk or Devon. Across north-western Europe, wide hedges have been found to support more woodland plants than thin ones, almost certainly reflecting the greater microclimatic stability within them (Litza *et al.* 2022). The wider a hedge is, the more resilient it will be to climate change, and in particular to extreme weather events.

Philip Wilson (2019) lists 234 vascular plants characteristic of ancient hedges. Among them are Primrose, Bluebell, Barren Strawberry, Wood Anemone, Wood-sorrel, Common Dog-violet, Ramsons and Early-purple Orchid, along with many ferns. In woodland relict hedges one would expect to find an especially good complement of such plants, but they are frequent too in many hedges that are clearly human-made, such as West Country banked hedges. Indeed, the presence of one or a few of these woodland plants is likely to be indicative at best that a hedge is ancient – other sources of evidence should be sought before reaching any conclusion about its age and origin.

Likewise, just occasionally, hedges have plant communities that are remnants of flower-rich grassland, now lost from adjacent fields through agricultural change. These relics may be the only survivors in a landscape of once widespread habitats such as the species-rich hay meadows of the Pennine dales, and so have considerable conservation value (Wilson 2019).

BELOW LEFT: Tufted Vetch and Meadowsweet add colour to a hedge margin.

BELOW: The presence of Wood Anemone in this hedge indicates that it is probably ancient, but not conclusively so.

ABOVE: A coastal hedge bank in full flower, above Crackington Haven in north Cornwall.

Classification of herbaceous communities

According to Philip Wilson (2019), some herbaceous communities found within hedges can be fairly easily placed within the National Vegetation Classification, introduced in Chapter 7 (Rodwell 1991). The first such community I associate nostalgically with high summer, when the flowerheads of False Oat-grass click loudly on both sides of the car as I drive down narrow Devon lanes. It is the one labelled MG1 'False Oat-grass grasslands' and is typical of ungrazed situations such as road verges, tracksides and arable field margins. In northern England, MG3 'Sweet Vernal-grass–Wood Crane's-bill grassland' can occur in hedge margins as remnants of upland hay meadows. Meanwhile hedges near west-facing coasts can include fragments of MC9 'Red Fescue–Yorkshire-fog grassland', MC12 'Red Fescue–Bluebell maritime bluebell community' and H8c 'Heather–Western Gorse heath'.

From a practical perspective, Countryside Survey data have been used to produce a more useful classification of hedge-base and marginal floras, to guide restoration work (Critchley *et al.* 2013). This identifies six broad types of herbaceous vegetation:

- Woodland herbs
- Species-rich grassland or semi-improved grassland
- Rank grassy vegetation
- Species-poor pasture
- Disturbed arable and sparse vegetation
- Sparse vegetation

In my experience this classification works reasonably well, although I would give Nettle- and Goosegrass-dominated vegetation a separate class of its own given how commonplace it is, rather than subsuming it within disturbed arable. Other plants that frequently dominate hedge floras are Bracken and Ivy, and these too probably deserve their own categories rather than being lumped with woodland herbs. Also, the classification does not obviously encompass vegetation that grows on damp ditch sides. Our farm, like many others in the West Country, has many hedges where the ditches and margins are dominated in June by Hemlock Water-dropwort, a luxuriant umbellifer of tall-herb fens. It finds our hedges and climate in north Devon so much to its liking I have seen it growing well even on the top of tall hedge banks, its roots no doubt carried there with soil cast up from the bottom during bank maintenance work.

LEFT: Hemlock Water-dropwort dominates many damp West Country hedge ditches and margins for a short time each year, in June.

LEFT: Tapered Drone Flies (*Eristalis pertinax*) – hoverflies – and the muscid *Graphomya maculata* on a Hemlock Water-dropwort flowerhead.

Hedge flora condition

Sadly, the overall picture of hedge herbaceous flora quality is one of decline, reflecting ever more agricultural intensification. Across Britain, between 1978 and 2007 when the last Countryside Survey was done, there was a marked decline in the diversity of plants at the base of hedges, falling from 16.1 species per plot in 1978 to 14.0 in 2007 (Carey *et al.* 2008). The most noticeable trends were for grass cover to increase at the expense of forbs, for species associated with high soil fertility to become more dominant, for plants of open, disturbed ground to decrease, and for plants tolerant of shade to increase.

ABOVE: Nettles and Goosegrass dominate this hedge margin – an all-too-common sight.

BELOW: A terribly thin hedge bottom with nothing but Goosegrass seedlings and a few blades of grass. Fertiliser pellets were being sprayed direct into the hedge base as I was there.

Explanations for these changes are not difficult to deduce: fields have been ploughed too close to their edges; fertilisers have been spread or run off into hedge bottoms; basal vegetation has been inadvertently or deliberately sprayed with herbicides; intensive stocking has led to the replacement of palatable plants with a much narrower range of unpalatable, usually coarse, ones; and, as more and more hedges have been allowed to develop into lines of trees, so shade levels have increased to the detriment of flowers of open ground.

All too often, I'm sorry to say, these days hedge bases and margins support little more than swathes of Nettles and Goosegrass, the result

Professor Good's survey of hedge flora in Dorset

A unique opportunity to assess long-term changes in hedge flora at a county level has been made possible by the foresight of Professor Ronald Good. Between 1931 and 1939 he recorded the plants of over 1,000 hedges in Dorset as part of a large-scale botanical survey (Good 1937). For each of the hedges, which varied in length from 57m to 1,040m, he counted the number of herb and fern species found in approximately one hour. Between 60 and 70 years later, in the spring and summer of 2001, Nick Button from the Dorset Environmental Records Centre revisited a random sample of 357 of the hedges and again counted the number of plants present, using the same methods.

When Ronald Good's and Nick Button's data were compared by Joanna Staley and colleagues at UKCEH, they found that the number of herbaceous plant species per hedge had increased from an average of 23 to 26, those hedges that started out with a lower number gaining more than those already relatively species-rich (Staley *et al.* 2013). However, they also found that differences in species-richness between hedges became less marked over time – in other words the hedges had become more homogeneous. Overall, across all hedges, the total number of species recorded decreased.

The main drivers of change appeared to be firstly an increase in the number of competitive plants that respond well to artificially high levels of nitrogen, and secondly a decrease in traditional management practices leading to an increase in shade-tolerant herbs. Unexpectedly, the changes attributed to nutrient enrichment did not appear to be related to adjacent land use, so presumably must have been down to atmospheric nitrogen deposition.

Alder Buckthorn is one of several native shrubs that declined in frequency in Dorset hedges between the 1930s and 2000s.

ABOVE: Grape-hyacinth, possibly the rare native *Muscari neglectum*, on a field margin in south Lincolnshire.

of fertilisers either running off from adjacent fields or being directly spread into them, poor and wasteful agricultural practices. Their legacy can last for decades if not centuries and be very difficult to rectify. Atmospheric nitrogen deposition does not help.

Rare herbs of hedges

Very few plants in Britain are restricted to hedges, the Plymouth Pear being a rare exception. Nearly all hedgerow plants are generalists capable of surviving in more than one habitat type. Nevertheless, there are quite a number of rare herbs closely associated with hedges, as described by Philip Wilson (2019). Here are some examples:

- Starved Wood-sedge: an Endangered species of light dry soils. One of only two surviving populations occurs on the southern slopes of the Mendips in Somerset, on a shaded hedge bank.
- Deptford Pink: another Endangered plant. The largest remaining populations, those in south Devon and Kent, are associated with hedges.
- Copse-bindweed and Field Cow-wheat: both are considered Vulnerable to extinction and are now largely restricted to hedge bottoms and green lanes.
- Grape-hyacinth: a further Vulnerable species, it is a rarity of grassy field banks and hedge bottoms in Breckland and elsewhere in East Anglia.
- Western Ramping-fumitory: endemic to Britain, this is virtually confined to west Cornwall, where it is locally present in hedges.
- Spiked Star-of-Bethlehem (Bath Asparagus): a frequent hedge species in a very small area around Bristol and Bath.

The fauna of hedge bases, banks and edges

This is where Robins and Grey Partridges nest, Hedgehogs feed, Common Lizards and Slow-worms shelter, newts and Common Toads spend the summer or hibernate, and a multitude of caterpillars feed. It's where the predators of many crop pests, like ground beetles and spiders, spend the winter and build up their numbers before moving out into the fields. Try searching grass tussocks growing at the base of a hedge in the winter – you may be surprised, as I was when I first tried it, at the number of insects and spiders you see.

Copse-bindweed

Not to be confused with the very similar Black-bindweed (*Fallopia convolvulus*) or with Hedge Bindweed (*Calystegia sepium*), both of which are common garden weeds, Copse-bindweed (*Fallopia dumetorum*) is a rare climbing annual of hedges, thickets and woodland edges on well-drained soils, its long trailing and winding stems growing up to 3 metres in length. Following a substantial range contraction since the 1950s, the herb is now largely restricted to hedge bottoms and green lanes (Wilson 2019), although it is also found in woodland. It is rated Vulnerable to extinction in the UK. Erratic in appearance, it sometimes occurs in quantity following felling, thinning or coppicing; lack of active management of hedges, especially of periodic rejuvenation by laying or coppicing, is likely to threaten population survival. Another threat comes from artificially raised nutrient levels, whether due to fertiliser drift from adjacent land or atmospheric nitrogen deposition.

Copse-bindweed, a rarity closely associated with hedges.

The diversity of invertebrates associated with hedge bases, banks and margins is huge, covering as it does everything from worms, woodlice, spiders, harvestmen, mites, slugs and snails through to a wide range of insects. Altogether, far too big a subject to cover in any depth in a book of this length, so I will restrict myself to remarking on pollinators, butterflies and bee-flies.

About 80% of our native plant species rely on insects to pollinate their flowers. Thus healthy pollinator populations are vital in most terrestrial habitats, hedges included, and not just for the biodiversity but ultimately for important ecosystem services too, including soil protection, flood control and carbon sequestration. Without pollinators many hedgerow herbs would not only be unable to exchange genetic material, collectively they would also produce fewer of the seeds that so many wild birds depend on for food and their numbers would in turn fall. With fewer birds, dispersal of seed attached to feather or foot will be restricted, and a vicious cycle of decline may set in.

RIGHT: A solitary bee (*Andrena* species) covered with Dandelion pollen.

Many still think only of bees when pollinators are mentioned, but increasingly the importance of other insects, especially flies, is being recognised. Crudely, pollinators can be divided into two sorts, those that prefer flowers with short corolla tubes and those with longer mouthparts that like flowers with deep corolla tubes. Butterflies, moths and bees mainly fall into the latter category, favouring hedge flowers like Bugle, knapweeds, vetches and woundworts. On the other hand, flies, including hoverflies, long-horned beetles and wasps prefer flatter flowers – umbellifers like Hogweed, Wild Angelica and Wild Carrot are among their favourites. Some plants, such as thistles, buttercups and Hemp-agrimony, attract members of both sets of insects. Thus, for healthy pollinator populations, hedge management needs to favour a broad range of different types of plant, not just those that are good for bees. More information on this can be found in Chapter 12, which covers the role hedges can play in boosting populations of crop pollinators.

On fine high-summer days, hedges with plentiful grasses and flowers at their base can be graced by colonies of hundreds of butterflies, particularly 'browns' and skippers. Among the browns, the one most closely associated with hedges is the Gatekeeper, as might be expected from its other common name, the Hedge Brown. Females lay their eggs on a range of bents, fescues and meadow-grasses, at the base of bushes, choosing tallish plants growing in sunny, sheltered spots for their caterpillars to feed on. The adults nectar at a variety of flowers,

brambles and thistles being strong favourites. Meadow Browns and Ringlets are less strongly associated with edge habitats, but they too can be common along hedges with broad grassy and flower-rich margins, the Ringlet especially in damper places. In fact, hedges provide a lifeline for these two butterflies in intensive farmland. Adults seek food and shelter in the lee of hedges on windy days, while the caterpillars feed on much the same sorts of grasses favoured by Gatekeepers. The Speckled Wood is the fourth brown often to be seen along hedges, particularly in southern Britain, preferring taller ones. While adults defend patches of sunlight in dappled shade, they may often be seen basking on hedge foliage or feeding on bramble

ABOVE: A Wasp Beetle (a wasp-mimicking longhorn beetle) feeding on a Hogweed flower.

LEFT: Hogweed is exceptionally attractive to many flies, beetles and wasps, and should be welcomed.

RIGHT: Meadow Brown on Hemp-agrimony.

flowers. Their larvae too feed on a range of grasses, including coarser ones like Cock's-foot and Yorkshire-fog.

Turning to the skippers, both Small Skipper and Large Skipper are commonly to be seen flitting along hedges, seeking mates or flowers. Their caterpillars too feed on grasses. The Small Skipper specialises on Yorkshire-fog, preferring tall, sheltered clumps, while the Large Skipper has a strong preference for Cock's-foot, being particularly attracted to large tussocks growing on open ground. The adults of both species visit a wide range of flowers including clovers, knapweeds, thistles and brambles.

RIGHT: Large Skipper. The caterpillars feed on grasses, favouring large tussocks of Cock's-foot.

Other butterflies have caterpillars that are well known for feeding mainly or exclusively on Common Nettles – namely Red Admiral, Peacock, Small Tortoiseshell and Comma, species that are among the best known and most attractive of all our butterflies. But they are fussy in their choice of suitable nettle patches – it's no good well-intentioned gardeners or farmers simply leaving a nettle patch in a shady corner (Thomas and Lewington 2010). This fussiness, together with the sheer abundance of nettles and the impact of insecticides, no doubt explains why it is often hard to find any caterpillars at all when walking along hedges, even though their margins are dominated by the foodplant. When selecting suitable nettles for egg-laying, Small Tortoiseshells favour young tender plants growing in full sunshine on the edges of large nettle beds, while the Peacock chooses large vigorous plants growing in sheltered situations, as alongside a tall hedge. The Red Admiral, a migratory butterfly, is less fussy than the previous two species and will lay its eggs on semi-shaded shoots while still ignoring those in full shade: to be chosen, nettles must be vigorous, with plenty of fresh green tips. The Comma, which uses Hop and trimmed elm too, deposits its eggs in sunny, sheltered places on the tenderest of

ABOVE TOP: Bramble flowers are among the best hedge flowers for insects, attracting many butterflies like this Red Admiral.

ABOVE: Small Tortoiseshell caterpillars devouring Common Nettle in a hedge margin.

ABOVE LEFT: Germander Speedwell is among the plants which favour disturbed soils.

ABOVE RIGHT: Bare patches on hedge banks such as this one made by a tractor tyre are important for much wildlife – like mosses, annual plants and nesting solitary bees. This one had a colony of Ashy Mining Bees.

leaves. Like the Red Admiral it lays its eggs singly; Small Tortoiseshell and Peacock lay theirs in large batches, leading to the conspicuous masses of young caterpillars under protective webs that many readers will be familiar with.

Sunny bank sides are a favoured habitat for invertebrates – vertebrates too. Here patches of bare earth are especially useful, perhaps caused by livestock – our bulls love rubbing their heads forcefully against hedge banks, raising clouds of dust in the process – perhaps by large farm or forestry vehicles scuffing the sides with their tyres, or perhaps just the result of wind and water erosion. Butterflies and lizards bask on the hot ground, even Adders before and after hibernation (Prestt 1971). Solitary bees excavate their nesting holes in the exposed soil. Short-lived plants like Hairy Bitter-cress, Common Field-speedwell, Red Dead-nettle and Foxglove find somewhere to grow, as do prostrate perennials like Daisy and Wild Strawberry. Where else in many farmland landscapes is there any bare ground suitable for colonisation by the many animals and plants that depend on it?

Most of our native ground-dwelling mammals make good use of hedge bottoms and margins. Strips of tussocky grasses may look rather boring to us when viewed casually from above, but they are not only excellent for overwintering invertebrates but also fine habitat for voles, mice and shrews – the reason why Barn Owls can often be seen hunting along hedges. Common Shrews and Pygmy Shrews feed on abundant invertebrates in the leaf litter, Bank Voles on a wide range

Bee-flies

As adults these flies are, without doubt, cute – with a round furry body, long legs and a long proboscis. They can often be seen in the spring hovering like tiny hummingbirds in front of Primrose, Bugle and other flowers along hedges. Their larvae, however, are not so adorable, eating the grubs of solitary bees.

Martin Drake (2013) describes a narrow country lane on the Devon–Somerset border with closely trimmed hedge banks where, each year, passing tractors and trailers shave off some of the vegetation to reveal intermittent stretches of bare sandy clay. These attract a vast colony of many thousands of Ashy Mining Bees. On a hot March day, while passing by on foot, Martin, a very accomplished entomologist, happened to notice vacated pupae (exuviae) of bee-flies sticking out of the entrances to bee burrows. He collected a few. When he got home, to his surprise and delight, he found about half belonged to the scarce Dotted Bee-fly, the others being from the expected and much commoner Dark-edged Bee-fly.

This chance discovery led Martin to revisit the site to count the numbers of exuviae of each species present. He worked out that the Ashy Mining Bee aggregation was host to over 100 larvae of each bee-fly, perhaps many more, even though he rarely saw adults of either species. Martin concluded that careless driving aids the survival of the Dotted Bee-fly.

A Dotted Bee-fly feeding from a Primrose.

Hedgehogs

Most aptly named, Hedgehogs forage, nest and hibernate along hedges, preferring wide ones with dense bases in which to breed (Hof and Bright 2010, Haigh *et al*. 2012). One reason why they stick so close to hedges – you might even say they hog them – is to reduce their vulnerability to Badgers and Foxes, both of which eat them. Populations of Hedgehogs and these predators are negatively correlated: the more Foxes and Badgers present, the fewer Hedgehogs.

In gardens, hedges are better than solid fences for Hedgehogs, since they provide not just cover and food but also allow movement between properties. As I write, the People's Trust for Endangered Species and British Hedgehog Preservation Society are running a campaign to encourage hedgehog-friendly gardens that include hedgehog-sized holes in the bases of fences. No need to do this with hedges!

The range of Hedgehogs appears to have been stable over recent years but evidence points to a substantial decline in farmland numbers, one so severe that they are now rated as Vulnerable to extinction. Happily, in urban areas Hedgehogs appear to be faring better, with populations stable if not increasing. Arable land is often now so Hedgehog-unfriendly that the animals preferentially seek out gardens and built environments within which to forage.

The loss of hedges and subsequent loss of habitat and landscape connectivity are heavily implicated in Hedgehog farmland decline, alongside the increased use and toxicity of pesticides and an increase in predator numbers. No doubt an underlying factor is the ongoing insidious and rapid decline in invertebrates, whether expressed in terms of species-richness, abundance or biomass.

Hedgehogs are indeed, as their name suggests, closely associated with hedges.

of plant matter, while the ubiquitous Wood Mouse is omnivorous. Harvest Mice build their tennis ball-sized nests among the stems of tussocky grasses in hedge margins – vital habitat for this tiny declining rodent, its very existence in intensive farmland now under threat. Weasels and Stoats follow hedge bottoms and margins, hunting all these small mammals. Moles tunnel in the soil, searching for worms and other invertebrate prey. In between bouts of myxomatosis and

ABOVE: Such tussocky grass margins may look rather dull but are good for overwintering invertebrates, small mammals and Barn Owls.

LEFT: Hedges provide a last refuge in intensive farmland for the declining Harvest Mouse.

rabbit viral haemorrhagic disease, Rabbits and their burrows can be abundant, in turn attracting Foxes. Omnivorous Badgers forage along hedge lines, occasionally excavating setts within them. Even Hazel Dormice, arboreal creatures when active, hibernate in damp depressions on the ground at the base of hedges, in small, compact woven nests.

Ditches

When present, these are the powerhouses of hedges, the rotting vegetation and mud producing vast amounts of invertebrate biomass, the engine behind many a food web. Robert Aquilina tells me of a study looking at water-filled farm ditches alongside hedges near Loddington in Leicestershire. Using floating emergence traps they caught 99,936 flies (Diptera) between April and August one year (2005), which, extrapolated, works out at 31 million flies per hectare. This compares favourably with results of my own, from emergence traps set in a wet woodland on our farm in 2019 – here I estimated that each hectare produced 38 million flies a year. These are much higher

RIGHT: Exceptionally fine fen vegetation in a farmland hedge ditch managed as a sediment trap. Mill Farm, Worcestershire.

numbers than would be expected from drier habitats – I estimated a figure of just 10 million flies per hectare in an oak–Ash–Bluebell wood on the farm, mostly tiny sciarids and gall-midges. Of course, most of the flies emerging from ditches are small too, mainly non-biting midges (Chironomidae), but still, against a background of rapidly falling numbers of insects in Britain, the numbers are deeply impressive and the diversity of species remarkable.

ABOVE: This Hawthorn hedge has grown alongside a deep reed-filled ditch or 'drain' between intensive arable fields in the Lincolnshire Fens.

All these flies provide food for larger invertebrates such as spiders and dragonflies, and for numerous birds like Chiffchaff, Chaffinch, Wren, Dunnock, tits, Swallow and House Martin, all of which feed their young on invertebrates. Bats too find open ditches and ponds to be rich feeding grounds. Research shows that unshaded farmland ponds produce 18 times more emergent aquatic insects and 25 times the biomass compared to unmanaged overgrown ones, and that bird numbers are strongly positively linked with the numbers of emerging insects (Lewis-Phillips *et al.* 2020). I expect the same to hold true for hedge ditches – those that are unshaded because the trees and shrubs are trimmed will yield many more insects and support many more birds and bats than those where the shrubs and trees have been allowed to spread, casting the ditches into heavy shade.

Among the best hedge ditches I have seen are those at Mill Farm in Worcestershire. Jonathan Boaz, the farmer, is a remarkably energetic and enthusiastic advocate of farming in ways that produce both plentiful food and wildlife. He has excavated wide ditches alongside hedges specifically to catch sediment and nutrients running off fields above. Some of these ditches have been colonised by exceptionally fine floras, with a variety of emergent and marginal plants typical of rich fens such as Reedmace (Bulrush), Common Club-rush, Greater Pond-sedge and Great Willowherb. Every few years Jonathan cleans the ditches, adding the rich organic sludge to large compost heaps prior to returning the nutrients and sediment to the fields as a soil improver and fertiliser. An excellent example of nutrient retention and recycling, in the process preventing pollution of watercourses and eventual loss of soil to the sea.

Most hedge ditches hold open water, whether standing or flowing, only intermittently, their purpose being to take unwanted water following heavy rainfall from field to stream and river. Still, their bottoms are often permanently damp and so provide habitat for many moisture-loving plants, molluscs, and a host of insects whose larvae need to be kept moist to survive. One snail I was surprised to find in the ditch that formed part of Hedge 2070 was a tiny bivalve, just 2–3mm across, the pea mussel *Pisidium personatum*.

On the sides of such ditches and in other moist places by hedges plants such as Field Horsetail, Common Figwort, Common Comfrey, Hemlock Water-dropwort, Water Mint, Hemp-agrimony, Common Valerian, Meadowsweet and Great Willowherb thrive. During dry summers, especially in the south-east of England, the only flowers available for pollinators within farmland settings are often those growing along the sides of ditches – they provide resources critical for the survival of these insects, a role that is becoming increasingly important with climate change.

BELOW: The Tiger Marsh Fly (*Helophilus pendulus*), a hoverfly with aquatic rat-tailed maggots that can inhabit hedge ditches. It is also known as the Footballer Hoverfly or the Sun Hoverfly.

Other frequent inhabitants of farm ditches that remain moist in summer are droneflies (hoverflies within the genus *Eristalis*) and the striped 'footballer' hoverflies (*Helophilus* species), all with rat-tailed maggots, adapted to a life in stagnant water through having long telescopic breathing tubes, the 'rat tails'. If there's permanent open water, then hedge ditches can support specialised water beetle faunas, as well as a range of dragonflies, damselflies and caddisflies.

On the vertebrate front, in addition to birds and bats, hedge ditches can be important for all three British species of newts, and for

Grass Snake

The longest of our three British snakes, growing to over a metre in length, and the only egg-laying one, Grass Snakes feed primarily on fish and amphibians. Hedges can be good habitat for them and provide vital connectivity between ponds, whether farm or garden. Their ditches supply amphibian prey in the form of newts, frogs and toads, while vegetated bases and bramble patches provide shelter and cover from predators such as Buzzards and Pheasants, the latter suspected to be major predators of juveniles.

Bank and ditch sides together with hedge margins, especially those that are south-facing with a varied vegetation structure, bare patches and both wet and dry places, allow Grass Snakes to thermoregulate effectively, providing the conditions they need either to bask or to cool off. Rabbit burrows and other underground cavities such as those among tree roots make suitable places in which to hibernate. Piles of rotting vegetation, together with compost and manure heaps, are used for breeding, fermentation supplying the heat needed to incubate the eggs, which are laid in batches. While searching for Hazel Dormouse nests one autumn, I was taken aback to find a young Grass Snake twined among bramble stems about 50cm off the ground – was it searching for dormice too?

Hedges can provide excellent habitat for Grass Snakes.

frogs and toads, if not for breeding then as places to seek their prey, to overwinter or to hunker down in dry summers. Water Shrews follow hedge ditches – although fully equipped to be aquatic, these small insectivorous mammals are often found in places well away from open water. In the banks of a big ditch alongside a hedge in Taunton, Somerset, I have even seen Water Vole burrows. Among the birds that frequent hedge ditches is the Moorhen, often roosting and occasionally (and unexpectedly) nesting in the shrubs above.

Concluding remarks

From a wildlife perspective, the bits at the bottom and edges of a hedge are every bit as important as the shrubs and trees, yet they attract far less interest and attention. Well-meant action to maintain or restore dense shrubby growth and standard trees seldom does much good for this herbaceous flora or its dependent wildlife. I have seen many a hedge where the shrub layer has been beautifully rejuvenated by laying or coppicing, yet the ground flora has been ignored and is rubbish. All too often, bottoms, banks and ditches are treated as entities entirely separate from the hedge, not part of it, an attitude reinforced by nearly all agri-environment support schemes funded from the public purse.

The sad fact is that while farmland hedges can have very fine shows of herbs and ferns, supporting a wealth of wildlife, the great majority are in a remarkably poor state of health in this respect. While in 2007 just under half (48%) of all farm hedges in Britain were in good structural condition as far as the woody bits went, this figure was reduced to just 12% when factors affecting herbaceous growth were added in, falling to a dismal 6% for arable land (Barr and Norton 2008). Often hedges lack even the narrowest of margins: they are cultivated tight to the woody stems. Many more support little in the way of herbs other than Nettle and Goosegrass, owing to artificially raised nutrient levels. In arable areas the situation is exacerbated still further by widespread contamination with pesticides, whether fungicide, herbicide or insecticide, taking a heavy toll not just on plants but on animal life too. Nearly as bad, grassland hedge bottoms and margins suffer from intensive stocking, the sheep and cattle eating or trampling flowers out of existence, leaving just bare ground or coarse unpalatable grasses. Altogether, it's a dire situation.

Holistic thinking is desperately needed, recognising not only that hedges are far more than lines of shrubs or trees, but also that the condition of basal and edge flora and fauna relies heavily on the management of adjacent land. Only then will we have a chance of seeing hedges reaching their full potential, as oases of nature within the sea of intensively managed fields that dominates our lowland landscapes.

ABOVE LEFT: Navelwort (Wall Pennywort) growing among the lichen-encrusted stones facing a Cornish hedge.

ABOVE: Diverse flora on a south Devon hedge bank, with Hart's-tongue fern, Navelwort (Wall Pennywort) and Red Campion.

LEFT: Primroses and Lesser Celandines grace a hedge bottom.

The multi-purpose hedge – traditional uses

chapter ten

Thorny Questions

In the Dark and Middle Ages,
If we trust to History's pages,
You might search the landscape round,
Not a hedge to be found.
Instead of little tidy squares,
Mine, and his, and yours, and theirs,
My field, his field, your field, their field,
All formed one enormous bare field.
How they knew without a hedge,
How far any land extended,
Which was middle, which was edge
Where the whole caboodle ended,
History that tells us so much
About the French wars and the Dutch,
Never says a word to show.
I would also like to know
In a land of hedge divested
Where on earth the hedge-sparrows nested,
And what did hedgehogs do about it?
Hedge for them means home and name;
What was their life like without it?
Were they simply – what a shame! –
Hogs, until their hedges came?

R. H. Charles, *Punch*, 28 November 1928

This delightful poem, which Ernie Pollard, Max Hooper and Norman Moore found to start their 1974 book *Hedges*, neatly captures just how very useful hedges can be. How on earth did we manage before them? But the poet did not cover the half of it. Boundary markers, yes, and wildlife habitat too, but so much more besides.

For most of their long history they have been indispensable for keeping cattle or sheep in or out of fields, for shelter and shade, and as a source of firewood. Alongside this they have at times been managed to provide forage for farm animals, either as browse or tree hay, or for ornamental purposes, even for fruit and nuts for us humans. In this chapter I explore these traditional uses. Subsequent chapters will cover the many new functions for which we value hedges such as carbon capture, soil conservation, crop pollination, pest control, removing pollutants from air or water and as a source of renewable energy.

OPPOSITE PAGE: Walking our Devon cattle along a narrow country lane bordered by high hedges.

Boundary markers

Once, before the time of accurate maps and the pinpoint precision of satellite systems, hedges and walls were the best means of marking out ownership and occupancy of land. Estates, farms, fields, gardens and all manner of other landholdings were bounded with hedges for all to see. After all, hedges lasted longer than wooden fences, even dry-stone walls – given just a little help, they were even self-repairing. Without them, would it have been possible to implement most of the Enclosure Acts? Most likely not.

But now that we have Global Positioning Systems (GPS), hedges are not nearly as important as boundary markers. We don't really need physical markers on the ground at all to establish our rights in law; and if we do require a physical barrier to prevent trespass or for our security, then walls, barbed wire, steel mesh or panel fences are more effective and easier to maintain.

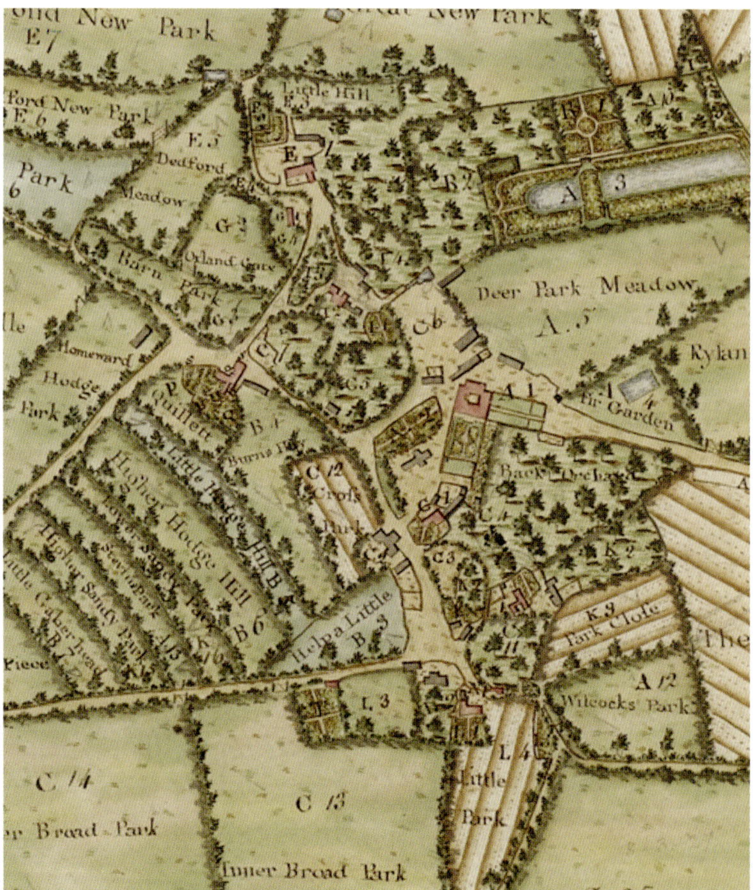

RIGHT: Hedges have long served to mark property and field boundaries. Detail from the Calmady estate maps of Longdown and Down Thomas, Devon, 1788–1789.

Livestock control

Barbed wire was first patented in Ohio in 1867 and revolutionised the management of cattle: it's so much easier to erect and maintain a barbed wire fence than it is to plant and look after a hedge. Stock netting has followed, enabling the efficient control of sheep and pigs. Now, it's fair to say, wire has made most hedges redundant in terms of livestock control. When hedges become gappy, farmers nearly always react by erecting wire fences alongside them. As we have found to our cost on our own farm, it takes a very good hedge indeed to keep a bull away from heifers signalling their readiness to mate. Apologising to a neighbour after one of their cows produces a brown calf rather than the expected black and white one is awkward to say the least.

The redundancy of modern hedges as functional fences is brought home forcefully by the almost universal condition attached to any grant given for new hedge planting on a livestock farm, namely that

BELOW: Wire fences have now made most farm hedges redundant as livestock barriers.

it should be protected by permanent wire fences. The same applies to hedges just laid or coppiced. High stocking rates and reduced labour availability have destroyed our belief that hedges can be relied on to control farm animals, on their own.

Electric fences can be used to afford short-term protection to hedges while they are becoming established, or just after they have been laid or coppiced, but require regular maintenance and don't afford the peace of mind of a good wire stock fence. Revolutionary 'no fence' systems may be the way forward, where cattle are fitted with collars that give an audible warning as they approach an invisible or permeable boundary they must not cross – if they get too close, they get a mild electric shock. The boundaries need have no physical presence on the ground; they are set by drawing lines on a map on a mobile phone.

Firewood

The unauthorised taking of firewood from hedges was once a serious offence. Oliver Rackham (1986) repeats an Elizabethan ruling of 1567 from Felsted in Essex:

> *Any person breaking any hedge or stealing wood be put next Sunday or holyday in the stocks for 2 hours at least, and the wood be placed before them, signifying the cause of the punishment.*

Rackham goes on to say that by 1600 at Ingatestone, again in Essex, hedge breakers were being whipped until they 'bleed well' and receivers of stolen wood were spending all Sunday in the stocks. The 1590s and 1600s were terrible years of cold and poverty.

Two hundred years later William Marshall, in his 1796 *Rural Economy of the West of England*, wrote:

> *Many farms have no other woodland, nor supply of fuel, than what their fences furnish; yet are amply supplied with this; besides, perhaps, an overplus of poles, cord wood, faggots and the bark of oak, for sale. Hedgewood is looked up to as a crop; and is profitable as such; beside the benefits received from the mounds [banks] and stubs [coppice stools], as fences.*

Much the same was true in the bocage country of Normandy. Here, hedges were called the 'peasants' woodland' because they were the main source of fuel for the poor, woods being reserved for the rich, for timber and for hunting.

ABOVE LEFT: Part of the firewood crop from a coppiced 17-year-old Grey Willow hedge.

ABOVE: Faggots cut from hedges for use as fuel for cooking ovens. Normandy, 2010.

Hedges continued to be an important source of fuel within living memory. Ross Dickinson recounts to me how, in the 1930s in Hertfordshire, after his grandfather and father had finished laying a hedge and cut out of it what logs they needed, they would let the village know that the remaining stems and brash could be taken for free. By next morning, the site would be completely clear. In a similar vein, Roger Parris, veteran and expert hedge layer, remembers that when he and his father were laying hedges on their farm near Honiton in East Devon, once they had taken what firewood they wanted, others, including local bakeries, took the rest for free and nothing was wasted. A far cry from today's bonfires. I remember visiting a farm in Devon in the 1980s where the elderly farmers were still making faggots, bundles of sticks, from their hedges with which to heat their bread oven and the kitchen 'copper' that was used for boiling water.

Today other sources of heat energy are available and, for most, affordable – the 2022/23 cost-of-living crisis arising mainly from the Russian invasion of Ukraine excepting – and managing and harvesting hedges for firewood has become a minority occupation. Still, an appreciable number of farmers and smallholders do choose to heat their properties, in part as least, with firewood taken from their hedges. We have two log stoves fuelled largely by sticks from our farm hedges, extracted during laying or coppicing. Each of those sticks has already warmed me several times, as the adage goes, first when I cut it out of the hedge, next when I sawed it to length, thirdly when I carried it to the store and finally when, after seasoning, I heave a basketful to the stove.

In recent years there has been a welcome resurgence of interest in managing hedges for fuel, using new techniques and machinery. It is one of the few ways that hedges can be of direct and obvious economic benefit to farmers, by reducing expensive heating bills, even generating cash through the sale of firewood or woodchip. More on this in Chapter 14.

Fruit and nuts

Hedges have always been appreciated as a source of fruit and, in Britain before Grey Squirrels were introduced, for Hazel nuts. Who does not enjoy meandering along a country lane and picking blackberries from hedges? Or perhaps collecting crab apples for jelly? Or gathering sloes for sloe gin – although I have to say this is far too sweet for my taste, a waste of good gin. Still, if you want good crops of edible fruits like apples or pears, it is probably better to plant the trees as an orchard rather than in hedge lines, where they are likely to grow more slowly and be more difficult to harvest.

Only Bullace and Damson, two varieties of the same subspecies of plum (*Prunus domestica insititia*), have been planted widely within hedges for their fruits, such hedges being a distinctive landscape feature in some parts of England, as described in Chapter 2. They form dense bushes, often difficult to distinguish from Blackthorn although usually less thorny, so in the days when stock-proofing was important they made good dual-purpose hedging plants.

RIGHT: Most Hazel nuts in England and Wales are taken by Grey Squirrels long before we can collect them.

Not easily told apart, their stones are the best distinguishing feature, being deeply furrowed in Damson, smoother in Bullace. Otherwise, Bullace fruits tend to be smaller and rounder than the oval ones of Damson. The fruits of both are fleshy but astringent, best used in jam-making. It is said that Damsons take a long time to produce good crops, an old rhyme going:

He who plants plums
Plants for his sons.
He who plants damsons
Plants for his grandsons.

ABOVE: Bullace, a type of plum, often planted in hedges for its fruit.

Windbreaks, shelter and shade

A few years back I offered to help improve the English translation of Georg Müller's epic work on the field boundaries of Europe. The initial translation had been done by someone with German as their first language, and I must admit to chuckling when I came across a lengthy section entitled 'Breaking wind'. The number of ways that hedges can break wind is, it appears, many. I was sorely tempted not to correct the English, but Georg was covering a serious subject so I relented and did. Hedges, serving as windbreaks, shelterbelts and sunshades, have great economic value for protecting crops and farm animals from exposure to harsh elements, conserving soils, boosting yields and improving welfare.

Physics and practical experience show that a hedge can reduce wind speed by more than a fifth over a distance 8–12 times its height downwind and 2–4 times upwind (Wolton *et al.* 2014b). Thus a hedge 7m high will affect crop yields and animal wellbeing within a band 14–28m upwind and 56–84m downwind, a substantial area. On occasion, the sheltered zone can extend as far as 20 times hedge height for single rows of trees, or even 30 times for multiple rows (Biber 1988). It is worth noting that hedges are more effective than walls, because solid structures result in substantial turbulence downwind. The best hedges for wind-breaking are those that allow 40% of the wind to pass through them, making for slower, smoother, more even air flows. Blocks of woodland, by contrast, provide poor shelter to crops and livestock in adjacent fields; they are too solid.

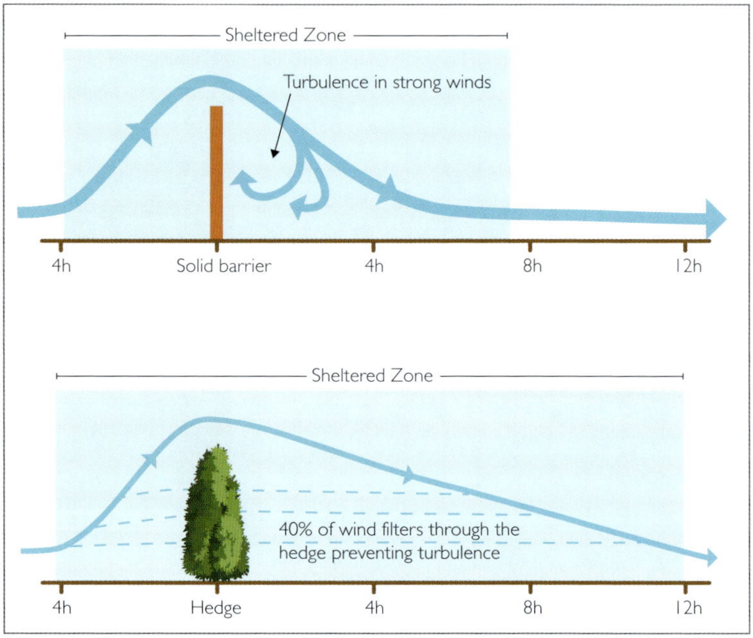

RIGHT: Hedges are more effective windbreaks than solid barriers like walls. A hedge can reduce wind speed by more than 20% over a distance 8–12 times its height downwind, and 2–4 times upwind. (h, hedge height.)

Astonishingly, hedges can even have a considerable impact on local climate. Comparisons of hedged and unhedged landscapes in Brittany found hedge networks to decrease wind speed in more or less flat farmland by 30–50%, as well as increasing daytime temperatures, reducing night-time temperatures and increasing humidity (Guyot and Seguin 1976). A remarkable piece of research concluded that hedges across the Danish part of the Jutland peninsula, which at its narrowest part is about 50km wide, reduce the speed of the prevailing westerly wind by, on average, nearly 10km per hour, the measurements being taken in places well away from hedges or other wind-breaking structures (Jensen 1954, reported in Pollard *et al.* 1974). Makes me wonder what impact the dense hedge networks in the part of Devon where I live are having on the local climate. Without the hedges, would it be a drier, windier place, hotter in summer and cooler in winter? One thing is for sure – it would be much bleaker and duller.

For crops, reducing wind speed not only lessens physical damage such as the flattening of cereals ('lodging') and shredding of orchard blossom, but also reduces water loss from soil and leaf through evaporation and transpiration. Night dews and humid conditions last longer into the day too, and daytime temperatures are raised. All this leads to faster growth. The crops likely to benefit most are those growing on well-drained soils in dry areas, those near the coast and

threatened by salt spray, and broadleaved, water-demanding ones like potatoes, beans, sugar beet, strawberries and many vegetables. In Kent, hop gardens are in places protected by tall poplar hedges, just as many vineyards in France are protected by lines of cypresses.

ABOVE: This hedge in Jutland, Denmark was carefully designed and planted to serve as a windbreak to protect crops and prevent the light sandy soil blowing away.

Increases in yields resulting from windbreaks can be substantial. Across Europe generally, sugar beet yields may be expected to increase by 11–12%, wheat by 6–26%, maize by 10–15%, grass by 27–67%, potatoes by 9–17%, apples by 16–75% and pears by 121% (Biber 1988). Winter wheat, barley and hay (mixed grasses and legumes) are more responsive to protection than spring wheat, oats and maize (Kort 1988). In Britain, at present, there are rather few places characterised by windbreak hedges planted to protect crops, partly because water stress has not been a major problem and partly because soils are mainly heavy and not easily blown away. However, with climate change leading to an increase in the number and severity of droughts and gales we may expect to see this change. Soil loss due to wind erosion is a real issue in arable places with sandy or peaty soils such as in the Fenlands of Cambridgeshire and Lincolnshire, and the Brecklands of Norfolk and Suffolk, and has been for decades. Where crops require irrigation and water supplies are short, windbreaks are likely to be especially beneficial: a good one can reduce evapotranspiration by 20–40% over a distance of 12 times windbreak height, substantially lessening the need for watering (Solagro 2000).

Of course, there will be times when hedges act negatively on crops. On our farm, with its wet land and high rainfall, they reduce the quality of the hay two or three windrows deep into the field because it does not dry so well; elsewhere, hedges delay the autumn ripening of crops. More fundamentally, the shade cast by hedges and the competition between trees and crops for moisture are likely to depress yields in field headlands and margins within a distance up to three times the height of the hedge. Still, most of the figures given in the preceding paragraph take this into account and the clear consensus is that, on balance, hedges result in a net increase in yield – any decrease close to the hedge is more than offset by the increase further away. In part this reflects the fact that yields in headlands and margins tend to be depressed anyhow due to soil compaction, physical damage caused by turning machinery, increased pest damage (from Rabbits, for example) and reduced nutrient status (Kuemmel 2003).

There's no denying, though, that if the hedges were removed entirely and the ground so freed up were cultivated, yields and profit for some crops like cereals would often increase despite the damage and stress to plants resulting from unimpeded wind – at least in the short term. But removing hedges on such narrow grounds is to

BELOW: The fine sandy or peaty soils of the Fenlands of eastern England are prone to windblow and loss in the absence of hedges.

overlook the wealth of benefits they can bring to farm businesses and to society, not to mention to wildlife, many of which cannot easily be costed. It also makes no sense – just the opposite – if farms are organic or grow crops that rely on natural populations of pollinators or predators of crop pests (Chapter 12).

Turning from crops to livestock, the sight of lambs sheltering in the lee of a hedge during gales and snowstorms is a common one – refuge from wind and driven snow is especially important for newborn lamb survival, at least 30% of lamb deaths each year being attributed to hypothermia. Less well known is the substantial value of summer shade for cattle. In our temperate climate, faster-growing and more commercial breeds are distressed by high summer temperatures just as they are by severe winter cold; even hardy native breeds avoid intense direct sun by seeking shade where possible. Cattle and sheep standing or lying beneath parkland or hedgerow trees is a bucolic scene of high summer often seen in the works of British painters such as Thomas Gainsborough (1727–1788), John Constable (1776–1837) and Samuel Palmer (1805–1881). And earlier, in the 17th century, Claud Lorrain and many landscape painters of the Dutch 'Golden Age' such as Aelbert Cuyp had pursued a long tradition of depicting such scenes.

When they suffer from heat stress, cattle close down their digestive and reproductive systems and redirect the blood towards cooling

LEFT: One of our Devon cows licks dry her newborn calf, tucked away in the base of a hedge for safety.

RIGHT: Calves shelter in the shade of hedgerow trees on a hot summer's day.

the body. This means the gut membrane becomes more permeable, allowing toxins to escape into the bloodstream and triggering an inflammation cascade: heat-stressed dairy cattle suffer more from mastitis that those with access to shade. Heat stress also affects fertility, birth weights and milk yield, and depresses weight gain (Gregory 1995, Whistance 2018b).

Altogether, access to trees and shrubs, including those in hedges, is highly important for cattle health and productivity, and not just for the shade they cast. As I explain in the next section, they are an important source of nutrients lacking in intensively managed grassland, that is unless expensive mineral licks are supplied. And they provide the branches and trunks on which cattle and sheep love to rub, to remove moulting hair and fleece and sharp seeds, and to dislodge external parasites like ticks, which often carry diseases. As Lindsay Whistance (2018b) from the Organic Research Centre explains, when among trees, cattle are more social, less aggressive and less stressed – simply, they are happier. Being primarily animals of open forest, they are able to behave naturally. Many has been the time when we know a cow has calved but it has taken some time to find the calf to check that all is well because the mother has tucked it safely away at the base of a hedge, reducing stress both for both calf and dam (if not us).

Browse and tree hay

Few farmers and smallholders realise that hedges can be a valuable source of low-cost supplementary forage for their livestock. Given half a chance, cattle, sheep, goats and horses eagerly browse green leaves and twigs directly from hedges and will avidly consume dried leafy branches, known as 'tree hay', in the winter. In the summer, Paula and I have often watched our cattle reaching up to browse contentedly on fresh leaves in hedges, willows being especially favoured, even when they have access to abundant lush grass close by. If allowed to behave naturally, a rare occurrence in modern farming systems, cattle usually fill their guts with grasses and forbs at the beginning and end of each day – when not resting, the time in between is spent searching out sources of the additional nutrients they need, such as the leaves of trees and shrubs (Whistance 2018a).

In drought years hedges can be a life saver, the tree leaves remaining green long after the grass has dried up, while in long winters, with supplies of hay running low, freshly cut Holly used to be prized as a feedstock in places such as the New Forest and the Pennines. On our farm, a recent summer storm twisted a large leafy bough off an oak, and as it fell it got caught up in the lower limbs. It was partly obstructing a gate, so in October I took my chainsaw to it – and as I did so and the branches fell to the ground, the cattle herd wandered over out of curiosity. To my surprise the lead cow let out a bellow and then started avidly to tuck in to the dried brown leaves, followed by the rest of the herd.

BELOW: Reaching for a tasty leafy twig full of nutrients.

Browse is nutritious. Wych Elm, Rowan, Goat Willow, Aspen, Grey Alder and birches all exceed meadow hay and Red Clover in fat, carbohydrate and protein content, except that Red Clover exceeds Rowan for protein. A similar picture has been found when

ABOVE: Goats enjoying browsing a hedge.

comparing the nutritional value of tree leaves with that of Alfalfa (Lucerne) and rye grasses. Even the tannins in tree leaves have health benefits, preventing bloat, enhancing protein utilisation during digestion, and leading to better growth and wool and milk production. They also have antioxidant activity as well as helping animals to rid themselves of gastrointestinal parasites: sheep with heavy worm burdens shift their diets to tannin-rich plants, and by doing so reduce these burdens by up to a half (Kendall *et al.* 2019, Whistance 2018a).

Better still, the deep roots of trees and shrubs mean that browse is often a much better source of trace elements than grasses and herbs (Kendall *et al.* 2021, Whistance 2018a and personal communication). Different trees are important for different elements. For example, Goat Willow is especially good for zinc and cobalt, Ash for magnesium, and elms for iron. Leaf selenium concentrations, on the other hand, vary more between sites than between species – this is an element that is often deficient in grasslands – while iodine concentrations vary with season.

The leaves of most of our native trees and shrubs are edible and safe, with some notable exceptions such as Yew, Wild Privet and Box. However, palatability varies considerably. The order, according to Scottish Forestry's *Woodland Grazing Toolbox*, consulted in 2023, is as follows, the most palatable first:

1. Ash, Aspen, Elder and willows
2. Elms, Hazel, *Holly, oaks and Rowan
3. Blackthorn, Hawthorn, Sycamore and Wild Cherry
4. Beech, birches and *Scots Pine
5. Bird Cherry and *Common Juniper
6. Alder (even though it is perhaps the best forage of all for energy and protein)

* Holly, Scots Pine and Common Juniper, being evergreen, can be useful winter forage.

The leaves of brambles also provide valuable forage, having high levels of vitamins C and E and minerals, especially magnesium, as

well as tannins with anti-inflammatory and antibacterial properties, while blackberries, along with other edible berries, have high levels of micronutrients. Our sheep eagerly eat any Ivy lying on the ground after a hedge has been laid – presumably it too has particular nutritional or medicinal properties.

ABOVE: A Devon cow snacks on hedge foliage as we move the herd along our farm lane.

For medicinal purposes, animals voluntarily switch to less tasty browse to rectify nutritional imbalances. For example, sheep offered different varieties of willow sometimes choose those with a high salicylic acid content even though it is bitter. The reason why the sheep do this is not known, but it is telling that the acid is an effective painkiller – it is a key starting material for the manufacture of aspirin – as well as being an anti-inflammatory, antimicrobial and antifungal agent.

Hedgerow herbs too are useful. When we drive our cattle down a quiet country lane, they stop to munch eagerly on Hogweed and Wild Angelica, being difficult to move on until they have had their fill. The former, with its rough hairy leaves, is said to scour helminthic worms from the gut, while the latter is perhaps just delicious.

To make tree hay, leafy boughs are cut from hedges in the summer, bundled tightly and either hung outside to dry or stored green under cover, ready for use as supplementary winter forage for cattle, sheep and ponies. A labour-intensive business, tree hay harvesting is now largely an activity of the past, at least in Britain and Ireland. However, there's been a recent resurgence of interest amongst those producing high-end meat. Trials conducted by various proponents including Ted Green (2016), well known for his passionate advocacy for ancient

ABOVE: A 'shredded' oak. Normandy, 2010.

trees, show that even naive animals need little urging before they tuck eagerly into tree hay, often consuming not just the dried leaves but the twigs and bark too.

Traditionally, to produce tree hay, trees growing in hedges were pollarded, that is lopped above the height that cattle, horses or deer can reach, with the new shoots harvested each summer or so. Alternatively, the trees were allowed to grow tall and straight, just the side shoots and branches being regularly removed, a process known as shredding. An old photograph of a cabbage field near Exeter in Devon has in the background a tree, probably an elm, managed in this way, and while on a Hedgelink visit to Normandy in 2010 we saw shredding still being practised (Wolton *et al.* 2010). No doubt pruned side branches were also at times used for firewood, and the tall straight trunks for timber.

Sometimes during dry summers, when the grass was brown and the soil parched, trees were pollarded or shredded so livestock could feed on the green leaves as the branches fell, leaving the bare branches and twigs to be collected when seasoned to fire bread ovens or the like. It is ironic that, in past times, elms and Ash were especially favoured by those harvesting tree hay – precisely the two trees worst affected by disease these days, a most unfortunate coincidence.

Concluding remarks

The changes in recent decades in the ways we use and value hedges have opened up the opportunity to do things differently. There is a particular upside to hedges no longer needing to be stock-proof: they can be planted and managed much more flexibly, to meet the many and various needs of the present, including those urgent ones relating to the biodiversity and climate crises. Rather than being a loss of function to decry, it is liberating.

For instance, no longer do we need to follow the standard prescription, originally developed it is said to keep livestock off railways lines, of planting eight Hawthorn plants per metre in a staggered double row. Instead we can choose whatever planting mix and density we like, to fulfil our desired objectives. Neither do we have to worry obsessively about small gaps developing along a hedge – they can be filled by laying or new planting in due course. In the meantime, as we have seen, they may well provide the bare earth essential for many plants to get a foothold and for solitary bees and wasps to burrow in, all the while acting as passages between fields for species of open ground with poor dispersal abilities.

The rekindling of interest in the potential of hedges to supply wood fuel and supplementary nutrition for farm animals is particularly encouraging, since these are ways in which farmers can receive direct, tangible, economic benefits from their hedges, providing motivation to keep and manage them. It is also abundantly clear that when cattle, sheep and other livestock have access to hedges they have the chance for a healthier and happier life. This too brings economic benefits, as well as being ethically desirable. Such access must, of course, be kept within limits, or the hedge may suffer unduly.

LEFT: Cattle that have access to hedges are happy cattle!

Soil conservation, cleaner water, less flooding

chapter eleven

Have you ever noticed how hedges that contour hill slopes form terraces – capturing soil, preventing its loss to the sea? Or heard that hedges can remove pollutants from water, helping to keep our streams and rivers clean? Or perhaps wondered how it is that they can reduce the risk of our land and homes flooding? These are among the modern-day reasons we have for looking after hedges and planting new ones. They are 'nature-based solutions' to some major environmental problems, and powerful ones at that. I explore them in this chapter.

Soil conservation

Depending on field size, gradient and soil type, heavy rainfall can wash huge quantities of soil off bare land as the water hurries downhill. Yet soil is effectively a non-renewable resource, since it takes between 100 and 400 years to create just 1 centimetre's depth. The loss of more than a tonne per hectare per year should be considered irreversible within a time span of 50–100 years. In this context, reported losses of 86 tonnes per hectare per year for maize crops growing on slopes (Solagro 2000) are truly shocking.

Hedges are one of the most effective ways to retain soil on farmland. Look carefully at any well-established hedge contouring a slope, and the chances are that you will see the ground level is higher on the upside than the downside. You may think the hedge has been planted along some natural ridge, but in all probability the break of slope is the result of soil being trapped by the hedge, building up over decades or even centuries on the uphill side. Not surprising when one considers that hedges can remove 94% of the sediment from water streaming off fields (Borin *et al.* 2010).

OPPOSITE PAGE:
A new hedge bank built to delay storm water from the field above reaching the stream and so reduce flood risk to the town of Braunton (Devon) further down the catchment. It has yet to be planted with shrubs.

ABOVE: This recently coppiced hedge in the Usk Valley has, over the centuries, captured soil from the field above to form a terrace.

This positive impact of hedges on local topography and soil thickness has been nicely demonstrated in a modelling study based on a densely hedged undulating landscape in Brittany, where historical records show the hedges to have been established 1,200 years ago, around the year 800. Simulations reveal that over the centuries the hedges have had a profound impact on topography: the accumulation of soil above hedges and its erosion below them have created pronounced shelves of raised ground, in other words terraces, across formerly smooth hillsides. Overall, if the landscape had not been hedged, 74% of it would have *thinner* soil now than originally, while in fact 62% of it now has *thicker* soil than would have been present 1,200 years ago, a direct result of the hedges (Follain *et al.* 2006).

The efficient way hedges can conserve soil and control erosion is highly appreciated across much of the world. In the tropics, hedges – often, as remarked in Chapter 2, composed of stiff grasses like Vetiver or Napier Grass (also known as Elephant Grass) rather than shrubs – are widely planted across slopes. In Britain and Ireland, however, the usefulness of hedges in this respect is much under-recognised, despite crop yields being heavily influenced by soil depth as well as by soil health. This largely reflects the short-term approach to agriculture that has prevailed since the Second World War. Only now is the importance of deep, healthy soil starting once more to be truly appreciated. In southern Wales, I recently met a farmer who every so often transports soil from the hedge at the bottom of a steep pasture back to the top. A lot of effort, time and diesel, but worth it from his point of view: he recognises that maintaining the farm's soil reserves is critical if, as he hopes, future generations of his family are to have a chance of making a decent living from the land.

Hedges that contour slopes capture soil that would otherwise be lost simply by acting as a permeable physical barrier, slowing down surface water and giving the soil particles a chance to settle. Banks, tussocky grasses and dense tangles of stems all help make this process more efficient, and the more hedges there are traversing a slope, the better. As a hedge ages and soil accumulates on its upside, so it becomes ever more efficient at slowing water flow and so preventing soil loss.

Hedges can also prevent the loss of wind-blown loss of soil from farmland, as described in Chapter 10.

ABOVE TOP: All this eroding soil was trapped after just one heavy rainfall event by the netting placed across the slope of a maize field by the West Country Rivers Trust and Catchment Sensitive Farming project. Hedges do the same job.

ABOVE: The hedge bank in the foreground has checked the loss of soil from this maize field.

Preventing pollutants reaching watercourses

Our rivers are far from healthy and, even after decades of effort trying to clean them up, most remain heavily polluted. In England, in 2019, just 14% of rivers were in good ecological condition, assessed against the EU Water Framework Directive. For the rest, there is a big difference between the wildlife that should be present and that which actually is. In Wales and Scotland less than a half of rivers are in good ecological condition; in Northern Ireland it is less than a third. Across Britain and Ireland a major cause of this sorry situation is diffuse pollution from agriculture – the cumulative loss of fertilisers, pesticides and sediment from field to watercourse.

The best way to tackle this pollution is to eradicate or at least reduce it at source through better farming practices, but next best is to impede the movement of water from land to ditch, stream or river, giving a chance for it to be cleansed *en route*. This is where hedges

ABOVE: Buffer strips with trees and hedges can prevent a high proportion of agricultural pollutants from reaching watercourses.

come into play: those running alongside watercourses and those that contour slopes can be remarkably efficient in this respect. For example, in Poland research has shown that a hedge can absorb 85% of the nitrate leached from crops (Solagro 2000).

There's a great deal of research into how effective grassy buffer strips are at limiting the quantity of diffuse pollution reaching watercourses, but relatively little into the benefits of adding trees and shrubs to these strips. In Brittany, a tall oak hedge with a low grassy bank running between an arable field above and a pasture below removed four times the quantity of nitrates the bank would have done on its own (Caubel *et al.* 2003). A more detailed study, this time in north-east Italy, found that a newly established 6-metre-wide grass strip planted with two rows of Plane trees interspersed with Guelder-rose shrubs reduced total run-off of water into a ditch below by 78%, and cut losses of nitrogen by 74% and of phosphorus by about 80% (Borin *et al.* 2010). It also stopped 94% of suspended solids, that is water-borne sediment, reaching the ditch. A 20-year-old riparian grass strip 5 metres wide with a single line of Plane trees performed even better, removing nearly 100% of nitrates and dissolved phosphorus. This well-established woody buffer strip also reduced concentrations of herbicides by between 55% and 95% depending on the chemical properties of the active ingredients – terbuthylazine, alachlor, nicosulfuron, pendimethalin and linuron.

Just how effective a hedge is at intercepting pollutants will depend on a range of different factors, including its location, dimensions, soil structure, species composition, age and management (van Vooren *et al.* 2018). It's a complex matter and further research is needed, especially into the increased effectiveness of buffer strips when they contain trees and shrubs, under a wide range of conditions.

Well-sited hedges remove pollutants from water flowing across the land surface and percolating through soil in three ways. Firstly, they physically impede the surface flow of water, allowing the grasses, herbs and stems to trap the fine sediment to which most pollutants are bound (rather few are in solution), a process that banks improve. Secondly, a hedge's numerous and deep tree and shrubs roots greatly increase the permeability of the soil, making it much easier for water to soak in and penetrate deep underground. And thirdly, microbial activity and the active uptake by trees and shrubs of nitrates and phosphates into their foliage, stems and roots removes nutrients from water, further reducing the likelihood of them reaching streams and rivers. This last process is at its best in hedges with a wide variety of trees and shrubs, since different plants have different nutrient needs and growth patterns, optimising nutrient uptake and cycling throughout the year.

Do hedges reach a saturation point, beyond which they can absorb no more nutrients? For nitrates this is unlikely, since nitrogen is lost to the air as leaves and other biomass are decomposed by microbes. But for phosphates the story is different – here hedges, like other forms of buffer strip, can become saturated and start to leach phosphorus back into the water in the dormant season as organic matter starts to decompose. The answer is to harvest the wood, for example through regular coppicing, taking it well away from the watercourses the hedges are protecting – although research into the frequency with which this should be done has yet, it seems, to be done. When harvested in this way, a hedge planted to control diffuse pollution can at the same time be managed as a source of wood fuel, in addition to being a carbon sink, stabilising eroding banks and providing wildlife habitat. Likewise, contaminated sediments trapped by hedges and their ditches should be moved well away whenever the opportunity arises.

BELOW: Where hedges hold back water after heavy rain like this, they reduce peak flows in streams and rivers and consequently diminish the risk of flooding.

Reducing flood risk

As some communities have found to their cost, hedge removal from hillsides can have the unfortunate and unintended consequence of increasing flood risk to land and homes. This happened in the parish of Kenton in Devon, where the removal of banked hedges in the 1970s to create large fields led to frequent inundation of downstream settlements – the authorities had to build new terraces to replicate the job once done by the hedges. Elsewhere across the country, we see hedges being put back or planted to reduce flood risk: when contouring slopes or bordering watercourses, hedges provide a soft, natural solution of the sort now frequently preferred to building ever more costly hard structures like flood defence barriers.

Another example comes from the village of Hillfarrance, near Taunton in Somerset, which used to be regularly flooded by the Hillfarrance Brook, a tributary of the River Tone. To address this, George Pidgeon, accomplished hedger and stalwart of the Blackdown Hills Hedge Association, tells me that in about 2010 the Environment Agency installed a sluice gate across the brook upriver of the village to divert water during periods of high flow onto adjacent farmland. At the same time as the sluice was installed, new hedges, some with banks, some without, were made across the farmland, both to further divert the water and to slow its movement. Subsequently George says the village has not flooded, even during periods of exceptional rainfall such as the winter of 2013/14. A further example comes from the town of Braunton in north Devon, susceptible to severe flooding after heavy storms as water pours unhindered off fields above. Here, among other measures, new 'leaky' hedge banks have been made across slopes, to slow down the surface flow during cloud bursts and reduce the chances of the River Caen, which runs through the town, breaking its banks.

BELOW: A new hedge bank being made across a field above Braunton in north Devon to reduce the risk of the town flooding. Drainage pipes are inserted through the base of the bank so that storm water is released, but slowly.

As with soil conservation, much of the best research into the impacts of hedges on the frequency and severity of flooding comes from north-west France, led by the INRAE Brittany–Normandy research centre in Rennes. Philippe Merot (1999), for his PhD research, compared the hydrology of two similar 32ha catchments in Brittany, one with and the other without a hedge network. In the hedged catchment, the hedges were on banks and at a density of 10.6km per square kilometre, so as good as the better hedged parts of Devon, Cornwall and Ireland. Sixty per cent of the hedge length was perpendicular to the slope and there was a continuous bank bordering the stream in the valley bottom. As expected, Merot found that the evacuation of storm run-off took longer in the hedged catchment, and peak flows in the streams were reduced. Following a typical storm, run-off volume and peak flow were one and a half to two times lower in the streams draining the hedged catchment than in those draining the hedgeless catchment. In the winter in the hedged catchment about 5% of a storm's precipitation reached the watercourse as storm flow, while the equivalent figure varied between a few per cent and 15% for the catchment without hedges, depending on rainfall intensity. Merot concluded that even though hedges may have lost their traditional functions they now have an important role to play in reducing rapid surface run-off and so downstream flood risk, especially in areas of intensive farming such as those where maize is grown for silage or biomass.

Hedges act to reduce the risk of flooding downstream in three different ways, as shown in the diagram below. Firstly they physically slow up water running over the ground, giving it time to percolate into the ground – once below the surface it moves downslope much more slowly. Fortuitously, the accumulation over time of soil from fields above to form terraces means that hedges become ever more effective at this, allowing more and more water to penetrate the

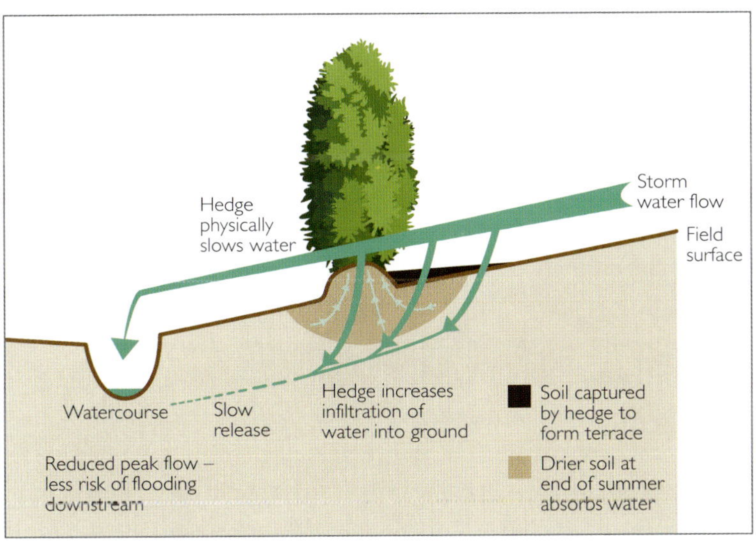

LEFT: How hedges reduce flood risk.

ground. Secondly, any surface water that does not infiltrate into the ground is released onto the land below at a slow rate, the hedges acting like long leaky dams. And thirdly, by the end of the summer the soil underneath and beside a hedge tends to be drier than that in open land, partly because the trees and shrubs take up and transpire more water than the crops, and partly because the hedge canopy intercepts a large proportion of rainfall before it reaches the ground, giving it the chance to evaporate. This soil dryness means that the ground below and beside hedges can absorb more rainfall early in the winter, again reducing the volume of surface water reaching watercourses below. The effect extends beyond the drip line of the shrubs and trees: in Brittany, a 15-metre-high oak hedge resulted in a zone of drier soil that extended 9 metres upslope and 6 metres downslope. In a dry year it took three months at the end of the summer, in a wet year one month, before the ground became saturated by autumnal and winter rains (Ghazavi *et al.* 2008), precious months of reduced flood risk.

The amount of rain that never reaches the ground because it evaporates from leaves, twigs and so forth is remarkably high: an investigation of two largely Hawthorn hedges in Wiltshire found that 57% of rain falling on the canopy never reached the ground beneath when the Hawthorn was in full leaf, a proportion that reduced only slightly, to 49%, after the leaves had fallen (Herbst *et al.* 2006). Have you ever sheltered beneath a tree in a heavy rain shower and wondered where the water goes?

Returning to the way hedges increase the downward movement of water into soil, they are particularly effective in this capacity where the surrounding farmland has been compacted through the movements of heavy agricultural machinery or through the treading of many sheep or cattle – an all-too-frequent occurrence in modern-day farmland. When compacted, the air spaces between the soil particles are pressed out, making the ground much less permeable, and much less amenable to life and plant growth.

The effectiveness of trees and hedges at reversing such compaction has been admirably demonstrated by research carried out at Pontbren in north Powys, Wales. Here, in a ground-breaking collaborative project, 10 farming families, between them managing 1,000 hectares, got together to improve the sustainability of their businesses, both economically and environmentally. One action they took was to plant strips of native trees, mainly birches and Alder but with some

Blackthorn, oaks and Ash, across sheep-grazed upland pastures, to provide shelter for animals and to yield wood products, notably woodchip for animal bedding. This afforded an ideal opportunity to investigate the impacts of grazing and tree-planting on the soil. Astonishingly, by the time the trees were just six or seven years old they had already increased the rate of infiltration of water into the ground by no less than 60 times compared to adjacent compacted sheep walks. The wider the tree belt, the greater the infiltration – that said, strips just 1 metre wide were nearly as effective as those 5 metres wide, showing that even young narrow hedges, for that is what they effectively were, can be very effective in such situations (Carroll *et al.* 2004). Further research at Pontbren revealed that 12-metre-square plots of native trees typical of hedges – Alder, Ash, birches, Hawthorn, Blackthorn, other *Prunus* species, Hazel and Rowan – planted in intensively managed sheep pasture with degraded soil reduced average run-off volumes by 78% and increased infiltration rates by 67 times (Marshall *et al.* 2013).

A recent study in Cumbria confirms that hedges can be very effective at reducing overland flow of water (Wallace *et al.* 2021). Water movement along a 17-year-old Hawthorn–Blackthorn hedge running down a slope between intensively managed rye grass fields was measured. The hedge had a 6-metre-wide grassy margin with Nettles, brambles and Blackthorn suckers. Impressively, the hedge increased the permeability of topsoil by 22–27 times compared to the adjacent fields. All three factors mentioned above – greater water infiltration, greater capacity of soils beneath hedges to absorb and store water, and increased loss of water through evaporation and transpiration – were instrumental in reducing surface water flow following high-rainfall events.

In Devon, elderly farming neighbours have told me that, after they removed hedges to enlarge fields, they were sometimes forced to carry out additional drainage because the ground where the hedges had been became very wet. This no doubt reflects the capacity of trees and shrubs when in leaf to take up much more water through their roots than grass and cereals, releasing it into the atmosphere through transpiration (Thomas *et al.* 2012). At times of high rainfall and high ground-water levels, the hedges were keeping the land workable. In past times, before large machines and diesel, farmers must surely have been aware of this, knowing that hedges planted in the right place could dry out waterlogged places.

Hedges that stretch across flood plains can reduce flooding too, by keeping the water on the land for longer, 'slowing the flow' as they say. Holding back storm water in this way helps prevent rivers downstream from overtopping their banks and flooding settlements. During our Hedgelink visit to Normandy in 2010 our hosts one day, Jean and Nicolle Colette, told us how the flood plain of La Sée at Tirepied was formerly crossed by some 40 parallel hedges that served to slow down the flow of water, enabling the valley floor to act as a slow-release reservoir. Local people believed that the removal of most of these hedges was responsible for the increased frequency with which a town further down the catchment flooded. When we had a look at one of the remaining hedges we found it to be wide and dense and could well believe that flood water would be slowed by its low earth bank and by the numerous stems and low branches of Hawthorn, Hazel, willows and brambles: trapped water-borne debris was evidence of this. Jean and Nicolle told us that, during flood episodes, water could be heard trickling through the hedges.

On the other hand, on a visit to the Netherlands in the 2010s I heard how, to the dismay of local conservationists, hedges were being removed from the floodplain of the Meuse (Maas) upstream of Amsterdam and Rotterdam, to reduce the flood risk to those cities. Here the imperative was to allow the flood plains to empty quickly so they had the capacity to store more water should there be further heavy rainfall further up the catchment, in France and Germany. So, where major river systems are concerned, while hedges across flood plains and hillsides in the higher reaches of catchments can be helpful to reduce flooding, lower down they may do the opposite, at least where major cities are located next to river mouths.

ABOVE TOP: Hedges across the flood plain of the River Thames near Oxford, May 2022. Such hedges can 'slow the flow', reducing the risk of downriver settlements flooding.

ABOVE: The Thames flood plain under water in 2007. The hedges shown in the top photo are in the centre foreground.

ABOVE: A new banked hedge made along a contour, in part to reduce flood risk further down the catchment. It was planted just four years earlier (see photo on page 336). Brook Manor, near Buckfastleigh on Dartmoor.

Concluding remarks

Conserving soil, preventing water pollution and reducing flood risk can be added to the long list of useful things hedges do. All are very topical matters, given the pressing need to foster deep and healthy soils – the basis of all terrestrial life and most of the food we eat – to improve the quality of our polluted streams and rivers, and to reduce the havoc and distress caused by climate change, including increased flooding. The awareness that hedges can be of real help in addressing these big issues is growing rapidly.

Planting, reinstating and looking after hedges that contour fields and border watercourses are without doubt good natural solutions, helping us move towards a more sustainable future for water and soil. If only those who in past times encouraged the removal of so many of our hedges had known this, some of our gravest problems would surely be less severe.

Crop pollination and pest control

chapter twelve

It is a common misconception that the very existence of hedges must reduce food production, mainly on the basis that they occupy land that could be used to grow crops, but also because they are thought to harbour pests, weeds and diseases. But, as we shall see, the balance of evidence suggests that hedges do just the opposite. Few weeds, pests and diseases spread from hedges into crops; rather, hedges boost productivity by increasing the richness and abundance of pollinator communities, and by enabling the predators and parasites of crop pests to survive and spread. In these days of mounting concern over pollinator decline, the impact of pesticides on non-target organisms including ourselves, and growing pesticide resistance, farmers and growers are increasingly recognising and embracing the help on offer from hedges.

LEFT: The Eyed Hawk-moth is among the many moths that are potential crop pollinators. Its caterpillars feed on a wide range of shrubs and trees commonly found in hedges, including Crab Apple, willows, Blackthorn and Wild Privet.

OPPOSITE PAGE: Many hoverflies, like this Black-thighed Flower Fly (*Syrphus vitripennis*), are effective both as predators of aphids (as larvae) and as pollinators (as adults).

The ways in which hedges influence crop pollination and pest control have many similarities, which is why research programmes often investigate the two topics together, and why it makes sense to cover them both in the same chapter here. There are even insects that can perform both roles, notably hoverflies: many of these are effective pollinators as adults while their larvae are voracious consumers of aphids, including some that are crop pests. When they fly into crops from hedges, the hoverflies may either be searching for nectar and pollen or looking for aphids near which to lay their eggs.

Hedges and crop pollination

Worldwide, a high proportion of human food plants rely on insect pollination. This holds true for Britain, although by land area and calories the bulk of the crops we grow, such as wheat, barley, oats and maize, are wind-pollinated. It is our fruits and vegetables that largely depend on insect pollination – our diets would be much impoverished if it were not for bees, flies and moths. No apples, strawberries, raspberries, cucumbers, courgettes, fresh tomatoes, broad beans or runner beans.

A few crops, including oilseed rape, are not wholly dependent on insect pollination but it helps: while this plant is largely self-fertile or wind-pollinated, seed set is enhanced by insect pollination, to an extent that varies between varieties. Overall, it is reckoned that some 13% of UK farm income is dependent on pollinators, but this is likely to be an underestimate since it does not take account of the fact that clovers and other legumes – plants vital both for livestock forage and for nitrogen fixation – rely on insect pollination.

Do hedges increase farmland pollinator diversity?

Hedged landscapes support communities of pollinators that are considerably more species-rich and abundant than those with crops alone, as might be expected given their much greater floristic and structural diversity. Thus when Michael Garratt and colleagues (2017) from the University of Reading counted bumblebees, hoverflies, honeybees and solitary bees in wheat fields along transects running parallel to hedges at 16 different sites in southern England, they found the number of these pollinators to decline significantly with distance away from the hedge: 10m from the hedges there were nine times fewer wild bees and half the number of hoverflies.

The diversity of insect pollinators

Many insects are pollinators, not just bees! They include true flies, butterflies, moths, beetles and wasps, as well as solitary bees, bumblebees and honeybees. Most flowers are visited by several species of pollinator, and most pollinators visit a range of flowers. A study of a British meadow found that by far the majority of flower visitors were true flies (Diptera) (Memmott 1999). The same holds true for hedge blossom: while solitary bees, followed by Western Honey Bees, are the most frequent visitors to Hawthorn flowers, dung flies (Scathophagidae) followed by hoverflies are for Blackthorn, hoverflies followed by Western Honey Bees and bumblebees for bramble, and wasps, houseflies (Muscidae), blowflies (Calliphoridae), flesh flies (Sarcophagidae) and parasitic flies (Tachinidae) for Ivy (Staley *et al.* 2019).

Of course, flower visitation is not quite the same as pollination – it may not result in seed set – and bees are well known to be effective pollinators. But it's worth remembering that bees are adapted to collect pollen to take back to their nest cells or broods – they don't want to pass it on to other flowers – while flies and other insects pick up and carry pollen from one flower to another incidentally. Honeybees and bumblebees are, from a flower's perspective, particularly adept pollen thieves, and so not necessarily the most effective pollinators even among bees. Indeed, research shows solitary bees to be more effective at apple pollination than honeybees even when present in much lower numbers (Weekers *et al.* 2022).

Even beetles, like this Garden Chafer on bramble, can be effective vectors of pollen from one flower to another.

In a similar vein, although hoverflies are effective pollinators, they are by no means the only flies (Diptera) up to the job. A study covering 33 British farms found that other families made up the bulk of flower-visiting flies (82% by number and 73% by species), non-hoverfly flies bearing an estimated 84% of total fly-carried pollen (Orford *et al.* 2015). So, while many studies on crop insect pollination focus on bees, and some on hoverflies, these by no means provide the full picture. Nocturnal insects like moths have been almost totally overlooked.

Many different types of true fly (Diptera) are effective pollinators, not just hoverflies. The Yellow Dung Fly seen here is covered with pollen from pussy willow.

Of course, the distance that pollinators attracted onto farmland by hedges move into fields or orchards will depend a lot on the quality of the hedges, the crop type and whether there are weeds in the crop. Does the hedge sport a good variety of flowering plants, both woody species and herbaceous ones? Does it provide plenty of opportunities for pollinator nesting, larval development or overwintering? Is it well connected to other hedges and semi-natural habitat? Is the crop one that is attractive to pollinators, like apples, oilseed rape or field beans? If not, does it grow alongside 'weeds' that are?

The benefits of hedges in boosting pollinator numbers are greatest in intensively farmed landscapes where there is little other semi-natural habitat to provide flowers or places suitable for immature stages like eggs and grubs (e.g. Alison *et al.* 2022). Even where flower-rich strips have been deliberately cultivated for pollinators, the additional presence of hedges boosts pollinator numbers – indeed, it's usually better to plant hedges than floral strips, given the choice (Donkersley 2019, Image *et al.* 2023). Hedge shrubs, trees, herbaceous margins, banks and ditches together provide a greater diversity of flowers over a longer period than floral strips, as well as more opportunities for nesting and larval development and better shelter from adverse weather. They are also more permanent and so provide safer nesting opportunities for ground-nesting bees, escaping cultivation and being relatively free of herbicides. Evidence suggests that floral strips on their own need to be several years old before they deliver substantive pollination benefits to adjacent crops. Pollinator populations need to be given time to build up, yet unlike hedges these strips are typically short-lived (Albrecht *et al.* 2020).

Bees are usually not only more abundant in hedges than in cultivated floral strips but their communities are more species-rich and contain more specialists and scarce species, at least where the hedges are of reasonable quality, containing a good range of trees and shrubs and herbaceous margins (Kremen *et al.* 2019). In intensive farmland, hedges are often preferentially used by bumblebees as nesting sites – the Garden Bumblebee and Common Carder Bee, long-tongued species that pollinate crops such as beans and red clover, preferring tussocky grass margins, whilst the Buff-tailed Bumblebee and Red-tailed Bumblebee, short-tongued species that pollinate more open flowers like oilseed rape and apples, prefer banks (Kells and Goulson 2003).

In line with well-established ecological principles, diverse bee communities are likely to be more stable and better able to withstand

LEFT: Even hive honeybees imported for pollination purposes benefit from hedges, since they provide food when the crop is not in flower.

shocks such as drought. They are also likely to be better at crop pollination, since most crops can be pollinated by a range of different insects – some of these will be more abundant one year than another, others fly at different times of the day or at different heights, or are able to cope with cloudy, cold or windy conditions, and so on. Ideally the only time when hedges should not provide much nectar and pollen is when the crops are in flower, so the pollinators have an incentive to move out into them, but this is a tough ask.

Do hedges concentrate or export pollinators, and what impact do they have on crop pollination?

So hedges have more species and a greater abundance of pollinators than crops alone, but do these pollinators actually move out into the crops, so increasing the likelihood of pollination, or do the hedges concentrate available pollinators within themselves, having a neutral or even negative effect on crop pollination? This is in many ways the all-important question, but it is only in the last decade or two that research has attempted to find the answer.

The most thorough investigation has been done a long way away, on another continent, in the Central Valley of California. Here the impact of hedges on the numbers of wild bees, honeybees and hoverflies in fields growing processing tomatoes was assessed by Lora Morandin and Claire Kremen (2013). The hedges had been planted

10 or more years beforehand, with a range of native shrubs providing flowers from early spring to late autumn. As we would expect, the abundance of wild bees and hoverflies was greater in the hedges than in unmanaged field edges composed largely of non-native grasses and non-native flowering herbaceous plants. Less expected was the discovery that uncommon species of bee were seven times more abundant in the hedges. Morandin and Kremen suggested that this preference by bees and hoverflies for hedges over herbaceous margins reflected not just the increased floral resources, but also the greater availability of nesting sites in the hedges: rodent burrows and cavities in ground debris for bumblebees (*Bombus*); hollows and cavities in dead wood and stems for hole-nesting species like leaf-cutting bees (*Megachile*), yellow-face bees (*Hylaeus*) and mason bees (*Osmia*). Importantly for the question posed in the heading above, they found that hedges did not concentrate the wild bees, honeybees or hoverflies away from cropped field areas. To the contrary, the presence of hedges increased the number of wild bees present in the fields – the hedges were net exporters of wild bees.

To test whether this export of bees into the fields would be likely to result in increased crop pollination, potted plants of canola, a plant very similar to oilseed rape, were placed in amongst the tomatoes at distances of up to 200m from the field edges and watched for pollinator visitors (Morandin *et al.* 2016). It was necessary to use canola plants because processing tomatoes, used for canning, drying, paste, puree, pulp, ketchup, tomato sauce and tomato juice, do not themselves need insect pollination for seed set. The results suggested that if the fields were growing canola rather than tomatoes, the presence of bordering hedges would increase seed yields by 21%: for a typical 16ha field, this translates into a 36% profit increase following hedge planting – provided, that is, no hive honeybees are introduced. The researchers note that this may over-egg the situation since placing a handful of canola plants in a field full of otherwise unattractive plants may concentrate pollinator activity on them, leading to an overestimation of pollinator services when scaled up to whole-field level. Nevertheless, the results do indicate that maintaining or planting hedges can, at least in certain circumstances, result in a substantial increase in profits as a result of better crop pollination, with further profit expected from improved pest control. The increase in canola pollination due to bees emanating from hedges extended for at least 200m into the fields.

In Europe several further studies have been carried out to explore whether or not hedges increase crop pollination. Firstly, research was carried out in an area of intensive farmland in Lower Saxony, Germany, to see if hoverflies spill over from hedges into crops of oilseed rape (Haenke *et al.* 2014). The hedges were mainly composed of Blackthorn, Hawthorn and Dog-rose. The results strongly suggested that spill-over was occurring, particularly where the hedges were connected with one another and with woodlands rather than being isolated, and where the proportion of oilseed rape in the landscape was low relative to the other main crop, winter wheat. The effect was most marked for hoverfly species with aphid-eating larvae. So it is probable that the presence of hedges enhanced both crop pollination and aphid control, especially the latter, by exporting hoverflies.

A related piece of research, this time in a part of the Swiss central plateau characterised by a small-scale mosaic of arable crops, grasslands and forest fragments, found that good-quality hedges, with at least five woody species every 10 metres and grassy margins at least 3 metres wide, increased the number of visits by pollinators to oilseed rape flowers by 4% (Sutter *et al.* 2018). However, this did not translate into an increase in yield, probably because the variety of oilseed rape planted was one with a particularly low reliance on insect pollination for seed set. The main determinant of yields was found to be the number of insecticide applications – no doubt these chemicals had an adverse impact on the pollinators. The results might have been markedly different in organic systems, or in landscapes with a lower proportion of the area occupied by semi-natural habitats other than hedges.

Another recent European study of note was conducted in an area of intensive agriculture in north-eastern Italy (Dainese *et al.* 2017). Radish plants were placed along the edges of winter wheat fields, and the numbers and types of pollinators visiting them counted, along with the numbers of seeds set. The most numerous pollinators were hoverflies, followed by wild bees (i.e. not introduced colonies of honeybees and bumblebees), with butterflies, other flies and beetles each making a minor contribution. The wheat fields used in the experiment had three different types of field margin – simple strips of herbaceous vegetation, simple hedges composed of a single storey of a few species of tree, and complex hedges with multiple storeys and a greater diversity of trees. Both hedge types also included strips of herbaceous vegetation. To the surprise of the researchers, the type of field margin had no impact on rates of pollinator visitation or on seed set. On the other hand, the cover of hedges in the landscape within a 1km radius had a strong effect, seed set increasing by 70% as the hedge cover increased from 1% to 6%. (Equivalent results were found for pest control, with parasitism of wheat aphids increasing from 12% to 18% as the cover of hedges increased – see later.) That the quality of individual field margins had no impact may well have been because the main pollinators were hoverflies, insects that typically move through landscapes far more freely than bees because they are not tied to returning to a precise place to provision nest cells or to feed larvae. The take-home message from this research is that environmental improvement schemes should focus just as much, if not more, on the extent and connectivity of hedges in a landscape context as on the quality of individual hedges.

The presence of hedges has also been found to increase pollination rates of strawberries. Potted plants were placed in arable fields in Lower Saxony, Germany, and pollinator visits and pollination rates recorded (Castle *et al.* 2019). The results were striking – the weight of strawberries produced by plants placed near isolated hedges was 29% lower than where the hedges were connected to woodlands, and 32% lower where the field margins were just grassy. Ninety per cent of strawberries produced by plants near connected hedges were classified as marketable, compared to 75% of those near isolated hedges and just 48% near grassy margins. Whether these effects remain true for whole fields of strawberries was not tested, but the results are promising all the same – if fields are connected by hedges to woodland, then pollinator populations are increased, leading to improvements in fruit yields and quality. In this instance the main flower visitors were dance flies (Empididae, 45%), pollen beetles (*Meligethes* species, 29%), hoverflies (12%), bees (mainly bumblebees and honeybees, 5%) and the Red-brown Longhorn Beetle (*Stictoleptura rubra*, 4%).

Hedges may not, though, always be a silver bullet for increasing crop pollination. They have not been found, for example, to increase seed set in sunflower fields, again in California's Central Valley (Sardiñas and Kremen 2015). This is probably partly because the bees, in California at least, that are the main pollinators of sunflowers are specialists that do not use hedge resources to any great extent, and partly because the sunflower fields were very large relative to hedge area and crop pollination needs very high.

ABOVE: Dance flies (empids) are frequent flower visitors commonly found in hedges, and may be important pollinators of strawberries.

Whether or not useful pollinator spill-over occurs from hedges into crops depends on many factors, including the crop, its extent, hedge quality, local pollinator communities and the landscape context. Overall, though, the studies that have been done all either demonstrate that hedges are likely to export pollinators into adjacent crops or are neutral in this regard – none has suggested that hedges concentrate pollinators to the detriment of crop pollination.

How far into fields can hedges improve pollination rates?

Naturally, the answer to this question will depend on the pollinators involved and the crop. A review looking across all the available research on this topic, covering both floral strips and hedges, found that, as a generalisation, crop pollination falls off rapidly with distance from the field edge, from about 50% at 10m to 20% at 20m (Albrecht *et al.* 2020). However, the positive benefit of pollinators emanating from hedges can extend much further than this – as mentioned above, canola plants placed 200m away from field edges within tomato fields were significantly better pollinated when the field edges were occupied by hedges than by herbaceous strips (Morandin *et al.* 2016), while research done in canola fields in Alberta, Canada, founded

RIGHT: Drone hoverflies are excellent honeybee mimics. Here a Eurasian Drone Fly (*Eristalis arbustorum*) feeds from a kale flower, picking up pollen on its hairy legs and body in the process.

that seed set was significantly increased within 750m of uncultivated land (Morandin and Winston 2006). This is well within the distance that some bumblebees can travel from their nests when foraging: by capturing individuals and looking at genetic markers, it has been found that the Buff-tailed Bumblebee can forage at least 625m from its nest sites, although the Common Carder Bee was found to move only half this distance (Darvill *et al.* 2004). Subsequent research has shown that Buff-tailed Bumblebee workers can regularly fly even further from their colonies, 1.5km, to visit rich pollen resources, in this case a field of Borage in Hertfordshire (Osborne *et al.* 2008). Even solitary bees can forage substantial distances away from their nests, if the ability of 16 species to find their way back after being translocated is anything to go by: depending on the species, they could do this over distances of between 150m and 600m through farmland in Germany (Gathmann and Tscharntke 2002). Hoverflies and some other flies, as well as some other non-bee pollinators like moths and butterflies, are likely to be able to travel considerably further into fields, since they are not bound by the necessity to return to nest sites. Indeed, hoverflies have been found to distribute themselves across farmed landscape much more freely than bees (Jauker *et al.* 2009).

An interesting aside is that where landscapes have a high proportion of mass-flowering crops like oilseed rape, they may attract pollinators away from hedges, reducing pollination rates of the shrubs and trees and so reducing food supplies for overwintering and migrant birds. Another is that the presence of weeds in fields may well help to sustain pollinator populations across intensively farmed landscapes

Hedges, bumblebees, apples and pears

Monsieur Legent by an ancient pear tree in his orchard in Tirepied, Normandy, explaining how important hedges are to the bumblebees pollinating his crop.

In May 2010, together with Hedgelink friends, I visited an ancient apple and pear orchard in Normandy (Wolton *et al.* 2010). The orchard was full of gnarled, twisted old trees, the pasture grazed by tethered Normandy cattle, white with black blotches. A Cirl Bunting sang from a nearby hedgerow tree. It was a bucolic scene, and the orchard was the pride and joy of the elderly owners, Monsieur and Madame Legent. With little prompting from us, M. Legent extolled the importance of the bordering hedges for his fruit crop. It is here, he explained, that bumblebees, the main pollinators of their trees, have their breeding nests and where they overwinter, in amongst the dense vegetation and roots at the hedge base. Here too is where they gather much of the pollen and nectar needed to raise their broods when the apples and pears are not in flower.

Bumblebees have a long season, from when the queens emerge in the early spring through to late summer or the autumn – and they need to be able to find food for themselves and their broods throughout this time. The spring blossom of hedge trees and shrubs and summer-flowering herbs in their margins and ditches are often vital resources in farmed landscapes, whereas any flowering crop typically meets their needs for just a few short weeks. On our farm in Devon, queen bumblebees, solitary bees and hoverflies emerging from their winter quarters fly first to pussy willow catkins in our hedges to refuel. From then on, right the way through to the autumn, there are hedge flowers available for them, whether it be May blossom, Bugle or Common Knapweed in the margins, or Ivy in the short days of late September.

Our visit to the Legents' orchard in Normandy culminated in being seated around the table in their farmhouse kitchen sampling the cider, calvados and pommeau. They were right to be so proud of their orchard and its produce – it was terrific!

(Nicholls and Altieri 2013). A third – not related to pollination, it's true, but still interesting – is that hedges can at times boost populations of earthworms in adjacent farmland, so improving soil structure and functioning (Holden *et al.* 2019).

Hedges and pest control

Do hedges help keep crop pests under control?

It is often assumed that natural enemies of crop pests, their predators and parasites, spread out from hedges into fields. But how strong is the evidence for this? And if they do so, is it to such an extent that yields are increased and the need for insecticides decreased? Probably. But so much depends on the circumstances. Altogether it's a thoroughly intractable issue, so many are the possible interactions and permutations. The volume of research is mountainous, with many hundreds of papers published on the subject, possibly thousands. Yet we still don't know the answer for sure, even for major crops like wheat.

So it's not surprising that farmers struggle to know whether their hedges are genuinely useful, cost-effective sources of natural enemies. Are they worth looking after solely on this account? Are environmentalists being over-optimistic when they extol the value of hedges for pest control? The balance of evidence suggests that hedges are indeed of value in this way (Wolton *et al.* 2014b, Holland 2019, Albrecht *et al.* 2020), yet after decades of research there remains no strong consensus among the scientific community on the matter, other than in organic, pesticide-free systems. It's fortunate that there are plenty of other good, proven reasons for farmers to value their hedges.

RIGHT: A hoverfly larva (*Scaeva* species, probably *S. pyrastri*) feeding on aphids on kale flowers.

The diversity of natural enemies

Many insects, spiders and other invertebrates are predatory or parasitic on crop pests. The main ones are:

Parasitic wasps (Hymenoptera). The diversity of these is staggering – in Britain and Ireland there are some 6,400 known species. They can be abundant too, as anyone who has operated a Malaise trap or used a sweep net will know. Stand by a hedge for a few minutes in the spring, summer or autumn and you are almost certain to see several of the larger ones; the more numerous smaller ones are likely to escape notice. Collectively the wasps concerned are known as the Parasitica, the main groups being ichneumons, braconids and chalcids. Technically they are parasitoids rather than parasites, since they result in the death of their hosts – which include among their number many invertebrates, including crop pests such as aphids. Indeed, they are probably the most important of all natural enemies. As an example of their sheer diversity, the six most important pests of oilseed rape in Europe, a pollen beetle, four weevils and a midge, are hosts to at least 80 species of parasitoid, mostly parasitic Hymenoptera (Ulber *et al.* 2010).

Many parasitic wasps are very small, less than 3mm long, and are dispersed far and wide by the wind. The extraordinarily delicate members of the family Mymaridae are really tiny, some just 1mm long as adults, and known as fairyflies or fairy wasps. Their larvae are so small they live in the eggs of other insects, devouring them from the inside.

Most parasitic wasps are, as far as is known, not host-specific – so although some search for aphids, for example, they may not be too worried about precisely which species they lay their eggs on. Some are parasitic on other parasitic wasps, so-called hyperparasites. The interrelationships between parasitic wasps and hosts leads to complex communities developing – that is, in the absence of insecticides.

Parasitic wasps, more correctly called parasitoids because they kill their hosts, are probably the most numerous and effective of all natural enemies of crop pests. This particularly large species, an ichneumon, *Heteropelma megarthrum*, kills the caterpillars of various moths.

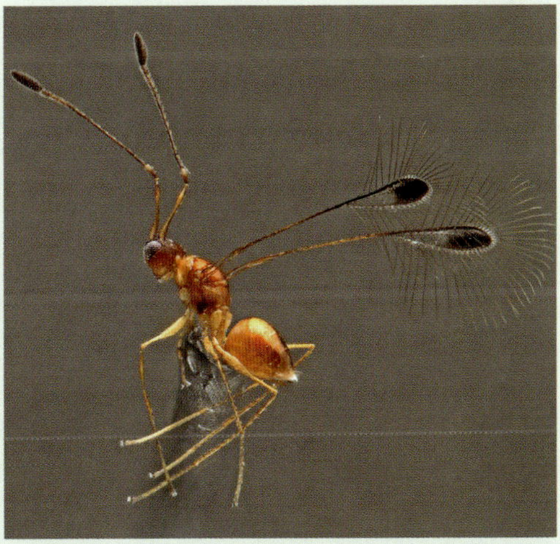

This minute wasp, *Mymar pulchellum*, with a body length of less than 1mm, is aptly known as the Beautiful Fairyfly. It is probably a parasitoid of myrid bugs – many of which are crop pests – laying its own eggs in the bug eggs, its larvae killing the eggs.

Flies (Diptera). Many hoverflies have larvae that are avid consumers of aphids and related bugs. I have watched them suck out the juices from aphids on brassicas in my vegetable garden with remarkable rapidity. As adults, hoverflies are usually largish and brightly coloured, often black and yellow, so they are a popular choice for those researching natural enemies. Other, drabber, flies are much overlooked, even though many of the 7,000-plus fly species found in Britain and Ireland are predatory on other insects too, either as adults or as larvae, or both. These include dung flies (Scathophagidae), dance flies (Empididae) and long-legged flies (Dolichopodidae). Yet other flies are parasitoids, including the big-headed flies (Pipunculidae) which lay their eggs on leafhoppers, planthoppers and other bugs, and the tachinids (Tachinidae). These last ones, typically black bristly flies, are of particular importance, killing a wide range of invertebrates but especially caterpillars. They are well known as effective biocontrol agents.

Tachinid flies are important parasitoids of crop pests. They are characteristically bristly flies, like this one, *Pales pavida*. It lays its eggs on or near a wide range of caterpillars, the larvae killing their hosts.

Beetles (Coleoptera). Many of the 4,000 or so species of beetle found in Britain and Ireland are potential predators of crop pests. The beetles concerned belong mainly to three families: the ground beetles (Carabidae) that feed as both adults and larvae on a range of

A Violet Ground Beetle (*Carabus violaceus*) eating a slug.

invertebrates including slugs, moth caterpillars like cutworms, and cereal aphids; the rove beetles (Staphylinidae), a very diverse and numerous family with about 1,000 species in Britain many of which are predatory; and the ladybirds (Coccinellidae), which are well known by gardeners to feed both as adults and as larvae on aphids and other pests. Most of these beetles are good fliers so can potentially disperse far into crops from hedges, although, as their name implies, many ground beetles prefer to run – if any are found well away from field edges, it's probably because their larvae can winter underground.

Bugs (Hemiptera). Although most of the 1,830 true bugs found in Britain and Ireland are plant feeders, including many major pests such as aphids, scale insects, mealybugs and psyllids (jumping plant lice), some are predatory. These include the rather inappropriately named flower bugs (Anthocoridae), a few plant bugs (Miridae) and damsel bugs (Nabidae). Most feed on aphids and other soft-bodied insects, along with mites and leaf-mining caterpillars, although the damsel bugs are more aggressive and prey upon a wide variety of insects and spiders.

Lacewings (Neuroptera). The larvae of all the 70 or so species of this order in Britain and Ireland, with their fearsome-looking pincers, are predatory and known to consume large numbers of aphids. The adults, with their net-veined wings, are predatory too, but rarely seem to feed. A few lacewings, including some of the larger green ones, are common in hedges. Many overwinter as adults in leaf litter.

Spiders and harvestmen (Arachnida). Collectively, these are well known to be efficient predators of crop pests, wolf spiders (Lycosidae) and money spiders (Linyphiidae) especially so. The former are fairly large and fast-moving spiders that hunt mainly at ground level, while the 280 British and Irish species of money spider mostly build webs and can be so numerous that they are capable of catching up to a third of the aphids flying through wheat crops. The majority of money spiders are extremely small and able to disperse by drifting through the air on strands of silk, an activity called ballooning. In this way they can penetrate far into fields.

Other natural enemies. Includes earwigs, ants, predatory wasps, centipedes and predatory mites.

Lacewing larva: a predator of crop aphids.

Money spiders like this can be major predators of crop pests like aphids. Most are very small.

Are hedges significant sources of crop pests and diseases?

Rather than being helpful by exporting natural enemies into crops, hedges could make matters worse for farmers by spreading weeds and pests. No doubt this does sometimes occur – I have seen seeds from stands of Creeping Thistles on hedge margins blowing far into fields, and don't doubt that docks and Common Nettles can spread their seeds into adjacent headlands, nor that the underground stems of the grass Common Couch can invade the soil. Yet one seldom sees these plants in arable crops even if abundant just a few metres away in hedges: modern mainstream farming methods are very effective at controlling perennial weeds, based as they are on non-stop production, regular cultivation and frequent herbicide application. It is only in organic or regenerative systems that they may be a problem – and any concerns here are likely to be trumped by the benefits hedges deliver for pest control, pollination and so forth.

The great majority of crop weeds are annuals or colonisers of disturbed ground, and so do not prosper in hedge bases or margins. You won't, for instance, see Black-grass (Twitch), a major pest of cereals in Europe with a high level of resistance to herbicides, in hedges unless they have been treated badly. Spraying herbicide onto hedge margins or into the bottom of hedges – such bad practice – is only likely to create a favourable seedbed for annual weeds and increase their numbers.

Turning to insect pests, it's true that plant-eating ones like psyllids and aphids can be abundant in hedges, especially in marginal herbaceous vegetation, and that some of these can be crop pests. For example, the cereal aphid *Sitobion avenae* can use alternative hosts in hedges such as willows, while the Black Bean Aphid (*Aphis fabae*) overwinters on Spindle in hedges. Several other arable crop pests can overwinter in hedges, such as the Rape Winter Stem Weevil (*Ceutorhynchus picitarsis*), Cabbage Seed Weevil (*C. assimilis*), various flea beetles and a number of potato leafhoppers (Holland 2019). In general, though, the great majority of the many herbivorous insects found in hedges are of little or no concern as crop pests, and the few that could be harmful are far outnumbered by potential natural enemies (Précigout and Robert 2022).

Likewise, hedges may act as reservoirs for crop diseases or their vectors, but it appears rarely do so in practice. The best-known example is of Barberry acting as a host for the Black-stem Wheat Rust (*Puccinia graminis*), although this is not considered a major problem at

LEFT: Rabbits frequently excavate their burrows in the bottom of a hedge and feed on adjacent crops, causing a local reduction in yield.

present, partly because, in Britain and Ireland at least, Barberry is a rare shrub in hedges outside gardens, secondly because modern wheat varieties are more resistant to the rust, and thirdly because it can be controlled effectively with fungicides (see Chapter 5).

Among vertebrates, only Rabbits can be a significant problem, living and breeding in burrows in the base of hedges. Travelling by train through East Anglia as a boy I recall seeing large numbers of Rabbits in young cereal crops close to hedges, and the green shoots being much sparser as a result. In the past these Rabbits were sometimes of value, so much so that in the 1920s and 1930s many farmers paid their rent with them. It was once thought that seed-eating birds like finches and buntings that shelter, roost and nest in hedges might have a harmful effect on crops, but research does not support this, at least not in Europe. In any case, numbers of these birds are for the most part so low now that they are hardly likely to be a problem, however voracious they might be.

Overall, available evidence suggests that crop yields are rarely reduced as a result of hedges increasing weeds, pests and diseases; typically, crops are unaffected in this way. We can safely dispel farmers' understandable concerns on this score, except perhaps where Rabbits are concerned.

Hedges, Badgers and bovine tuberculosis

Bovine TB (bTB) is a disease that causes much suffering in cattle and can infect humans through unpasteurised milk. In the UK, in the late 2010s, it resulted in the compulsory slaughter of over 30,000 cattle a year at an annual cost to the taxpayer and industry of around £150 million. Moreover, outbreaks can be traumatic and have a severe impact on the mental health and welfare of farmers and farming communities. Eradication of the disease is therefore a government priority – but one contentious part of the strategy to do so has been to cull Badgers, a significant vector of the bacterium.

In 2007 Fiona Mathews and colleagues at the University of Oxford published an analysis of whether landscape features have any impact on outbreak frequency. The study covered 120 dairy farms in two areas, one in northern Devon and the other on the Herefordshire–Gloucestershire border, each covering 1,000 square kilometres. Half the farms had recent bTB outbreaks and half had not. After controlling for herd size, distance to other farms with recent breakdowns and so forth, Fiona found that hedge density, the number of hedge gaps and the number of ungrazed field boundaries were the main variables explaining outbreak frequency. Hedge-poor farms, with a low density of hedges and poor connectivity, were at significantly greater risk of bTB than hedge-rich ones. An increase in hedge length of 1km per 100ha was associated with a decrease in the risk of breakdowns of 12.5%, equating to an annual reduction of 251 infected herds in the West of England (based on 2004 figures). So the more hedges a farm has, the wider they are and the better connected, the lower the chances of contracting bTB, and by an appreciable margin.

The explanation for this effect is not known, but it's most likely that the presence of hedges reduces Badger–cattle transmission because a higher proportion of contaminated grass is kept out of the reach of cattle. Badgers preferentially use hedges as movement corridors and for their latrines, so where cattle are excluded from these areas by either hedge growth or fences, contact with the disease is reduced.

The presence of hedges has been shown to reduce the risk of dairy herds contracting bovine tuberculosis, probably through reducing contact with Badger urine and faeces.

How do hedges help natural enemies of crop pests to survive on farmland?

Hedges help natural enemies to survive, reproduce and increase in number in several ways. Firstly they provide prey or alternative hosts when the crop is absent, something that is especially useful in the autumn and early spring, allowing individuals to fatten up before and after overwintering. They also facilitate rapid spring population growth of fast-breeding species like money spiders, so more are available to move out into the crops. Pests typically complete their life cycles faster than do their predators or parasitoids, so if no alternative sources of food are available, natural enemies will be unlikely to be able to build up their populations fast enough to act as effective biocontrol agents.

Secondly, hedges provide the nectar and pollen that many adult insects like hoverflies, lacewings and ladybirds require. Unless cereal crops are weedy, they will provide no nectar at all, leaving many natural enemies dependent on nearby hedges and other semi-natural habitats for their continued survival on the farm. Nectar from hedge blossom and herbs like umbellifers have been found to be especially important for many short-lived parasitoid wasps, increasing their longevity and fecundity (Précigout and Robert 2022).

BELOW: The flowers of Hogweed and other umbellifers are an excellent source of nectar for many natural enemies. This flowerhead boasts four species of hoverfly, as well as two members of the housefly family (Muscidae).

RIGHT: Hollow stems of plants like Hogweed (shown here) and brambles provide places for insects to overwinter safely.

Probably just as important, hedges provide shelter from unfavourable weather such as strong wind or rain, especially crucial after fields have been cropped. They also, of course, provide relative safety from insecticides and other biocides, preventing local extinction. And finally, in a similar vein, they provide stable sheltered places for both dormancy during hot dry periods and for overwintering: many invertebrates favour undisturbed soil, leaf litter or the cover and hollow stems provided by tall herbs, brambles and the like for these purposes.

Do natural enemies of crop pests spread out from hedges into adjacent crops?

There is plentiful evidence that hedges support a high diversity and abundance of natural enemies. But what evidence is there that hedges increase the numbers of natural enemies in adjacent cropped land? Good evidence to back up this belief, widely held among environmentalists, is hard to come by. There are just a few conclusive studies, and they are limited in their general applicability.

In the study in north-east Italy mentioned earlier in the context of pollination, just as seed set of potted radishes increased by 70%, so the parasitism rate of wheat aphids increased from 12% to 18% as the cover of hedges in the local landscape increased from 1% to 6% (Dainese *et al.* 2017). Ladybirds showed much the same response, but not ground beetles, rove beetles or hunting spiders. In another study, this time in southern England, hedges were found to be an important source of the money spiders in adjacent wheat fields, numbers falling by 80% between hedge edge and 50m into the fields (Garratt *et al.* 2017). Similarly, rove beetles were more frequent next to hedges than 50m away, although here the numbers fell by just 25% over this distance, either because the beetles are more mobile than the spiders or because more can overwinter successfully in open fields. As with the Italian study, carabid beetle numbers were not found to be influenced by hedge proximity, perhaps because most individuals were of species able to overwinter successfully underground in open land. Other research has shown that some carabids do in fact move out from field margins, including hedges, into cereal crops, for example *Anchomenus dorsalis*, *Bembidion lampros* and *Demetrias atricapillus*. Indeed, a few *A. dorsalis* marked in field boundaries were recaptured up to 200m into adjacent crops later in the season, while significantly more *D. atricapillus* were found in fields surrounded by hedges than in fields with post-and-wire boundaries (Coombes and Sotherton 1986).

Usually the spread of natural enemies from hedges into crops is inferred rather than proven. For example, in Lower Saxony (Germany) hoverflies with aphid-eating larvae were more abundant some 6m into oilseed rape and wheat fields adjacent to hedges but not

LEFT: Some ground beetles, such as *Anchomenus dorsalis*, overwinter in the base of hedges and move out into fields during the spring as the crop grows. They are predators of cutworms and other crop pests.

when adjacent to woodlands, leading the researchers to surmise that spill-over from hedges into crops was occurring (Haenke *et al.* 2014).

In another study, this time in south-east France, numbers of overwintering Codling Moth larvae, a major pest of apple and pear orchards, were found to be reduced in those parts of orchards close to windbreak hedges (Ricci *et al.* 2011). The researchers conjectured that this was because populations of predators and parasites were more stable in the hedges than in the orchards, enabling effective biological control, but again this was not proven. However, it is known that anthocorid bugs, effective predators and control agents of the Pear Sucker, a psyllid that's a major pest of pears, cannot overwinter inside orchards (Cross *et al.* 2012). Instead, they use Nettle, willows and Hazel in nearby hedges for this purpose. This led the researchers to recommend that these plants be encouraged, to increase the numbers of anthocorids and those of other natural enemies of the Pear Sucker such as earwigs, ladybirds and spiders.

Sticking with orchards, research conducted in eight organic cider apple orchards in Asturias, north-west Spain, found many native plants in surrounding hedges to be attractive to adult hoverflies and capable of providing pollen and nectar throughout the spring and summer when the apple trees were not in flower (Miñarro and Prida 2013). Since hoverfly larvae are known to be significant predators of important aphid pests of apple trees in orchards such as the Rosy Apple Aphid, the researchers reasonably concluded that the presence of hedges and their flowers is likely to be important for the natural control of the aphids – but again, this has not been proved.

Can hedges save farmers money on insecticides?

The key question, at least in non-organic farms and orchards, is whether natural enemies emanating from hedges can ever be so effective at reducing pest numbers that these are kept under the threshold for cost-effective insecticide application. Despite the multitude of research papers published on natural enemies and hedges, this question has hardly ever been addressed. It is only Lora Morandin, Rachel Long and Claire Kremen (2014) working in intensively farmed parts of California, who have really explored it. In fields of tomatoes in the Sacramento Valley, not only did they find that hedges increased bee pollination rates of potted canola plants up to 200m into the fields, but also that the hedges increased the

abundance of ladybirds and decreased the abundance of aphids on the tomatoes over the same distance. Levels of wasp parasitism of stink bug eggs also increased, but over the shorter distance of up to 100m – the stink bug concerned, *Euschistus conspersus*, is a major pest of tomatoes and other fruits in the region (in Britain and Ireland we call stink bugs shield bugs). Crucially, just one out of eight fields adjacent to hedges reached threshold pest levels requiring insecticide application, compared to four out of eight next to 'weedy' field edges. The hedges had been deliberately planted with a range of native shrubs that provided flowers throughout the growing season. As such, they were likely to be of particular benefit to parasitoid wasps, reliant as they are upon nectar sources for energy as adults.

ABOVE: Nearby hedges increase the numbers of pollinators and natural enemies in orchards and so are likely to increase crop yields.

Building on these results, the Californian research team were able to estimate that it would take 16 years to break even from insecticide savings on the typical costs of planting a 300m hedge. For crops reliant on insect pollination, the added benefit of enhanced pollination by native bees reduced the payback time to seven years (Morandin *et al.* 2016). They concluded that planting hedges is an economically viable action to take, at least in the Sacramento Valley. However the results of the economic analysis are not readily transferrable to Britain and Ireland because, in contrast to most of ours, the Californian hedges did not require regular maintenance, nor did they take up space

ABOVE: A tiny parasitic wasp, a braconid of the subfamily Aphidiinae, among its host aphids on a kale plant.

that would otherwise be cropped. Still, these figures are the best we have, and they are indicative of what can be achieved.

Given the lack of research evidence, it is somewhat a matter of faith as to whether non-organic farmers can rely on hedges for cost-effective pest control. For those that choose to do so, they may be pleased with the results they experience, even if cause and effect are not fully proven. When members of Hedgelink visited East Lothian in 2010 we met two local farmers who told us that since they had planted new hedges and made beetle banks across their cereal fields neither had had to use insecticides. Beetle banks are made by raising strips of soil about 2 metres wide through arable fields. They are sown with tussocky grasses like Cock's-foot and Yorkshire-fog to serve as overwintering sites for the natural enemies of crop pests, and as such resemble and perform the same biological functions as many hedge bottoms, low banks and margins.

Concluding remarks

In a complex subject fraught with uncertainties, there are two things we can be reasonably certain of. The first is that hedges rarely spread weeds, pests and diseases into crops, especially where production is reliant on pesticides including herbicides. The second is that hedges attract pollinators and the natural enemies of crop pests into farmland landscapes, particularly where there is little other semi-natural habitat present. Basically, where there are more hedges, there are more pollinators and natural enemies.

We can be a little less sure about whether that increased diversity and abundance leads to more beneficial insects and spiders in the crops themselves. However, the balance of evidence suggests this is so – both pollinators and natural enemies do at least sometimes spill out from hedges into crops in sufficient numbers to increase yields. At worst their effects are neutral – hedges do not, or at least rarely, concentrate beneficial invertebrates, drawing them away from crops. We can be rather more positive about all this than were Pollard, Hooper and Moore in 1974, who when considering natural enemies concluded that there is no clear case to be argued either for or against hedges. Of course, a great deal depends on crop type, hedge quality and connectivity, landscape configuration and any insecticide use.

A significant increase in pollination may be achieved up to at least 200m into fields, perhaps up to 750m where bumblebees are involved and even further with hoverflies. Similarly, significant pest control may be achieved up to 200m, perhaps more if flies or money spiders, both easily dispersed by the wind, are the main agents. Taking the distance of 200m into fields as a reasonable working figure for both pollination and pest control, then fields should be no larger than about 16ha (40 acres) for hedges to exert some effect right into the centre, assuming square fields with hedges on all four sides. In a pastural Devon context, that's a large field; in an arable Lincolnshire context, a small one. By coincidence, writing in 1987, Dowdeswell says that from the point of view of efficient use of large agricultural vehicles like tractors and combine harvesters, the optimal size of fields is in the order of 18–20ha (45–50 acres). Whether this remains true to this day, where machinery is even bigger and increasingly satellite-guided, I don't know. Regardless, this field size is far larger than that which would be needed to achieve the suggested optimal hedge density for biodiversity of 10km per square kilometre (Staley *et al.* 2020, 2023) – that would require fields to be on average about 4ha in size.

The million-dollar question is, are hedges a cost-effective option for improving pollination or pest control? Are environmentalists being over-optimistic when they answer yes? In truth, we just don't really know – not for any crop, landscape or farm type. Difficult though it may be, more research is needed here. The best currently available comes from the tomato fields of California, where research suggests that new hedges can pay for themselves within seven years – but this is hardly likely to convince British or Irish farmers faced with the need to make hard economic decisions that may make or break their business. As Pollard, Hooper and Moore said 50 years ago, 'the case for preserving hedges must rest on other factors such as their value for landscape and wildlife'. This remains true today – unless, that is, you are an organic farmer or one who follows other sound agroecological practices like Integrated Pest Management (IPM). Here, planting new hedges and maintaining existing ones to enhance populations of beneficial invertebrates is surely a low-risk option with a good chance of success. What is there to lose?

Carbon capture and storage

chapter thirteen

While woodlands and forest trees are well known to be effective at removing the greenhouse gas carbon dioxide from the air, hedges are too, taking carbon into their stems and branches, into their roots and into the soil beneath them. They too help reduce human-made climate change. Even though hedges may not be game changers in this respect, they can still be an important player, as recognised by the UK's Climate Change Committee. In its 2019 *Net Zero* report, this committee called for a 40% increase in the extent of the UK's hedges, as part of the nation's contribution to reducing global warming. Consequently, and with climate change starting to bite and become a deeply worrying reality, there's now a great deal of interest in the role hedges can play.

OPPOSITE PAGE: With thoughtful management, hedges can both be dense and bushy at the base and have many mature trees, excellent for wildlife and carbon at the same time.

LEFT: Well-hedged landscapes like this one in Carmarthenshire capture substantial amounts of carbon dioxide from the air and store the carbon within biomass and the soil.

While most hedge carbon is recycled into the atmosphere within a few decades at most, some is sequestered away in the soil in an inert form where it can remain locked up for centuries, even perhaps millennia. Another climate-friendly aspect of hedges is their potential as a source of renewable energy, through their management and cropping for wood fuel to heat homes and businesses. That is the subject of the next chapter, while Chapter 15 addresses the tensions that can arise between managing hedges for carbon or wood fuel as opposed to for wildlife. Here, in this chapter, I explore and quantify the ways in which hedges accumulate and retain carbon, touching at the end on the extraordinary properties of biochar.

Where within a hedge is carbon stored?

Hedges store carbon both in their biomass and in the soil. Most of the biomass lies in the stems and roots of trees and shrubs, but an appreciable fraction occurs in non-woody parts like tree leaves and herbaceous margins. As this living material dies and decays, most of its carbon is recycled into the atmosphere within anything from a few days in the case of leaves, to a few decades in the case of woody stems.

The soil beneath and beside hedges can, however, store carbon for much longer. This soil organic carbon (SOC) occurs in two sorts, particulate organic matter (POM) and mineral-associated organic matter (MAOM), the latter known and much appreciated by growers as humus. POM comprises small organic fragments and is usually recycled in between 1 and 50 years, while MAOM consists of organic molecules bound to minerals and is much more stable, typically being stored for anything between decades and 1,000 years before it is recycled, assuming no soil disturbance. It is this form of SOC that is especially important in the context of helping to keep global warming at bay. As Becky Willson from Farm Carbon Toolkit puts it, POM is analogous to a current bank account, while MAOM is the savings account. Normally about one-third of SOC is POM and two-thirds MAOM, but this will vary considerably with soil type. Clay soils, with their fine mineral particles, have much greater capacity to store MAOM than do sandy soils. However, few soil studies involving hedges distinguish between the two types of SOC – it's one of several matters relating to hedge carbon on which we need more information.

We know that cultivation typically releases carbon dioxide into the atmosphere, so it is not surprising that the soil beneath and beside

LEFT: The soil beneath hedges stores substantially more carbon than the soil of arable land.

hedges, being undisturbed, contains more SOC than that within arable fields or grassland leys. Sofia Biffi and colleagues from the University of Leeds compared SOC stocks within the top 50cm of soil beneath trimmed, predominantly Hawthorn, hedges on five dairy farms in the Eden Valley, Cumbria, with stocks under adjacent grassland intensively managed for silage (Biffi *et al.* 2022). They found that SOC was on average 31.3% higher under the hedges than in the fields. The relative proportion of SOC increased with hedge age, from 3.3% for hedges 2–4 years old, to 57.2% for older ones. Other researchers have likewise found that SOC is appreciably greater beneath and immediately beside hedges than under arable or intensive grassland (e.g. Viaud and Kunnemann 2021). A study near Tadcaster in North Yorkshire found that SOC under hedges was nearly double that in adjacent arable land (Holden *et al.* 2019), but this would seem to be an exceptionally high figure and perhaps reflects the shallow depth, just 2–7cm, used for sampling: a review of all published studies within the temperate climatic zone reported an average increase of 32% (Drexler *et al.* 2021).

How much carbon is stored in hedges?

Matthew Axe is one of very the few people who has ever meticulously taken hedges to pieces both above and below ground to find out exactly how much carbon they contain in their biomass (Axe *et al.* 2017). For his PhD research in 2013 he studied three hedges, each at least 140 years old, on a farm in Gloucestershire, two composed of Hawthorn and the other a Hawthorn and Blackthorn mix. They had been laid some 12–18 years beforehand and had subsequently been regularly trimmed by flail to keep them short and bushy, with a width of between 2.6m and 4.2m across the canopy. Three 1m lengths were sampled in each hedge, one trimmed to a height of 1.9m, the others to 2.5m and 3.5m. Matthew carefully cut out all the stems and branches more than 10cm above the ground within each 1m length, regardless of whether they came from shrubs rooted within the sample lengths or not. He also collected hung-up dead wood and all woody litter on the ground surface. To sample below-ground biomass, he used a 3-tonne excavator to pull up the stumps and to dig up the soil, creating a hole as wide as the hedge canopy and 1m deep or until a rocky bed was reached. A pressure washer was used to find and clean all the stumps and roots, even down to tiny ones. All this biomass was dried and weighed so the carbon content could be assessed.

Across the samples, Matthew found that the average ratio of below-ground biomass to above-ground biomass was approximately one to one. As would be expected, above ground-biomass increased as the hedges became both wider and taller: the 1.9m high by 2.6m wide hedges had on average 10 tonnes of carbon per kilometre (tC/km), and the 3.5m × 4.2m ones 14tC/km. Above-ground biomass increased faster with increasing width than with increasing height. SOC averaged 44tC/km, so one and a half to two times the amount of carbon in total biomass, above and below ground.

Mary Crossland from the Organic Research Centre has also gathered empirical data on the carbon stored in hedges. Working at Elm Farm in Berkshire, she assessed the carbon stored in three hedges, one dominated by Blackthorn, another by Hawthorn and the third by Hazel (Crossland 2015). Each was judged ready for coppicing, having at least 15 years' regrowth since they were last cut to the ground (about 40 years' worth in the case of the Blackthorn and Hawthorn hedges), so they were much larger than Matthew's hedges. One winter all the above-ground woody growth in a 15m section of each hedge was extracted, chipped, dried and weighed. The amount of carbon in

ABOVE: One of the hedges Matthew Axe thoroughly sampled in 2013 in Gloucestershire, to measure the amount of carbon stored both above and below ground.

this woody material was found to vary considerably between hedges, reflecting differences in biomass. The Blackthorn hedge had the most at 46tC/km, followed by the Hawthorn one at 28tC/km, and the Hazel one least at just 18tC/km. All these numbers are, as one would expect, greater than the 10–14tC/km recorded by Matthew Axe from his shorter, trimmed hedges. Mary did not measure below-ground biomass held in roots, but from published work on coppices she estimated this to be one-third of that held above ground. She did, though, sample SOC, down to a depth of 50cm. Here the difference between hedges was less pronounced, the Blackthorn one estimated to have 90tC/km SOC, the Hawthorn one 59tC/km and the Hazel one 68tC/km: all higher than the 44tC/km found by Matthew even though he sampled to twice the depth.

The Leeds University study of trimmed thorn hedges in Cumbria has provided further information on the carbon stored within British hedges, for both above-ground biomass and SOC (Biffi *et al.* 2022, 2023). Their hedges, being smaller in volume than those looked at by Matthew, contained, as expected, less above-ground biomass and SOC than Matthew's, and much less than Mary Crossland's.

A fourth piece of research, this time in Ireland, throws further light on the carbon content of hedges. Kevin Black and colleagues took 40 samples from eight hedges in County Carlow and County Wexford (Black *et al.* 2023). For each hedge they weighed all above-ground and

below-ground biomass (to a depth of 1m) within five 6m stretches. Here too the hedges were dominated by Blackthorn or Hawthorn, although one contained a line of Ash trees coppiced 10 years beforehand. The eight hedges chosen for sampling were varied in width, height and structure, in contrast to those selected in the other three studies. The researchers found that the main determinant of the amount of biomass was the width of hedges and whether or not they were top-trimmed, although the hedge with the line of Ash had most of all. The ratio of below-ground to above-ground biomass was less than Matthew Axe found, ranging from an average of almost half for hedges that were kept less than 4m wide by trimming every 1–2 years to not much more than a quarter for those that were not top-trimmed, only side-trimmed once every 4–5 years. Wide and high hedges have more above-ground biomass relative to below-ground biomass than thin and short ones.

ABOVE: An experimental coppiced Hazel–Field Maple hedge at the Organic Research Centre's Elm Farm in Berkshire.

The results of all four studies are summarised in the table opposite (top). The main message from these figures is that the amount of carbon stored by hedges, especially above ground, varies considerably with hedge width and management. As I explain later, different tree and shrub species also have a big influence on biomass.

How rapidly do hedges absorb, or sequester, carbon from the atmosphere?

Mary Crossland estimated not just the carbon stored in the three hedges she looked at, but also the rate at which they sequestered carbon from the atmosphere, by calculating their annual growth rate. Her findings are summarised in the table opposite (below). When I laid a predominantly Grey Willow hedge with nine years' growth on our Devon farm and weighed all the extracted stems and branches, I estimated an above-ground sequestration rate of 1.45tC/km/year, within the range of Mary's figures. However, the team from Leeds University, working on younger hedges (12–37 years old since planting) in Cumbria composed mainly of Hawthorn with some Blackthorn, found a much lower above-ground biomass sequestration rate of just 0.06tC/km/year (Biffi *et al.* 2023). These hedges were kept

Measurements of carbon stored in hedges in Britain and Ireland

Hedge species	Management	Above-ground biomass	Below-ground biomass	Soil organic carbon (SOC)	Total	Reference
Hawthorn (two hedges), Hawthorn and Blackthorn (one hedge)	Trimmed to between 2.6m and 3.5m high	10–14	11	44 (to 1m depth)	65–69	Axe *et al.* 2017
Blackthorn	Coppiced with 40+ years' growth	46	15*	90 (to 50cm depth)	151	Crossland 2015
Hawthorn	Coppiced with 40+ years' growth	28	9*	59 (to 50cm depth)	96	
Hazel	Coppiced with 15 years' growth	18	6*	68 (to 50cm depth)	92	
Hawthorn with some Blackthorn	Trimmed to 2.10m tall and 1.75m wide	7**	—	28*** (to 50cm depth)	—	Biffi *et al.* 2022, 2023
Hawthorn and/or Blackthorn (8 hedges, one with a line of 10-year-old coppiced Ash trees)	'Regular' hedges – tops and sides trimmed frequently, 0.4–4.0m wide	14	6	—	—	Black *et al.* 2023
	'Irregular' hedges – sides only trimmed, and only occasionally, 4.4–11.5m wide	42	11	—	—	

Units are tonnes of carbon per kilometre (tC/km).
* Based on studies carried out in coppiced woodland.
** Derived from the figure of 40tC/ha.
*** Derived from the figure of 164tC/ha.

Estimated annual rates of carbon sequestration over a 100-year timeframe for three coppiced hedges in Berkshire (Crossland 2015)

Hedge species	Above-ground biomass	Below-ground biomass	Soil organic carbon (SOC)	Total
Blackthorn	2.1	0.39	0.74	3.22
Hawthorn	0.54	0.15	2.97	3.66
Hazel	0.50	0.20	0.40	1.10

Note that when sampled the Blackthorn and Hawthorn hedges had 40+ years' growth, the Hazel one just 15 years', and that the hedges were of different widths – it is the range of sequestration rates recorded across the hedges that is of interest, not the differences between species. Units are tonnes of carbon per kilometre per year (tC/km/year).

much shorter and thinner than the hedges Mary Crossland and I investigated, by trimming every 1–2 years, action which prevented new shoots from contributing to the carbon stock. Also, growth rates of Hawthorn and Blackthorn may well have been slower in Cumbria than in Berkshire or Devon.

The SOC sequestration rate of 0.22tC/km/year in the Cumbrian hedges (Biffi *et al.* 2022) was also less than the rates recorded by Mary for the three hedges she studied. Nevertheless, the Cumbria sequestration rates remain higher than suggested by studies based on models using data obtained from woodland or short-rotation coppice (SRC) (Falloon *et al.* 2004). This difference may be explained by the fact that the trees and shrubs in hedges differ considerably, both structurally and in management, from those in woodland and SRC settings. If a hedge has been trimmed regularly to a height of 2–3 metres it is likely to have a much greater density of stems than any woodland undergrowth or SRC, while hedgerow trees, especially standards, typically have large spreading crowns unlike those in woodlands: these differences will be reflected in SOC. Consequently, figures based on data taken from woodland or SRC are likely to underestimate the rate at which carbon is captured by hedges.

As hedges grow into lines of trees, and as any standards mature, so they capture and store more and more carbon for as long as they remain healthy. Over the winter of 2011/12 I was involved in a project to assess the practicality and cost-effectiveness of managing hedges and small woodlands in south-west England as a sustainable source of fuel. This work included full above-ground biomass sampling of two contrasting hedges in the Blackdown Hills, which span the Devon–Somerset border (Wolton 2012). As the hedges were coppiced we weighed all the stems, branches and twigs arising within a 30m length, leaving behind just a few stems for laying. The first hedge consisted mainly of Hazel (60%) with elms (30%), plus small amounts of thorns (5%) and Sycamore (5%). This hedge is shown in the upper photo opposite; it was at just the right stage for laying, with 25 years' growth and 4m high. The second hedge (shown in the lower photo) was much bigger, approaching a line of mature trees, with some growth over 8m high and at least 38 years old. Here the main species were Beech and Ash (40% each), with some Hazel (15%) and a little thorn (5%). The weight of above-ground biomass was strikingly different in these two hedges, the first one producing 40kg per metre, the second 340kg per metre – more than an eight-fold difference.

When measured, the above-ground biomass of the hedge in the lower photo was found to be over eight times that of the one in the upper photo.

The point here is that as hedges change from being bushy to becoming lines of trees, the amount of above-ground biomass they contain increases sharply. So while Mary Crossland's hedges, which were ripe for coppicing, held 3–4 times the amount of above-ground biomass held by Matthew Axe's classical trimmed hedges, hedges with a good proportion of mature trees like oaks, Beech, Ash or Sycamore are likely to contain still more biomass, indeed many times more. What's more, they are likely to sequester carbon at a faster and faster rate as long as they remain healthy. A report collating information on the growth rates of hundreds of tropical and temperate trees revealed that for most species the rate of biomass accumulation increases continuously with tree size (Stephenson *et al.* 2014). Large mature trees, when free from high levels of competition from near neighbours, do not act simply as senescent carbon reservoirs but actively fix larger amounts of carbon than smaller trees. In just one year, a single big tree can add the same amount of carbon to a forest (or hedge) as is contained in an entire mid-sized tree.

How do hedges influence the amount of carbon stored in farmland soils?

As leaves and twigs fall onto the ground they start to decay and are dragged beneath the surface by worms and other invertebrates; this, together with the death and decay of roots, leads to the build-up of soil organic carbon (SOC) beneath hedges. Normally this effect is limited to the soils under and close to hedges but where they contain tall trees it may extend for a considerable distance into adjacent fields. In Belgium, SOC was observed to be greater up to 30m into fields beside lines of tall poplar trees, presumed to be the result of wind-blown leaves being incorporated into the soil (Pardon *et al.* 2017). Earthworms are usually less abundant in acid soils than calcareous ones and as a result acid soils have less SOC in deep soil horizons; the worms are also less abundant in cultivated and disturbed soils.

Hedges also build up SOC in farmed landscapes by capturing soil eroding from slopes above them. As noted in Chapter 11, contouring hedges typically result in terrace formation, soil building up on the higher side. In the bocage landscape of Brittany, with its banked hedges, the topsoil above contouring hedges starts to become thicker as far as 60m upslope (Walter *et al.* 2003). As

the topsoil becomes thicker, so the amount of SOC it contains increases. So large is this effect that in parts of Brittany where the farmland has a very dense networks of hedges, 200 metres per hectare, the hedges may be responsible for 38% of the all the SOC stored in the landscape. Where the density is lower, at 50m/ha, this proportion falls to 13% – still a significant amount. It's tempting to apply these figures across Britain, where there is currently on average 42m of hedge per hectare, excluding the uplands (Staley *et al.* 2023), but the Brittany estimates can only be safely applied to those places such as Devon and Cornwall where the topography resembles the Brittany bocage, the hedges are wide and banked, and livestock farms predominate.

ABOVE: In a densely hedged landscape, hedges can be responsible for as much as a third of all the carbon stored in the soil. Near Truro in Cornwall.

Tree roots penetrate much further into the ground than do those of most herbaceous plants, including grasses, and accordingly it's been found that a considerable fraction of SOC beneath hedges occurs at a depth greater than 30cm, the limit of sampling in most studies even though tree roots can reach down at least 4 metres given the chance. This really matters, since SOC tends to be more stable the deeper it lies, as microbial activity lessens; it is deep SOC that is of particular importance in the context of climate change.

Can hedges continue to increase their carbon stores indefinitely?

Unfortunately not: the capacity of hedges to sequester and store carbon is finite, in terms of both biomass and SOC. Sooner or later an equilibrium must be reached. For biomass, as trees and shrubs are laid or coppiced most of their carbon is quickly recycled back into the atmosphere; similarly every time a hedge is trimmed, for example with a flail cutter, most of the carbon in the cut material is rapidly returned to the atmosphere. On the other hand, if hedges are allowed to mature without any cutting, their woody material may not be recycled for decades or centuries, depending on the species. Oak trees, for example, may continue to get larger for centuries while on the other hand Hazel bushes do not normally increase in biomass, at least above ground, when more than about 20 years old – by then, gains from new growth are fully offset by the death and decay of stems. For Hawthorn and Blackthorn hedges, Black and colleagues (2023) reckon that these reach peak biomass stocks, both above and below ground, about 30–40 years after planting.

Even SOC will reach equilibrium eventually, but this may take a long time. Work done on woodland soils at the agricultural research station at Rothamsted, Hertfordshire, suggests that SOC equilibrium may not be achieved for some 776 years (Robertson *et al.* 2012); others

BELOW: Hedgerow trees continue to capture carbon at a faster and faster rate for decades or even centuries until they become senescent.

have suggested that a much shorter period of a few decades is more likely, but this may reflect shallow sampling of the soil profile. SOC may be lost quickly from surface layers when hedges are removed: in Belgium there was no trace of hedge-generated SOC in the top 23cm of soil under grass strips from which hedges had been removed 60 years previously (Van Den Berge 2021).

Just how important are hedges for carbon storage and capture?

Hedges are so variable, particularly in terms of their height, width, species content and management, that it's very difficult to give any precise figures in answer to this question – but we do have enough information to know they are among our most important habitats in this respect. Taking storage first, data suggest that the overall amount of carbon stored in a trimmed Hawthorn hedge is about 40% of that in a 100-year-old native woodland, hectare for hectare, and for the SOC fraction the figure is 65% (Gregg *et al.* 2021). However, for line-of-trees hedges, particularly if composed of species like Beech, oaks, Sycamore or willows, we may reasonably expect hedges to store just as much carbon as native woodlands – and indeed a review of nine relevant temperate-zone studies covering 83 sites confirms this to be so even for hedges on a long coppice cycle, let alone those that are lines of trees (Drexler *et al.* 2021). Given that trees often grow bigger in hedges than in woodlands, because they are free to spread their branches, line-of-trees hedges are likely to store considerably more carbon than woodlands for a given basal area. The presence of herbaceous growth at the base of hedges and in their margins will add further to their carbon stocks. So it's safe to say, given how much land they cover – as I estimate below, in England about a quarter of the area of broadleaved woodland – that hedges are of considerable importance for carbon storage.

In terms of carbon capture rates, the figures reveal that coppiced hedges with at least 15 years' growth are capable of sequestering carbon just as fast as 100-year-old native woodlands, hectare for hectare (Drexler *et al.* 2021). More impressive still, where line-of-trees hedges are concerned, sequestration rates are likely to be higher than in woodland. This is because the trees within them often grow faster than trees in woodland, experiencing less competition for light, water and nutrients, and perhaps benefiting from fertilisers applied

ABOVE: Appropriately managed, hedges can capture and store as least as much carbon as broadleaved woodland, area for area.

to adjacent crop land. So hedges that are managed under a coppice rotation, let alone those that have developed into lines of trees, both *store* at least as much carbon as well-established woodlands and *capture* carbon at the same rate if not faster, area for area.

Better still, hedges that are managed under a coppice rotation can be used to produce renewable energy in the form of wood fuel (see Chapter 14). If this is used for heating as a substitute for fossil fuels, then even though hedges may sequester more and more carbon the larger they get, modelling shows that overall, for climate change mitigation, regular cropping on a coppice cycle is the better option (Crossland 2015).

If the Climate Change Committee recommendation that UK hedges be increased in extent by 40% by 2050 is taken to mean that hedge *length* should be increased by 40% by new planting, then it has been estimated, based on both SOC (in the top 50cm of soil) and total biomass (assuming below-ground biomass is equal to above-ground biomass), that doing this would result in between 1.5% and 4.5% of annual agricultural carbon dioxide emissions being offset over a 40-year time span (Biffi *et al.* 2023). However, these figures are based on predominantly Hawthorn hedges growing in the north of England that are kept short and thin (1.5m wide) by regular trimming. Hedges planted with a mix of species including fast-growing ones in more southerly locations and allowed to grow freely over this timeframe will offset a much higher proportion of agricultural emissions.

On the whole, though, it is easier for farmers to increase the extent of hedges on their land by letting existing ones get *wider* than by planting new ones. In this context, Matthew Axe tells me that if all the trimmed hedges in Great Britain were allowed to increase in width from 1.5m to 3.5m, then a further 26 million tonnes of carbon dioxide would be removed from the atmosphere, which is nearly five times the total amount of carbon dioxide emitted by UK agriculture in 2020 (Defra 2022a). Likewise in Ireland, recent research confirms that allowing hedges to increase in width would increase the amount of carbon stored (Black *et al.* 2023). As it is, models suggest that hedge carbon stocks in County Wexford and County Waterford are currently actually decreasing due to hedge removal and an increase in flailing and coppicing. Across Britain and Ireland, if hedges are permitted to grow taller at the same time as being allowed to become wider, then the amount of carbon dioxide removed from the atmosphere will be even greater.

Another comparison that can be made is between the sequestration rates of hedges and emissions from cars. A typical passenger vehicle emits about 4.6 tonnes of carbon dioxide per year. If we take the average amount of carbon sequestered annually by 1 kilometre of hedge ready for coppicing to be 3 tonnes, which is the equivalent

LEFT: Letting hedges get wider is an easy way of increasing the amount of carbon stored on farmland. Here Blackthorn suckers are being allowed to grow out from the original hedge line.

of 13.6 tonnes of carbon dioxide, then each year every kilometre of hedge will absorb the emissions of three cars. Or half a tractor. Scaling these figures up to a UK level, since in December 2022 there were 32.9 million cars on the road and allowing for the 3–4% that are electric or plug-in hybrid, the 814,000km of rural hedge in the UK could, if managed appropriately, soak up the emissions of 7.7% of cars than run on fossil fuels. Not an amount to be sniffed at.

Biochar – a remarkable substance

When wood is heated with little oxygen present, volatile gasses are driven off, leaving behind charcoal of one type or another, a process known as pyrolysis. The precise form of charcoal produced depends on the temperature to which the wood is heated, on the amount of oxygen present and on the speed of manufacture. We all know the type used in barbecues, the light black stuff often sold as briquettes that burns cleanly at a high temperature to cook our sausages. Less well known is the type called biochar that is used as a soil conditioner, even though it has quite remarkable properties.

Although pyrolysis leads to the loss of about half the carbon in wood to the atmosphere, the remainder, which is typically in the form of 80% pure carbon, is remarkably stable. When incorporated into the soil it can remain unchanged for hundreds or even thousands of years – witness the charcoal from ancient fire sites remaining visible to archaeologists millennia after it was created. In effect, the unstable carbon in biomass, material that would normally decompose rapidly and release carbon dioxide, is converted into a very long-lasting stable form. So it is not surprising that there is now a great deal of interest in converting biomass into biochar as a means of effectively permanently locking up carbon and helping to reduce climate change.

Biochar made from wood is not only a fine carbon store, it is also highly porous and has a very high surface-area-to-volume ratio. This means it can absorb large quantities of water and nutrients, a property that makes it ideally suited for use as a soil improver, especially in places where the soil is light and free-draining and those that are drought-prone. It acts as a sponge, soaking up the water and nutrients, which are then available for use by plants. Conversely, it can be used to improve the aeration and drainage of heavy soils. The list of benefits goes on. Since it tends to be alkaline, it can also be used to improve the fertility of acid soils.

LEFT: Making biochar from hedge brash using a simple kiln.

And it's an excellent habitat for soil microorganisms, including mycorrhizae. Given all these almost miraculous properties and few, if any, downsides it is not surprising that biochar has been widely used in the tropics to improve soil fertility for a long time, perhaps for 2,000 years in the Amazon basin. But it seems that most of us living in temperate zones are only now catching on to its potential. In Britain and Ireland we can surely expect to see it used more and more, especially in areas with dry, sandy soils prone to desiccation and nutrient loss through leaching.

So where do hedges fit in? Well, their woody biomass can be converted into biochar just as any other wood can. And it can be done quite simply. In Dorset I have seen the brash from a coppiced hedge being burnt in a simple dish-shaped kiln to produce biochar. Skill was required to feed the brash into the kiln at just the right rate to ensure the end product was charcoal, not ash, but it worked. When the time was right, to prevent complete combustion, the fire was doused with water, leaving behind soggy black fragments, the biochar. This can be air-dried to reduce its weight or left as it is. If it is to act as a slow-release fertiliser, then biochar must be enriched by adding manure or compost before it is added to soil, but again this can be done simply enough.

What better way to make use of the branches and twigs arising from hedge laying or coppicing, or even trimming?

Concluding remarks

The amount of carbon stored by a hedge and the rate at which it removes carbon dioxide from the atmosphere will vary hugely according to the species of tree and shrub present and the hedge's dimensions and structure. Other factors like growth rate and stage (shrubby or line-of-trees), and soil type, depth and drainage, as well as local climate, will have a big bearing too. Given all this, we need to sample a lot more hedges before we can say with accuracy just how much carbon in total is stored or sequestered by British and Irish hedges. For all this uncertainty, we do have enough good evidence to be confident that, area for area, hedges are able to absorb and store just as much carbon from the atmosphere as native broadleaved woodlands, if not more. Across Britain and Ireland there can be no doubt that the establishment and maintenance of hedges is an important and significant measure we can take in our countries' efforts to help reduce the rate of global warming, while allowing agricultural production to continue and providing a wealth of other benefits to farmers, society and the environment.

If we take the average canopy width of hedges to be 4 metres, taking into consideration line-of-trees hedges but ignoring marginal herbaceous growth, then the area occupied by hedges, both rural and urban, in Great Britain is about 300,000ha. That's about a quarter of the area of broadleaved woodland (1.3 million hectares). (In Chapter 8, I noted that altogether trees outside woodlands make up 30% of all the native broadleaved cover in Britain.) If herbaceous margins are included, together with adjacent cropped land where soil carbon has been increased due to hedge presence, then the area of land over which hedges act to increase carbon stock will be considerably greater.

In the Republic of Ireland, according to Ireland's National Forest Inventory (Forest Service 2017), hedges are even more important relative to broadleaved woodland. Here the area of such woodland (223,000ha) is actually *less* than the area covered by 'hedges' (282,000ha). The definition of hedges used by the inventory is broader than that in Britain, including as it does all lines of trees and shrubs 20 metres or less across the canopy; even so Neil Foulkes tells me that the area of hedges, as we normally think of them, probably does exceed the area of broadleaved woodland in Ireland.

Farmers and other hedge managers can increase the carbon storage capacity of their hedges simply by letting those that already exist become wider, taller or both. Compared to most other options available for increasing carbon stocks on farm, this is an easy thing to do, a quick win. Reflecting this, as I write, such action is currently being strongly promoted in Britain by the National Farmers Union. Otherwise, planting new hedges is an excellent way to increase carbon stocks, the wider the better. As we have seen, new hedges bring many other benefits too – for wildlife, for livestock, for soil conservation, for water quality, reducing flood risk and so on. From a carbon perspective, planting on arable land is likely to have the greatest benefits, but doing so on intensive grassland is still highly desirable. There's one place, though, where hedges should not be planted, and that is on organic soils such as peat – here the physical disturbance and increased microbial and mycorrhizal activity is likely to lead to net losses of soil carbon that continue for decades, rather than the desired gains.

Finally, biochar production is a rising star in the context of hedges and climate change. Given its simplicity to produce and extraordinary ability to lock up carbon long term, as well as its value as a soil improver, surely it will gain in popularity? Watch this space …

BELOW: This wide banked hedge would, if allowed to grow taller, sequester and store large amounts of carbon.

KOMATSU

Hedges for wood fuel

chapter fourteen

As I write, on a fine but icy January morning, the trees and bushes outside are frosted white and I'm grateful for the cosy heat from the wood stove in the next-door room, fuelled as it is with logs and sticks from our hedges. This winter, oil and gas prices are sky-high because of the Russian invasion of Ukraine, but our household is fortunate: thanks to our farm hedges we have been able to avoid putting on the oil-fired central heating almost entirely. Quite a few others I know are doing the same thing.

Not long ago hedges used to be highly prized for firewood in Britain and Ireland, as described in Chapter 10, but the arrival in the middle of the last century of cheap alternatives like oil and gas largely put paid to that. Now, in a reversal of fortunes, more and more farms once more meet some, or even all, of their heat energy needs from their hedges. For this exciting revival we are indebted to rural communities across the Channel, whose interest in managing hedges for wood fuel never fully lapsed. Enterprising individuals have developed new ways of cropping and processing hedge wood, making the most of modern machinery and technology. Over the last decade or two, here on this side of the Channel, we have learnt much from them and have been able to build up our own know-how.

If hedges were once again to be managed and cropped for wood fuel at scale, it would be a major boost for their conservation. Far too often hedges are perceived as being no more than a cost burden, of no real economic worth. However, cropping them for wood fuel is a tangible, direct way that farm businesses can save money or even generate an income. It's a good reason to look after hedges, more readily quantified than other benefits such as increasing crop pollination or reducing crop pests. Big wins are to be had all round if we get it right: there's a very strong case to be made for 'hedge farming', to use a term aptly coined by Dorset farmer and firewood

OPPOSITE PAGE: Coppicing a hedge for woodchip, using tree shears.

RIGHT: Excavator with fork attachment being used to remove coppiced material during experimental work to find the most cost-effective ways of cropping hedges for wood fuel, at the Organic Research Centre's Elm Farm in Berkshire.

merchant Ross Dickinson. The potential is huge, not just to improve farm profitability but at the same time to enhance hedge health, all the while providing renewable, climate-friendly energy.

This chapter covers the resurgence of interest in hedges as a source of wood fuel. I shall start, though, with sections on sustainability and impacts on human health, since much concern has been expressed on both accounts. Only then will I turn to the practicalities and feasibility of managing hedges for fuel, whether as logs or as woodchips, and to the experiences of some of the farmers leading the way. Finally, I give a nod to the other uses to which woody material extracted from hedges can be put, in agriculture and horticulture.

How green is hedge wood fuel?

There are those who say that burning wood is no better than burning fossil fuels, on the grounds that both processes release greenhouse gases, mainly carbon dioxide, into the atmosphere. They have a point, but the premise is not necessarily true. Sure, if forests are being felled at a rate faster than they are regrowing, then at best there's still a short-term loss of carbon to the atmosphere, so it's not a truly green process. Burning wood is only climate-friendly if there's no net loss of biomass at any point of time – that is, the trees in a given area must extract carbon dioxide from the atmosphere at least as fast as it is being returned there through felling and combustion. Given the urgency of the climate crisis and the need within a decade or two at most to stop and reverse the accumulation of carbon dioxide in the atmosphere, waiting for growth to catch up with felling is not an option.

A second prerequisite for burning wood to be green is that it should be displacing the use of fossil fuels – it's no good at all if it's just adding to overall energy consumption, since no reductions in greenhouses gases will result. And a third is that the amount of non-renewable energy used in the processing and transport should be small in comparison to the renewable energy generated and used for heating.

Fortunately, hedges can meet these three requirements, often easily. At a farm level, assuming hedges are managed for wood fuel on a 20-year coppice cycle, as is typical, if just one in 20 hedges is coppiced in any one year, the remaining 19, if allowed to grow freely, should accumulate enough carbon that year to offset the amount released by combustion. If, alternatively, a line-of-trees hedge is being felled, then a greater length of hedge elsewhere on the farm will have to be allowed to grow unchecked to achieve a carbon balance. I know that on our farm the canopy cover of trees and shrubs is becoming ever greater even though much of our heating comes from our hedges – most farms have plenty of scope to be net accumulators of hedge carbon, provided those managing them are not obsessed with being neat and tidy (more on this later). We should not forget either that the ongoing accumulation of carbon in the soil beneath and beside hedges is likely to make for favourable carbon balance sheets.

The second prerequisite, that hedge fuel should displace the use of fossil fuels, will be met if, for example, a farmhouse is being converted from dependency on fossil fuels to biomass, or if new builds have biomass boilers or stoves installed where heating would otherwise be by oil or gas. There have been horror stories of people who have installed biomass boilers with a capacity greater than that needed and then left windows open so they can claim subsidies, but hopefully that is now behind us.

As for the third prerequisite, that the amount of renewable energy generated should greatly exceed the amount of fossil fuel used in its production, that too is easily met where hedge wood fuel is concerned. Because it's the most cost-effective way, hedge woodchip and logs are nearly always used on the farm where they are produced or within the local vicinity, so little energy is required in transport – we are not talking here about moving lorryloads of wood across the country to biomass power plants, or even importing the material by ship from overseas. Figures from Normandy show that for every one unit of fossil fuel used to produce woodchip from hedges, 44 units of renewable energy are returned – a very favourable ratio and one even higher than for short-rotation coppice (1:33) or the tall perennial grass *Miscanthus* (1:27). When I laid 90 metres of hedge on our farm to produce 2.8 tonnes of seasoned firewood logs and sticks for our stoves, I used just 10 litres of petrol for my chainsaw and 3 litres of diesel for tractor trips carrying the wood, giving a return of 89 units of renewable energy for every unit of fossil fuel used.

Another concern about the green credentials of hedge wood fuel is that management on a coppice cycle, as is usual practice, may result in hedges that are less dense than those that are regularly trimmed, and so be less good as wildlife habitat. There is a grain of truth in this. In northern France I have seen hedges that have been coppiced many times that are now threadbare. However, it need not be so, provided hedges are allowed to spread a little as they grow, brambles and other rambling plants are tolerated, and grazing is controlled. Ross Dickinson made the decision when he took on Thane Farm in Dorset to manage nearly all its hedges on a 20-year coppice cycle without any flail trimming at all. Now, 40 years later, the hedges are a joy to behold – thick, species-rich and luxuriant. Each time he coppices a hedge he takes it back to its bare bones, paring away all

BELOW: Coppiced hedge with four years' regrowth managed by Ross Dickinson in the Marshwood Vale in west Dorset. A photo of this hedge freshly coppiced is shown in the box on page 283.

the side growth, but rapidly, within four years, the hedges once more provide excellent habitat, good enough even for Hazel Dormice. Since he manages the hedges on a well-staggered rotation, coppicing just one or two a year, there is ample habitat continuity across the farm. Elsewhere, if coppice regrowth is starting to look thin, then there is always the option of side-trimming to thicken a hedge up, with little loss of wood fuel crop.

Health concerns from burning wood

Wood smoke can be damaging to human health. It contains a range of air pollutants, tiny particles that have a diameter of less than 2.5 micrometres (PM2.5) being of particular concern. These can cause severe respiratory and cardiovascular problems. In the UK, some 38% of all such atmospheric particles of anthropogenic origin are said to come from wood burners and open fires. So there's no doubting it's an important issue, and regulations have recently been introduced around selling wood fuel and the manufacture of wood burners, to safeguard people. To reduce the health risk, the advice is only to burn logs that have a moisture content of 20% or less, to use wood stoves or boilers that pass emission standards, and to keep them and their flues well maintained. Also, it's best to avoid open fireplaces and burning wood in densely built-up areas.

Once legal requirements have been followed, it's a matter of personal choice as to whether to burn hedge wood for heat or not. True, it has much value as a form of cheap renewable energy but, however careful you are, there will always be at least a slight risk to human health. My family is comfortable with this risk, as indeed are many others; indeed, we enjoy our two wood stoves. Our farmhouse is isolated from other dwellings so there's no risk to the health of people living around us. Managing hedges in part for firewood helps to keep them in great condition for wildlife – it's a powerful incentive to care for them.

Burning wet wood not only leads to unnecessary pollution but makes absolutely no sense from an energy point of view. For example, a green log with a moisture content of 60% will when burnt give out only 30% of the heat compared to a seasoned one at 20% moisture content. This explains why Ash is considered the queen of all firewoods: when green, its 33% moisture content is much lower than that of most other species (Hazel 44%, Hawthorn, oaks, Beech and Sycamore 47%, elms 58%, willows 60%). Even so, Ash logs should be seasoned until their moisture content is 20% or less before burning. Then their calorific value is very much the same as for any other hardwood: weight for weight and at the same moisture content nearly all woody species give a very similar energy output. A kilogram of seasoned willow will give out the same amount of heat as a kilogram of seasoned oak. Conversely, being less dense, a cubic metre of seasoned willow will give out considerably less heat than a cubic metre of seasoned oak – which is why one should buy seasoned firewood by weight rather than volume, given the chance. It is important, of course, to check that logs are indeed seasoned – either with a moisture meter or, when experienced, by their look, feel and the sound they make when tapped together.

The different types of hedge wood fuel

Wood for heat is usually taken from hedges either as logs or as woodchip. Far more people use logs because you don't need any specialist machinery to produce them beyond a chainsaw, and stoves are much less expensive than biomass boilers. Woodchips, on the other hand, are more efficient since every last twig can be utilised and they can be produced rapidly and cost-effectively by large machines without any man-handling needed (Westaway *et al.* 2016, Wolton *et al.* 2016, Smith *et al.* 2023). On the downside, and it's a big one, woodchip biomass boilers are very expensive and the whole associated setup, including hoppers, hot-water storage tanks and a plethora of pipework, needs a lot of space – like an outhouse or barn.

Between these two main options falls that of using a machine that is usually called a branch logger but sometimes a chunker. This looks like a normal hand-fed chipper but instead produces mini-logs or short lengths of twigs – these are suitable for use in stoves or ovens when heat is required rapidly, or for kindling. They are very handy machines, bridging the gap between chip and log production, and reducing wastage.

Logs

Logs are usually taken from hedges as they are being laid or coppiced, so one can accomplish two things at once, both rejuvenate the hedge and replenish the winter firewood store. As I write, in the early 2020s, there's an additional ready source of logs on most farms, and that's

LEFT: Branch loggers convert small stem wood into mini-logs and kindling.

LEFT: Hedge logs. The larger ones will need splitting.

diseased hedgerow Ash trees. We have just had to fell a dozen or more Ashes growing beside a minor road that runs through the farm, all diseased. Already some had fallen in gales, requiring us to grab chainsaws to clear the highway.

There are new ways to reduce the expenditure of human energy on log production. The first of these is the use of a wood processor, machines which cut and split stems into logs of the required size. But hedge stems are often gnarled and twisted, the smaller ones in particular, so can be difficult to feed into these machines – they are really designed for nice straight poles sourced from plantations. Alternatively, powered saw benches and hydraulic log splitters can be used, although these require more manual handling than wood processors. Then there's tree shears, remarkable pieces of kit that are rapidly and deservedly becoming popular. In effect, they are powerful hydraulic secateurs or loppers mounted on excavators, capable of quickly cutting through large stems, 25cm or more in diameter. Better still, before they sever the stems they grab them, so once they have been cut the excavator can lift them in their entirety out of the hedge and place them in a neat row beside it – a real blessing, especially if the hedge is fenced or if it is on a high bank. They are very handy too when it comes to safely cutting back or taking down trees beside roads or those that have cables running through or near them.

I hired in a contractor with tree shears to coppice a tall hedge on our farm that was approaching a line of trees. It was some 200m long and I thought it would take him all day. But by lunch time he'd finished, and I had to find further stretches of hedge nearby that could do with management. All this work, achieved in just six or seven hours, would have taken two or three weeks for a team of two men to do with chainsaws, first taking the fences down and then dragging the cut material out of the hedges by hand. True, we

RIGHT: Tree shears resemble powerful secateurs or loppers and are used for gripping and felling trees much bigger than the branches shown here!

still had to extract stems of the right size for cutting into logs from the rows of trees laid out in the fields, removing side growth with chainsaw and billhook, but we would have had to do this anyhow. Another task we did was to go along the hedges cutting back the stumps closer to the ground. Tree shears rarely cut neatly – they tend to split and fracture remaining lengths of stem, to a degree dependent on the make and model used. So it's best to leave 40cm or so standing which can then be reduced to normal coppicing height by chainsaw, leaving a clean cut, an operation which does not take long. This will help to ensure the regrowth is stable. Three years on, the tree-sheared hedges on our farm have regrown well and are now once more thick and bushy.

ABOVE: Hedges can yield large amounts of firewood. This one was coppiced in a single morning by an excavator with tree shears.

One drawback of taking only logs from hedges is that all the small branches and twigs are wasted, at least as far as wood fuel is concerned. Twigs are just as good a source of heat as stem wood, or very nearly so – at the same moisture content, a kilogram of twigs has much the same calorific value as a kilogram of heartwood. When combusted, bark does produce a little more ash than logs, but modern wood boilers and stoves are so efficient this slight increase in ash is not an issue. Burning twigs and small branches in bonfires is not just bad for the climate, it is wasteful too.

ABOVE: Excavator with tree shears removing an unwanted tree from a tall hedge for firewood. Depending on the land, this is often best done in the autumn, while the ground remains firm enough to take heavy machinery.

In the winter of 2011/12, I set out to find out just how much energy is wasted if only logs are taken from a hedge for wood fuel. A wide, banked hedge containing a good number of different trees and shrubs but dominated by Grey Willow was ready to lay, having last been laid nine years beforehand and not trimmed since. After I'd laid a decent stretch that turned out to be 46m long, I sorted out all the removed woody material into five size categories – the smallest being thin branches and twigs less than 2.5cm in diameter and the largest stems more than 10cm in diameter, with the intermediate categories each spanning a 2.5cm range. The smaller branches and twigs I tied into bundles. Then, with the aid of a borrowed tripod, block and tackle, I weighed the lot green. Laborious work made more challenging by the season being exceptionally wet, the field quickly turning into a muddy morass as I trampled back and forth over the working area. But worth it. In all, the extracted material weighed 2,689kg. Of this, 41% fell into the smallest category and a further 30% into the next one (2.5–5.0cm). Few will bother to extract branches smaller than 5cm for use as firewood (unless they have a branch logger), so 71% of this hedge's biomass would have been wasted by most. If one decided only to use the bigger branches, as many do, those larger than 7.5cm in diameter, then the wastage would have been 91%.

These wastage figures appear typical. In East Devon, on the Cotley Estate, a Hazel–elm hedge produced just under 4 tonnes of material from a 30-metre stretch when it was laid (Wolton 2012). Of this, 92% was smaller than 7.5cm in diameter, within a percentage point of the predominantly willow hedge I looked at. In a third experiment, this one carried out on the National Trust's Wimpole Estate in Cambridgeshire, Simon Damant tells me that 76% of the 5 tonnes of material removed from a Field Maple–thorn hedge after laying was less than 5cm in diameter. In all three experiments, a proportion of usable firewood stems was retained in the hedge for laying, but this amounted to only a low fraction, about 10%, of hedge biomass.

If hedges are allowed to grow bigger before being harvested, to the point where coppicing is more appropriate than laying, a higher proportion of the woody material is suitable for log production. Our work on another hedge on the Cotley Estate, this one consisting mainly of Beech and Ash stems ripe for coppicing, produced 34 tonnes of woody material from a 30-metre stretch, of which just 38% was less than 7.5cm in diameter. So, if you want logs, the more mature a hedge the greater the proportion of usable biomass. Unless, that is, you have a use or market for mini-logs, making it sensible to run small stem material through a branch logger.

BELOW: The author weighing different sizes of material taken from a hedge after it had been laid, to find out the relative proportion of twigs and small branches to stems large enough to use as logs for the stove.

Just how much firewood can a hedge produce?

Ross Dickinson is a forward-thinking farmer and firewood merchant who is truly passionate about his hedges. Amongst other things, he is a strong and effective advocate for managing hedges on a coppice rotation rather than by regular flailing, to produce economically viable crops of firewood and provide rural employment opportunities. To my mind, he's a fine 'hedge farmer', one of the best. Both of his family's two farms are in the green rolling countryside of the Marshwood Vale in Dorset. He manages most of the hedges, which stretch for 19km, on a 15- to 20-year coppice rotation, only flailing those alongside roads. These coppiced hedges are in fine heart, in part because Ross is content to allow them to thicken out into the fields on either side as they grow, so they are dense and bushy throughout most of the coppicing cycle. Ross, supported by his family including his son Euan, markets firewood from these hedges and small farm copses together with hedge wood and woodland thinnings from neighbouring farms, supplemented by bought-in timber. Altogether, he sells some 175 tonnes a year. The stacks of cut and split logs air drying in his sheds are prodigious (I was quite proud of my own hedge log heaps until I saw his – now they seem puny). Half of this wood is sold in feed bags at the farm gate using an honesty-box system, the other half is delivered by pickup to customers within a 6km radius.

In the winter of 2017/18 Ross decided to carry out a ground-breaking experiment to determine just how much firewood of different sorts a single hedge could produce, and whether this firewood could be marketed profitably. He chose a 220m mixed-species hedge typical of the farm, with a 25% cover of Sycamore, a good firewood species since it grows quickly and produces clean straight stems and branches. He

Ross Dickinson beside his impressive pile of logs, many taken from hedges.

and Euan coppiced the hedge with chainsaws, leaving occasional standards, and sorted and processed the extracted material into four products: (1) round and split logs 20cm long, suitable for sale, (2) 'ugly sticks', thin, often twisted, logs that the public dislike buying, (3) small-diameter stems with their side branches removed, used to produce mini-logs 5cm long, and (4) brash for kindling twigs; the latter two products being supplied in net bags filled from a branch logger. Only some of the brash was used to make kindling, that from thorns, for example, being discarded and disposed of in bonfires. The ugly sticks were made good use of by Ross's family in their eight wood stoves, although some were sold to camping businesses, including upmarket 'glamping', for use on open fires. The mini-logs were sold to those with small top-loading log burners, for example in shepherd's huts or boats.

When all these products were weighed, the hedge was found to have produced 9 tonnes of saleable logs, 6 tonnes of ugly sticks, 99 net bags of mini-logs weighing 2.5 tonnes, and 260 net bags of kindling twigs weighing just under 4 tonnes. The income from the sale of all these products exceeded the costs of coppicing, sorting, processing and delivery – labour costs were covered, and a small profit generated. Including in the balance sheet the savings realised by not flailing the hedge every year increased the profit, as would receiving an environmental grant for the coppicing. Altogether, a respectable profit of 73% on expenditure could be achieved. And the equivalent of 8,500 litres of heating oil saved.

Ross concluded that, at a farm level, taking hedges away from an annual flail-trimming regime in favour of a coppicing cycle is economically worthwhile provided the regrowth is sold for firewood. A living wage can be made by the coppicer, creating local employment, and some income generated for the farmer – all with no need to buy or hire any equipment other than that which most farms already have to hand.

Ross Dickinson by his experimental hedge in the Marshwood Vale, Dorset, used to assess the economic viability of changing management from annual flailing to coppicing. The green net bags contain mini-logs produced by a branch logger.

Woodchips

It was on the Hedgelink visit to Normandy in May 2010 that I first came across hedges being used as a source of woodchips for heating (Wolton *et al.* 2010). It was quite an eye-opener. In the commune of Athis-de-l'Orne (now part of Athis-Val-de-Rouvre) we were taken to a large open-fronted shed that was full of woodchip, delivered by local farmers and largely sourced from hedges. This woodchip was being used to fuel a large biomass boiler heating the mayor's office, the school, a residential home for the elderly and six council homes. Our hosts explained that, faced with an unwelcome proposal for a nuclear waste storage facility, in 2006 the local community decided to demonstrate their green credentials by establishing a cooperative, Bois Bocage Énergie, to produce renewable energy from the farmed landscape. It worked, and the threat of the nuclear store was averted. At the time of our visit, there were 70 farmer producers with 110 individuals and 10 councils buying the chips. Most of the chips were being retained by the farmers themselves to heat their own farmhouses or as animal bedding, but about 20% was being taken to the central depot we saw for seasoning and sale. Fees paid by producers and users were sufficient to cover the costs of employing a project officer to coordinate the operation. Against a backdrop of large-scale hedge removal across Normandy, in this commune at least hedges were once more being valued by farmers and being actively managed: the future of the local bocage landscape looked brighter. Better still, considerable amounts of green energy were being used instead of fossil fuels or nuclear energy. The cooperative continues to function as I write.

BELOW: Hedge woodchip depot, Saint-Samson-de-Bonfossé, Normandy. The pale chips are fresh ones that are in the process of self-drying, the darker ones are ready for combustion.

ABOVE: Chipping trees coppiced from a hedge by tree shears using a tractor-mounted grab and a whole-tree chipper. St Amand, Parc Naturel Regional des Marais du Cotentin et du Bessin, Normandy.

Subsequently I have had further opportunities in mainland Europe to see hedges being used as a source of woodchip for heating, in Normandy again and in Germany. In the Parc Naturel Régional des Marais du Cotentin et du Bessin in Normandy another cooperative promotes the management of hedges through managing and harvesting them for woodchip. Here the mayor of Saint-Samson-de-Bonfossé proudly showed us a large woodchip boiler heating a new development of 10 sheltered housing units. Participating farmers who supply woodchips sign up to a management plan to ensure their hedges are managed sustainably and not over-exploited.

In the same regional park we were given a demonstration of hedge material being chipped. Before we arrived on site 80 metres of hedge had been coppiced by an excavator with tree shears, in just one and a half hours. We saw the felled trees, some with trunks 25cm in diameter, being picked up by tractor-mounted grab and fed whole into a large chipper towed by the same tractor. The chip was blown into a silage wagon pulled by a second tractor, to be taken to a farm shed for seasoning. It took one hour to chip all the trees. So, in just two and a half hours the hedge had been converted to 40 cubic metres of woodchip, enough to heat a typical three- to four-bedroom leaky farmhouse for a year. The cost of machinery hire and transport was much less than it would have cost to buy in the same amount of energy as oil. No manual work was involved at all. The farmer expressed considerable satisfaction.

ABOVE: David Johns collecting stems extracted from a hedge as it was being laid, for outdoor seasoning before use in a biomass boiler. Hillhead Farm, Devon.

In 2012 Nigel Adams and I went on a hedge road trip that took us from the Netherlands through Germany to Denmark. In Germany we were shown hedges that were being coppiced to produce woodchips for boilers and, in one case, in Kreis Steinfurt, for a small local power station. The processes used were very similar to those we saw in Normandy, and as we drove north towards Denmark we saw many other examples of coppiced hedges, few being trimmed in the typical British fashion.

Back in the UK, subsequent research carried out in the Tamar Valley and Blackdown Hills Areas of Outstanding Natural Beauty in south-west England (Wolton 2012, 2014b) and by the Organic Research Centre on their farms in Berkshire and Suffolk (Chambers *et al.* 2019) has confirmed that managing hedges to produce woodchips for heating is just as practical and economically viable on this side of the Channel as it is in mainland Europe. There are now quite a number of British farmers heating their houses and livestock units with woodchips harvested from hedges, in part at least.

For example, in 2013 David Johns, who farms in the South Hams in Devon, installed a biomass boiler to heat his traditional farmhouse and two adjacent properties. Nearly all the woodchip comes from the farm's hedges, which are managed on a laying or coppicing cycle, with trimming restricted to roadsides. Although expensive to install, within eight years the system had more than paid for itself, through receipt of a government renewal heat energy incentive and through savings on fuel bills.

Woodchip doubts and drying

Three objections are often raised whenever the idea of using hedge woodchip for wood fuel is mooted, despite the material being used successfully for this purpose on many farms on both sides of the Channel. These concerns relate to the quality of the woodchip. The first is that it has a lower energy value than that from forests – but, as we have already seen, the difference is marginal. The second is that hedge woodchips, having a higher proportion of bark than chips produced from the trunks of plantation trees, are more polluting – but modern boilers are so efficient that any increase in air pollution is insignificant. The third is that because small branch material is involved the chips produced are irregular in shape and size, so likely to cause blockages in boiler feed systems. While this could be a problem, it need not be a barrier, for there are several boilers and feed systems that can cope with irregular chips. In Britain it's the norm, in my experience, for farmers to feed their biomass boilers with markedly uneven chips, ranging from dust to palm-sized chunks, including pencil-shaped shards: provided they have the right boiler feed equipment, this is not a problem. Only when selling off farm is irregular chip size and shape likely to be an issue, but selling off farm will seldom be a competitive option outside cooperative ventures, because those producing chips from plantations will nearly always be able to undercut your price.

That brings me to methods of drying woodchip, another matter of worry to British farmers, understandably so given our high rainfall. There are two main methods, both of which are straightforward and work well even in our climate. The first, favoured by those on the continent, is to chip green and heap in a well-ventilated shed. Green woodchip piles have the convenient property of self-drying. Heat generated by fermentation evaporates off unwanted water, and within just 4–6 months the woodchip moisture content is 25–30% and the chips are fit for efficient combustion in boilers. Try putting your hand into even a small pile, a few wheelbarrows' worth, of fresh green woodchips a few days after it has been made and feel the heat! Provided there's not too much green leafy material present in the heap and it is not turned, it will not end up as compost, just nice dry woodchip ready for

BELOW: Woodchips produced by chipping seasoned hedge wood. Although of uneven size and shape they were being used without problem to fuel a farmhouse biomass boiler. Chips made from green wood tend to be more even-sized.

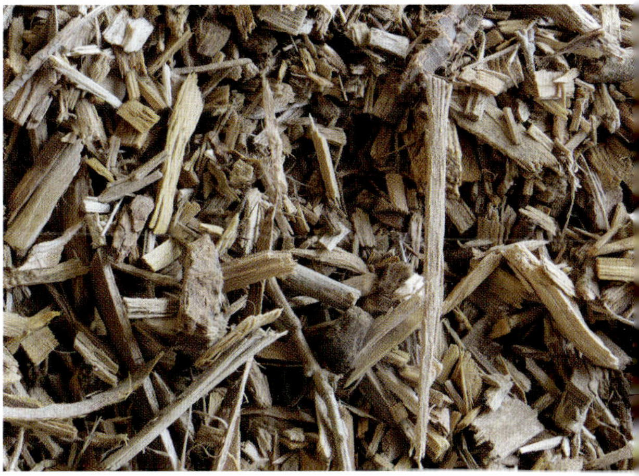

combustion. Neither is there any risk that heaps will burst into flames unless really huge – in Normandy we saw green woodchip piled metres deep in sheds leaning up against farmhouses. One drawback of this method is that about 15% of the energy content of the chips is lost in the fermentation process.

The other method is simply to pile cut stems and branches outside over the summer to air dry, chipping them as needed during the colder months. This method is preferred by most UK practitioners for a number of reasons. The first is that there's no need to have a covered storage space where fermentation can take place; instead, woodchips can be blown direct by the chipper from the outdoors into a boiler's feed hopper. Also, given a dry year, chips of lower moisture content can be produced, closer to 20%, increasing combustion efficiency, while no energy is lost through fermentation. Whichever way is chosen to dry chips, there's no need for any external heat source other than the sun. Some may use grain dryers or kilns, but this is not necessary and is likely to increase both carbon emissions and expense.

BELOW: Jonathan Boaz demonstrating the quality of the compost he produces by mixing together hedge woodchips, rotting matter gained from cleaning ditches and other organic farm materials. He spreads the compost on his fields at Mill Farm in Worcestershire, as a soil improver and fertiliser.

Other uses for hedge woodchips

Hedge woodchips, as with those sourced from other farm trees and bushes, can be used for purposes other than for heating. These include serving as bedding for housed livestock, as mulches to suppress weeds, and as soil conditioners. Woodchip can be composted if it is kept moist, turned and given long enough to decay, and then used to improve the soil in gardens or for horticulture. In fact, hedge woodchip is particularly good for this purpose in comparison to chips generated from larger-diameter material, because nutrients are concentrated in bark and buds. As research by the Organic Research Centre has shown, even fresh un-composted woodchip from small-diameter branches, known as ramial woodchip, improves soil fertility through increasing soil organic matter, water-holding capacity and nutrient levels (Westaway 2020). It is, however, likely to take longer to have a positive effect than composted woodchips.

In Chapter 9, when writing about ditch management, I mentioned how an enterprising farmer, Jonathan Boaz of Mill Farm in

Worcestershire, mixes the rich organic sludge taken from farm ditches into large compost heaps. He also adds woodchips derived from farm hedges, pollarded willows and other sources, first mixing them with green leafy material such as lucerne. When ready, the compost is spread onto his fields. In essence, he has found an effective way of recycling hedge fertility into crops, saving on the use of expensive artificial fertilisers.

What types of hedges are good for wood fuel?

Most hedges can produce respectable crops of wood fuel if managed right, but the best are those that are wide, growing on fertile soils and have tree species that accumulate biomass rapidly. Those located on high windswept places, or those on poor acid soils, are not likely to be worthwhile. The graph below compares the expected yields of different species (after Wolton 2014b). Hedges with a high proportion of willows like Grey Willow or Goat Willow are best of all. Although these trees produce wood with a low density and high water content when green they still accumulate carbon faster than all other native species, except perhaps poplars and Alder. But, as I find on our farm, willow stems and branches can sometimes be so twisted, bent and intertwined as to make extraction and processing a real pain, especially if a lot of manual handling and dragging is required. Sycamore, Beech and perhaps Hornbeam, even though they grow a little slower, may be a better option if planting a new hedge for

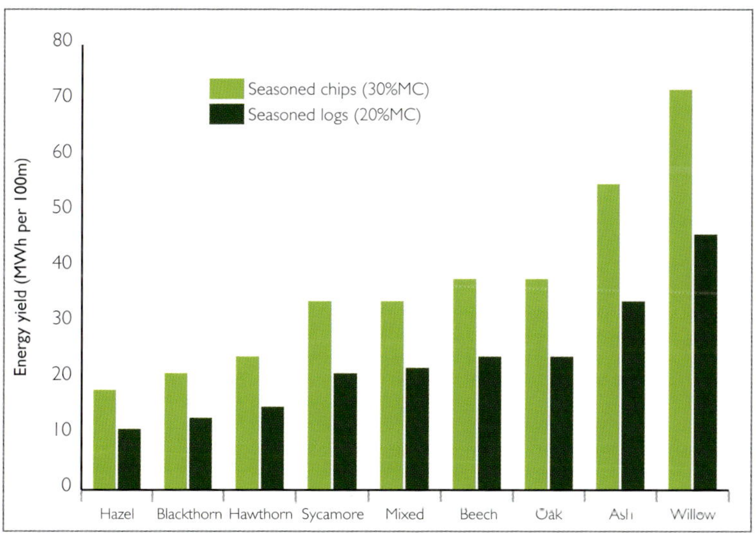

LEFT: The estimated wood fuel yields of hedges of different species when cropped at the optimal stage for coppicing, assuming an average width of 2m at the base and fertile soil. MWh, megawatt hour; MC, moisture content.

fuel, but beware these species are prone to heavy damage from Grey Squirrels which find their bark irresistible. The demise of Ash is deeply felt, as a tree that grows fast, has straight easily worked trunks and branches, and that is avoided by squirrels. Wild Cherry, another fast-growing species avoided by squirrels, is a possible alternative.

Cost comparison with other energy sources

Relative costs of different sources of energy are changing all the time. Even given this flux, for a farmer nothing compares to producing and combusting your own home-grown woodchips, assuming the use of tree shears and whole-tree chippers (e.g. Smith *et al.* 2023). Buying in heat energy, whether as gas (piped or LPG), oil, coal, electricity or wood (chips, pellets or logs), will almost certainly be more expensive. Even bulk buying in woodchip from commercial plantations, harvested by highly efficient forestry machines working at large scale, will almost always be a more costly option.

On the other hand, burning your own hedge-procured logs, if you cost in your own time or that of farm workers, will probably be on a par with other sources except for electricity, which is likely still to be more expensive. Indeed, buying in seasoned hardwood logs may be no more expensive than taking them from your own hedges. But if you are laying or coppicing your hedges anyhow for other reasons, such as to keep them in fine fettle for wildlife, then taking logs from them as you do so should financially be well worth your while, as well as being hugely satisfying and helping to keep you fit. If you are receiving grant-aid for rejuvenating or restoring your hedges, taking logs from them will almost certainly make financial sense.

Concluding remarks

I think it's safe to predict further growth in hedge farming for wood fuel. The drivers are threefold: farmers' desire for affordable energy, society's support for renewable energy, and national targets around carbon capture. In Chapter 13, I noted that managing hedges on a coppice rotation without top trimming is optimal for carbon sequestration – so too it is for wood fuel production, the two functions going hand in hand. Linked to this is the growing realisation that burning the brash arising from hedge management in bonfires is wasteful and not helpful from a climate perspective.

ABOVE: When I lay hedges on the farm, as here, I always extract as much firewood as I can to help keep our home warm.

There's considerable scope too for local communities to work together with farmers to manage hedges for wood fuel, in particular to benefit those in fuel poverty. This is being trialled, and how-to guidance is available for those wishing to set up such community groups (e.g. Dartmoor Circle 2012). Farmers benefit through getting their hedges laid or coppiced, while local communities get free firewood in return for voluntary labour, as well as a chance for social engagement and physical exercise.

There may be risks to hedges from over-exploitation, but if these do emerge they are unlikely to go unchallenged by the environmental movement, and can be guarded against by putting appropriate policies, incentives and regulations in place. The opportunities for improved hedge management, with fewer hedges either being tightly flail trimmed or neglected altogether, surely make it worth the risk.

As a final thought, the advantages we have over our forebears in terms of equipment are immense – chainsaws, hydraulic log splitters, excavators with tree shears and branch loggers – but there is one piece of equipment that has yet to be successfully developed, and that is a machine that can collect the woody debris that flies from the flail cutter. Some entrepreneurs have developed prototypes, but as far as I know no such machine has yet been developed to the stage of commercial production. We need one.

Hedge management

chapter fifteen

Hedges need to be managed, yet, truth be told, this is seldom straightforward. Just because they are common and largely created by humans, hedges are no easier to manage well than most other habitats. Allowed to grow unchecked, sooner or later they become gap-ridden, eventually derelict. To make matters more challenging still, the same happens when hedges are over-managed, as when closely trimmed year upon year – they literally go to pieces. Getting management right is a complex business, one that requires a good understanding of the ways hedges grow and mature. Both knowledge and skill are required to do a good job.

Over time, given the chance, lines of bushes turn into lines of trees, as nature intended but often to the frustration of our own objectives. That's not to say that unmanaged line-of-trees hedges have no part to play in the countryside or in built-up areas, they do. They support some wildlife that does not thrive in short bushy hedges, add structure and beauty to the scenery, and provide shelter and privacy. But most landscapes deliver best for wildlife, and for all the other many things hedges do for us, if the majority of the hedges are of the classic bushy form, ideally with standard trees studded along their length.

At least that was so until we started to worry about climate change and it became a national and moral imperative to lock away as much carbon as possible. That has thrown a spanner in the works – where should our priorities lie? Optimising hedges for wildlife or for carbon? Can an acceptable compromise be found? As I suggest in the section below on hedge shape, I believe one can.

Then again, a good hedge is more than a line of bushes, it should also have herbaceous margins sporting tussocky grasses or plentiful flowers, preferably both. Standard trees too. Banks and ditches add further value. All these components need to be considered. A holistic, integrated view of hedge management should be taken, but it rarely

OPPOSITE PAGE: Hedge laying in the Devon style, expertly done by Mike Reed at a Devon Rural Skills Trust hedge laying competition.

ABOVE: This oak, birch and willow hedge has become very sparse at the bottom, the result of no management for many years. The undergrowth has been shaded out and destroyed by cattle and sheep. The trees themselves are unlikely to develop high wildlife value – the hedge is ripe for restoration by coppicing and planting up gaps.

BELOW: This Beech hedge in Dartmoor National Park with its full-canopied mature trees full of holes, splits and other veteran features is excellent for wildlife and as a carbon store, as well as being a fine cultural and landscape feature. Eventually it will need replanting, but it is good for decades yet.

is. Far too often, management is thought about only in the context of the woody bits, even though we know that many widespread animal species require some combination of bush, tree, margin, bank and ditch to thrive, as described in Chapter 5. Such a rounded approach is by no means easy, though. For example, one way to delay the rate at which an unmanaged hedge becomes gappy is to allow it to spread out sideways into adjacent land. However, to do so is to put at risk any light-demanding plants in the hedge bottom and in its herbaceous margins and ditches, not to mention making future laying or coppicing more difficult.

Hedge managers must think about many things, but above all they need to be clear on objectives and to think at a landscape scale, as well as considering all structural components. As already emphasised in several chapters, no hedge should be managed in isolation from its neighbours, since there's no single perfect state for a hedge to be in. Only a hedge network can be perfect, even if we are not quite sure as yet what this might look like.

To manage a hedge, let alone a network of hedges, is exacting, time-consuming and often physically demanding. My neighbour and good friend Sally Vergette has just phoned, and in passing she mentioned that last summer's drought meant her sheep had burrowed deeper into the hedges than normal, creating gaps. These she is blocking

BELOW: Wildlife will not thrive on farms where all the hedges are trimmed hard every year.

ABOVE: A fine hedge for wildlife, thick and bushy, containing a mature Ash tree and a flowery margin with Red Campion and tussocky grasses.

with home-made hurdles of willow and Hazel, whereas before, on many a farm, rusty sheets of corrugated iron or old bedsteads would have served this purpose. But all the effort expended on getting management right is well worth it: the reward is thick, healthy hedges full of wildlife, providing shelter and shade for crops and livestock, conserving soil, and boosting pollinator populations.

Management cycle

No hedge can remain in the same condition for ever. The stems of shrubs and trees thicken and the patterning of their branches changes as they mature; banks erode, ditches infill and margins become overgrown. The upshot is that hedges must be taken through a cycle of management to remain healthy, a concept developed by hedgerow guru Nigel Adams. A leading light in both the National Hedgelaying Society and Hedgelink, over the last three decades he has been at the forefront of developing new ways of managing hedges, based on many years of practical experience. In the noughties, Nigel developed a 10-point scale of hedge management, now being called the Adams scale (Adams 2012). This has been hugely influential.

The Adams scale, shown graphically on pages 300–302, acknowledges that to remain in good heart the woody part of a hedge must periodically be *rejuvenated*, either by laying or coppicing, otherwise it will lose structure and become full of gaps. Nigel uses the word 'rejuvenated' rather than 'restored' advisedly, recognising that when a hedge is ready for laying or coppicing this is not because it has been damaged, mismanaged or neglected, but rather because it has gone through a natural process of growth and maturation and now needs to be reset. He explains that the word 'restore' is only appropriate where something has gone amiss and requires substantial corrective action. The rejuvenation approach is analogous to the management of coppice coupes in woodland, where stems are cut to the ground to be allowed to regrow to produce useful products of one sort or another, a cycle of management that has been practised for centuries. Restoration here is only necessary where the coppice stools have died or become moribund, perhaps due to deer browsing or disease.

With hedges, *restoration* is required when a significant number of trees and shrubs have died, leaving the hedge so thin that if cut down to the ground the regrowth would be insufficient to form a decent thick hedge. Gaps must be planted up. Such loss of hedge bushes and trees is usually the result of one of two things happening. The first is that the hedge has been allowed to develop into a line of trees, which owing to competition for resources have self-thinned. Here some trees may well have passed beyond the point at which they will regrow if felled: mature Beech is notably poor at surviving felling, failing to send up new shoots from the cut stumps. The second reason for loss of bushes and trees is that the hedge has been repeatedly trimmed, often incessantly, at the same height year upon year, reducing the vigour of woody plants to such an extent that they either die or are unable to regrow from cut stumps. Both processes are exacerbated by high levels of browsing or trampling by deer, sheep, cattle or other large mammals.

Switching attention briefly to other hedge components, restoration is also required if margins are lost or lose their flowers, perhaps because of cultivation, herbicides, heavy shade,

BELOW: A sad sight – all that's left of a banked hedge after years of being overrun by sheep and, doubtless, cattle. It is now desperately in need of serious restoration.

high grazing pressure or, most likely of all, because they have been swamped out by Nettles and Goosegrass. It's also necessary if banks have become badly eroded or ditches too dry, such as when infilled with sediment.

Returning to the woody part of the hedge, once familiar with the 10-point scale, hedge managers can, as Nigel puts it, 'read' any hedge. When the point at which a hedge lies has been determined, linked guidance presents appropriate management options. So, for example, if a hedge is at Point 7, that is with frequent healthy stems more than 4 metres high of the right thickness to lay, it is fit to be managed in one of three ways. It can be rejuvenated by laying, which will take it back to Point 4, or it can be to be reshaped back to Point 5, or it can be allowed to grow on for a few more years to the optimal stage for coppicing, Point 8. Which option is best to take will depend on management objectives, the state of other hedges in the network, the needs of any special species present, and so on. At times, it helps that the boundaries between options can be blurred, removing the need for black-and-white decisions to be taken. For instance, it is usually possible to coppice at least a few stems when laying a hedge without compromising the end result, and vice versa – when coppicing to lay a few stems, to improve habitat continuity. Hedge reading is usually best done in the winter after the leaves have fallen, because that is when it is easiest to see the hedge structure.

The duration of the hedge management cycle can vary greatly. If a hedge is on reasonably fertile soils and not trimmed at all, then it may be appropriate to lay every 10–15 years, or to coppice every 15–20 years. So much depends on objectives and local conditions. For most farmers the less often hedges have to be laid or coppiced the better, since these are expensive and time-consuming operations, the former particularly so. Long gone are the days when there was plentiful cheap farm labour available for this winter work. The answer is to trim. By doing so one can greatly extend the management cycle, perhaps to 50 years or more. As I shall explain later, trimming, done properly, can also be of great benefit to wildlife, although it may well be counterproductive if one's main goal is carbon capture or wood fuel production.

Even on our small farm, I could easily spend most of every winter doing nothing but laying hedges. As it is, I am saved from that by trimming a proportion every year. Fortunately our hedges, being wide and rich in different species of shrub and tree, are capable of

ABOVE: Beautifully restored hedges in East Lothian. Formerly very sparse, the remaining shrubs, mainly Hawthorn, were coppiced and the gaps planted up. Five years later they looked like this!

remaining in a healthy gap-free condition for decades, even if trimmed every year. One hedge has been close trimmed annually ever since we have been here, now over 30 years, to allow our neighbours to see out over the countryside from their farmhouse, yet it remains in reasonably good heart.

The same would not be true if the hedge consisted mainly of Hawthorn, as so many hedges elsewhere in the country do. This is a tree that only wants to grow from the top and quickly develops mop-headed growth when trimmed regularly, leaving the wind to whistle through the large gaps that develop beneath. Ash is another tree that does this, whereas Blackthorn, Hazel, Beech, Hornbeam, Field Maple, Holly and elms either constantly throw up new stems or suckers from the base or retain their lower branches for a long time in a living state. For this reason, among others, I would not recommend anyone to plant a new hedge with a high proportion of Hawthorn, even if characteristic of the local landscape. It is, quite frankly, one of the more difficult species to manage. In the days before wire, it used to be the hedging plant of preference because of its thorny stems and quick growth. Now, however, new hedges in livestock areas are nearly always fenced as a matter of course, so planting mainly thorns is pointless. Still, Hawthorn has its part to play because it produces large quantities of wildlife-friendly flowers and berries.

The 10 points of the hedge management cycle

Developed by Nigel Adams, adapted to include banks. Standard trees and herbaceous margins should be encouraged and maintained throughout. See also the illustration on page 302.

Point 1 Repeated close trimming has resulted in a short hedge with many gaps and sparse stems, the shrubs gnarled and often rotting. Any bank is usually badly eroded.

Action Coppice and replant any gaps. Restore the bank if necessary. Remove undesirable species like non-natives.

Point 2 Heavily trimmed and short, but not yet gappy or moribund. The shrubs are mop-headed with hard knuckles at the cut line. The stems are still too far apart to create a thick hedge after being allowed to grow up and then either laid or coppiced.

Action As above.

Point 3 Trimmed short yet the hedge remains dense. However, mop-headed growth is developing together with hard knuckles at the cut line and gaps at base. Stems are sufficiently frequent that when allowed to grow up the hedge can successfully be laid or coppiced.

Action Allow the hedge to grow up so it can be laid, coppiced or reshaped. The hedge may be trimmed as it grows up, but raise the cutting height each time.

Point 4 Recently laid, coppiced or planted.

Action In first few years trim the hedge often to create dense growth, initially low down, raising the cutting height 10cm or so each year until Point 5 reached.

Point 5 Dense with frequent healthy stems 1–3m high, above the ground or any bank.

Action Trim at intervals ranging from every year to once every five years but raise the cutting height 10–15cm each time, until Point 6 reached.

Alternatively, if carbon capture or wood fuel are the objectives, allow the hedge to grow up uncut until Point 8 is reached.

Point 6 Dense, with frequent healthy stems over 3m high. Hedge bottom may be starting to become thin.

Action Allow the hedge to grow up ready for laying (Point 7) or coppicing (Point 8).

Alternatively reshape with a circular saw or finger bar cutter to Point 5.

Point 7 Frequent healthy stems more than 4m high and of the right thickness to lay (typically 8–15cm in diameter).

Action Lay to return to Point 4, reshape with circular saw to Point 5, or allow to grow to Point 8 for coppicing.

Where there is a bank, cast up any soil that has slumped or eroded to ground level back up onto the bank top. Patch up any damage to bank facing.

Point 8 Tall hedge, typically 6–7m high and with some stems too large for laying (usually more than 15cm in diameter).

Action Coppice and plant up gaps to return to Point 4.

Manage any bank as in Point 7 above.

Point 9 Tall, often gappy hedge, typically over 7m high and with many Ash, Beech, oak, or Sycamore stems and with a sparse shrub layer.

Most stems are too large for laying and some may not coppice successfully. Any bank may be damaged from erosion or wind-rock and liable to collapse.

Action Fell or coppice and plant up gaps.

Restore bank as necessary.

Alternatively, if of particular value to wildlife, to livestock for shade or shelter, to the landscape or for carbon capture, allow to mature to Point 10.

Point 10 Mature line of trees, with spreading canopy. Unless the hedge has been allowed to spread out onto adjacent land, the shrub layer will often have been shaded out.

Action Retain as a line-of-trees hedge for as long as possible.

When the trees die or have to be felled, replant the hedge.

Restore any bank as necessary.

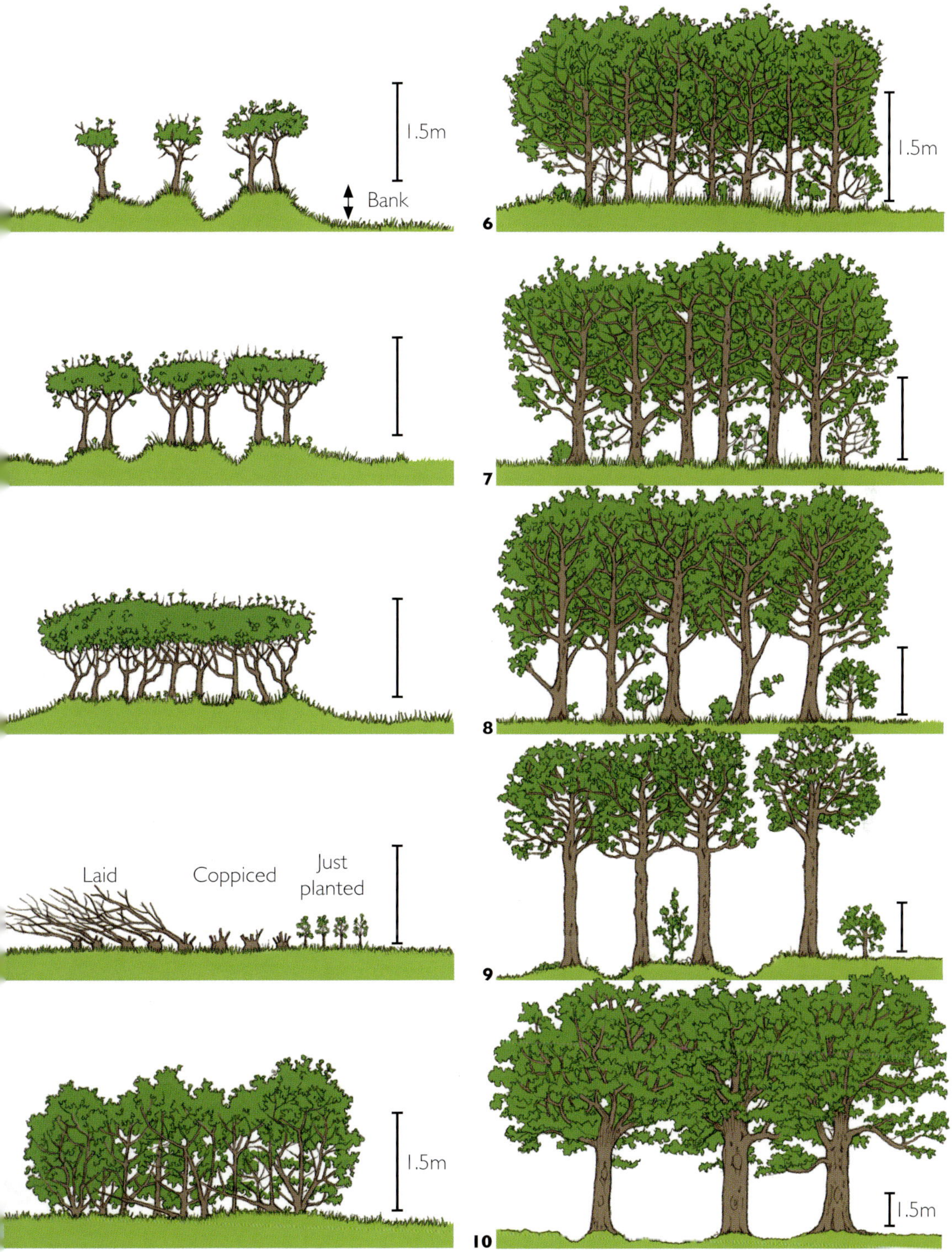
1.5m
Bank
6
1.5m
7
8
Laid
Coppiced
Just planted
9
1.5m
10
1.5m

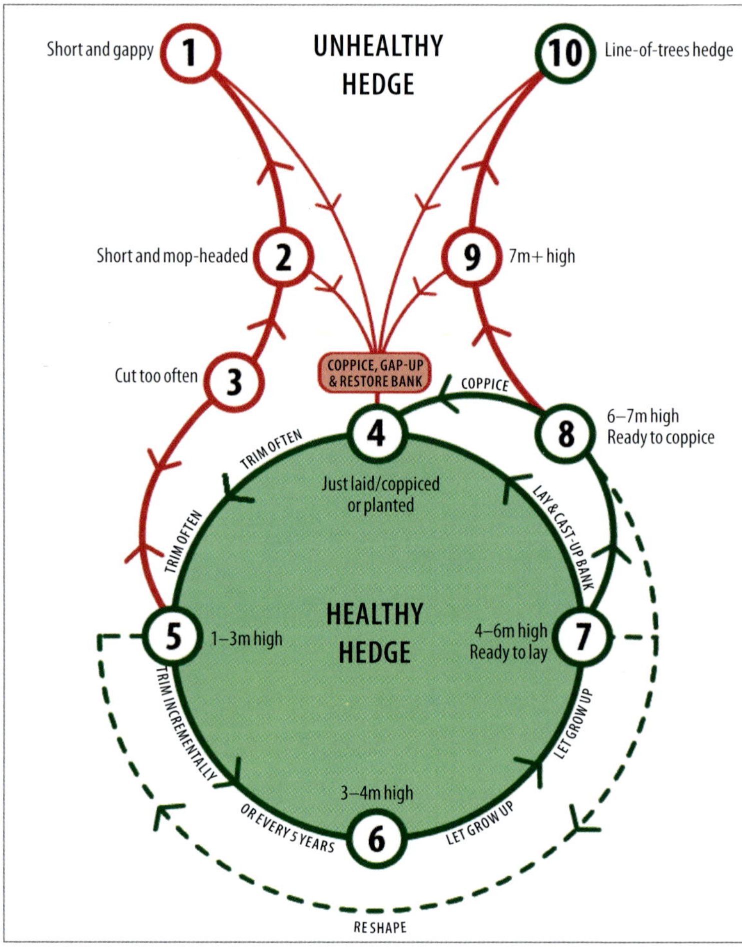

RIGHT: The hedge management cycle. Management should aim to keep hedge condition within the green circle, between points 4 and 7 or 8. If the hedge has entered into one of the states shown by the two red offshoots, it is in need of restoration, not rejuvenation, although line-of-trees hedges are often acceptable.

Trimming

No single hedge matter causes more controversy and angst than the use of the flail cutter. This is a tractor-mounted implement that contains a rapidly rotating shaft bearing rows of short, hinged, heavy-duty metal blades. They are very efficient at cutting through woody material, reducing it to fragments that fall to the ground. Concern about their use is understandable: flail cutters appear to be, and can be, brutal and destructive. Yet their use is the management method of choice for most farmers and has been for decades, ever since they were introduced about 70 years ago. They are even used by managers of nature reserves. I myself ask a neighbouring farmer, a hedge cutting contractor, to flail a few of ours every autumn. Why? To keep them thick and bushy, and great habitat for Hazel Dormice amongst much

LEFT TOP: The flail cutter is a powerful hedge management tool. Rapidly rotating metal blades cut shoots into little pieces that fall to the ground. This hedge, with three years' growth, is being cut in late September, while the ground is still firm enough to bear the weight of the tractor.

LEFT: This Hawthorn hedge has been tightly flail trimmed every year for many years and has lost its bottom growth. It needs restoration if it is to survive.

other wildlife. And, as I've already admitted, so that I don't have to devote most of my winter to hedge laying! The motivation of many other farmers is rather different – simply to keep the farm looking neat, tidy and well cared for, a matter I shall return to. Whatever the reason, used sensitively and in the right way, flail hedge trimmers are an important device in any hedge manager's toolbox. It's as well to remember too, as an elderly farmer recently reminded me, that if it had not been for the invention of flail cutters many more hedges would have been bulldozed out of existence in the 1960s, '70s and '80s. For many farmers, they reduced the cost of trimming to a level that was tolerable.

Regrettably though, poor hedge flailing practices remain rife. A remarkable but deeply disquieting study in Cornwall graphically demonstrates just how damaging flailing can be. Over an astonishing

34-year period between 1960 and 1994, in her home county of Cornwall, Sarah Carter (2008) collected evidence of the terrible harm caused by truly awful flailing along a mile (1.6km) of single-track country lane near Penzance, one that she walked back and forth nearly every day, occasionally twice, at times pushing a pram. To her horror, one sunny day in July 1972, she was faced with a tractor flailing the hedges. The destruction it wreaked she describes as a nightmare. For the next 12 years or so the county council continued to flail down the hedges on both sides of the lane, cutting them hard back to their banks not just once a year, but twice every summer, in May/June and again in July/August. Carter documented with despair the cumulative loss of wildlife from the hedges, year upon year. Where there had been abundant flowers, birds and insects, by the time the 12 years had passed the lane was a shadow of its former self. The number of plant species flowering in the hedges fell from 188 in 1971 to just 57 by 1985. The number of butterfly species dropped from 24 to a pitiful three. Eventually, the campaign she led to stop this destruction was successful, and the council reduced cutting to once a year, in late summer or autumn. Three years on, 14 of the lost butterflies had returned, and the number of plants flowering was 103, a partial recovery. Still, writing in 2019, Carter says the hedges remain a shadow of their former glory: while there may now be a good showing of spring flowers, the majority of summer and autumn flowering perennials have been lost and few birds, butterflies or bees are to be seen any more. Of course, not all this loss will have been at the hands of the flail, for we know there has been a general decline in farmland wildlife, but much of it will have been.

Thankfully, reflecting public concern, legislation to protect nesting birds and conditions attached to agricultural support measures, it's rare now to see hedges being cut more than once a year, at least outside gardens and parks, or to see flailing in the spring or summer, before the end of August, unless necessary for road safety. A big step forward.

The alternative to not trimming at all is to let hedges grow freely, managing them just by laying or coppicing once every 10–20 years – either that or, over a longer timeframe, giving them free reign to grow into line-of-trees hedges. The big downside of managing only by laying or coppicing without any trimming is that the interval between one laying/coppicing episode and the next is shortened, adding to the management burden. Nevertheless, an economic argument in favour of it can be made: the savings realised by not trimming may well outweigh the additional costs of labour incurred through having to lay or coppice more frequently. Another potential downside is that, without any trimming, hedges can become thin at the base and rather poor wildlife habitat. Ten years or so after laying or coppicing, a Beech or Hazel hedge that has been regularly trimmed will have a very different structure from one that's been allowed to grow unchecked. But for other hedge types thick bushy growth can persist throughout the management cycle without any trimming, as Ross Dickinson has shown in Dorset (see box on page 282).

The benefits of letting more than a few hedges in a landscape or network grow beyond the optimal stage for coppicing are less certain. While the trees within them many continue to capture carbon until they mature, and indeed do so at an ever faster rate, both hedge longevity and value as a wildlife habitat are likely to decline. Already about a third of the UK's hedges have been allowed to develop into line-of-trees hedges, largely not through any conscious design but due to neglect or forgetfulness. These hedges have a limited life span, at least in most agricultural contexts. Gaps, both along the length of the hedge and between ground and canopy, arise and expand as the shrub layer is shaded out and the trees age, thin and die. Banks erode and herbaceous margins become sparse or dominated by shade-tolerant plants, a situation often made worse in livestock areas by the

BELOW: Flail trimming the hedge on our farm in Devon where I recorded 2,070 species, to keep it thick and bushy.

ABOVE: This Alder hedge is now little more than a relic due to lack of care or management, although the cattle still find its shade and cover to their liking. The horizontal stem shows that the hedge was once laid along a bank that has now wholly eroded away.

incursion of sheep or cattle. The end result, sooner or later, is that much money and time has to be spent on the restoration of these unmanaged hedges, otherwise they will become relics and eventually be lost altogether.

Alternatives to flail hedge cutters

When we first took on our farm in the early 1990s, I rather naively bought a second-hand belt-driven finger bar cutter for use with our small tractor, to trim our hedges. Imagine your spread fingers being blades and your hands moving rapidly over one another, that's how they work. I quickly learnt that the old-fashioned machine I purchased was cumbersome and difficult to use and abandoned it to rust in a corner of our yard in favour of the much easier option of engaging our neighbour with his flail. However, modern hydraulic finger bar cutters, for use with tractors, loaders or excavators, are no doubt much easier to operate than mine was. Nevertheless, for reasons explained below, they are rarely used in agricultural settings, being the implement of choice only in places with high public visibility and access where a very clean cut is desirable to avoid the public outcry that can ensue from the use of flails. Larger models can cut material up to 6 centimetres in diameter and cover widths of over 2 metres on a single pass.

LEFT: Using a finger bar cutter to trim a hedge.

In many other northern European countries, cutting heads that consist of a series of circular saws are widely used, but they have not caught on in the UK. These 'shaping saws' can cleanly cut much larger material than either finger bars or flails, up to 20 centimetres or more in diameter, so can be used to reshape hedges with many years' growth. However, reshaping rarely seems a satisfactory alternative to felling stems right to ground level, that is coppicing, since the regrowth of most species occurs mainly at the cut line – resulting in top-heavy hedges with thin bases. I have seen old photographs of awesome single circular saw attachments, some with a blade a metre or more in diameter, mounted beside tractors with no guards – these must have been unbelievably dangerous. There's one such photograph, albeit with a modestly sized saw, in Pollard, Hooper and Moore's 1974 book on hedges. This reveals it was a two-person operation, one driving the tractor, the other turning handles and levers to position the saw correctly. These days it's all done by one person in the tractor cab using hydraulic controls, and the cutting heads typically have three or four smaller blades, although there are some models on the market that have eight and can cut swathes over 4 metres wide in a single pass.

The big disadvantage of both finger bar cutters and shaping saws, and the main reason they are seldom used by farmers in Britain and Ireland, is that the cut branches fall to the ground beside the hedge

ABOVE: A shaping saw, a tractor-mounted implement used to cut cleanly through stems and branches much bigger than finger bar or flail cutters can manage.

in a tangled heap and have to be dealt with. I remember seeing an elderly man walking along a Devon lane behind an ancient tractor with finger bar cutter attached. With a pitchfork he was casting the cut material up over the hedge into the field on the other side, and it was clearly hard work. They surely could not have managed more than a few hundred metres of hedge in such a way. Nowadays tractor-front fork and grab attachments are used to push cut material into bundles and lift them to a bonfire site or out-of-the-way place. But all this requires a lot more work and fuel than using a flail, and often a second tractor, so it's hardly surprising that neither finger bar nor shaping saw are popular today. Certainly it is not practical to use them to cut hedges alongside public roads – there's just too much traffic about.

Circular saws can be used to coppice hedges, that is to cut them down nearly to ground level so they can reshoot, but the lack of directional control of falling material remains a problem, and taller trees must be taken down in sections. Care also has to be taken not to hit stones or pieces of hidden metal. Far better to use tree shears to coppice hedges, powerful new machinery described in Chapter 14.

Flailing woes and elephants

Most of our native shrubs and trees produce their flowers on recent growth. So if a hedge is cut back hard in the autumn or winter, removing all the new twigs in the process, the following spring there will be few flowers. And if there are few flowers there will, of course, be few berries, fruits or nuts. This effect is obvious in the spring when travelling through farmland: tightly flailed hedges are devoid of the blossom of Blackthorn or Hawthorn, except for occasional patches where the implement has not followed indentations in the interest of achieving neat straight lines. In contrast, hedges uncut for a while, the longer the better, are often adorned with a white cloak of flowers.

Cutting at the same height year upon year is not good for the health of woody stems. They develop hard knobs or knuckles at the

ABOVE: Hedges that are cut hard back produce very little blossom and few berries the following year. This wide hedge has not been cut on the field side for a couple of years, allowing the Blackthorn and Gorse to flower.

cut line that reduce the vigour of stems, sometimes to such an extent that they die. While this may not matter so much in species that sucker like Blackthorn or elms, or which send up fresh shoots from the base like Hazel, it's bad news for others such as Hawthorn, Ash, Sycamore and birches. Remnant thorn hedges, reduced to scattered bushes, knuckle-topped and mop-headed, are all too common a sight in some arable landscapes. Often even these sorry vestiges continue to be flailed hard back every year for no reason other than for the sake of tidiness, sealing their fate.

The drawbacks of cutting repeatedly at the same height have long been recognised. In the middle of the 19th century, in *The Book of the Farm*, Henry Stephens (1851) wrote that 'hedges are woefully mismanaged in the cutting in many parts of the country'. That pithy statement remains true to this day. He goes on to say:

> *Without further consideration than saving the expense of a paling to guard a new-cut-down hedge, or in ignorance of the method of making a dead-hedge from the remains of an old live one, the stems of the old hedge are often cut over about 3½ feet* [1.07m] *high, to remain as the fence. The consequence is just what might have been anticipated from the known habits of the thorn – a quick growth of young twigs where the hedge was cut over; and the ultimate effect is, that of a young hedge standing 3½ feet above the ground upon bare stakes.*

Stephens follows this by observing that 'the only plan, therefore, to make an old hedge a valuable fence, is to cut it over near the ground, and form a dead-hedge with the part cut off to protect it.' In modern

ABOVE: This hedge is on its last legs due to mismanagement – the remaining Elder and Hawthorn bushes do not have long to live. They need to be coppiced and the gaps replanted.

parlance, that's 'coppice and protect regrowth', advice that holds good to this day. Dead hedges, constructed from cut branches, have long since fallen out of fashion – today we use wire fencing instead to protect young growth from livestock or deer.

Flailing shatters the ends of twigs and stems, especially when cutting through several years' growth. The damage caused is worse when flails have not been kept sharp or when the tractor is driven too fast. Understandably, many fear that this shattering is harmful to plant growth and survival through making the ingress of pathogens and rot more likely. No doubt this is true to some extent, but I have not come across any hard evidence in support. Indeed, it's remarkable that hedges that look terribly mistreated following flailing, with bare white wounds flashing on every branch and twig, can by the following summer look green and healthy. I have not seen any obvious signs of pathogens taking advantage of flail damage to infect trees: one does see fungal bodies on flailed ends and knuckle growth from time to time, but I suspect the organisms involved are, for the most part, not feeding on living tissues; either that, or the shrubs and trees are proficient at sealing off affected areas to prevent spread. An area where research is needed.

Forestalling the problems caused by close annual trimming is easy – simply introduce a rotation that leaves a decent proportion of hedges uncut each year, or raise the cutting height a little each time, perhaps just 10cm, a technique known as incremental growth trimming. This technique is another original idea proposed by Nigel Adams and has growing traction. It is favoured by many farmers because hedges can still be cut annually so they look neat and tidy – to many, a sign of good husbandry – yet bear blossom and fruit in significant quantities. More on this later.

Damage similar to that caused by flails has, it seems, a long history. A treatise written by Francis Blaikie in 1821 recounts the following:

> *Amongst the various methods of cutting hedges there is one of a very barbarous nature called (in Norfolk)* Buck-stalling. *The operation is performed by* cutting down *the tops of bushes about two feet* [60cm] *from the stubs, then slashing and shattering the stems to pieces, until they, in some degree, resemble the horns or head of an aged buck, from which resemblance I apprehend this ruinous custom derives its name.*

Blaikie says this practice was 'of great antiquity'. He lays emphasis on the words *cutting down*, having previously explained that only upward cuts made by billhook or slasher are clean, those made downwards causing splits and fractures. It would be fascinating to know now what the rationale for buck-stalling was – perhaps to encourage the growth of twigs lower down Hawthorn stems, helping to keep them bushier for longer? Perhaps someone would like to try it?

LEFT: Flail trimming can look brutal, but it's remarkable how quickly the frayed stems can recover. By the following summer you would scarcely have known this predominantly Grey Willow hedge had been cut.

One possible explanation for our native deciduous trees' and shrubs' extraordinary resilience to harsh treatment is that they are adapted to being eaten and bashed about by the Straight-tusked Elephant, a species that inhabited the forests of Europe during the Pleistocene, surviving until about 28,000 years ago on the Iberian peninsula. This giant was possibly the largest terrestrial mammal ever to have existed, up to twice the weight of an African Savanna Elephant. Elephants are messy browsers, trampling vegetation, tearing off branches with their trunks and stripping bark with their tusks or even their toenails – they don't bite cleanly through stems. The theory goes that our bushes and trees have evolved to cope with such damage, now replicated by the flail. Even the great Oliver Rackham considered this a possibility, and describes how he spent an instructive afternoon at Whipsnade Zoo watching Indian Elephants feeding on hedgerow trees (Rackham 2014).

How flail cutting affects flower and berry production

Several research studies have explored the impacts of trimming on flower and berry production. The most recent, and by some measure the most comprehensive, has been carried out by Joanna Staley and her team from the UK Centre for Ecology and Hydrology (UKCEH), funded by Defra (Staley *et al.* 2018a, 2019). On six sites across southern England they investigated the impact of cutting frequency (annually, biennially or triennially), and timing (early autumn or late winter), on the flower and berry production of Hawthorn, Blackthorn, bramble and Dog-rose. They also, for the first time, researched the impact of incremental growth trimming, the new technique mentioned earlier. The results were reasonably consistent across the six sites, but revealed a complex interplay between cutting frequency, timing and intensity. Because the matter is of such significance to hedge management and to wildlife, I've covered the results in some detail in the box on pages 314–315.

The main conclusions from the UKCEH research are as follows. Firstly, the longer one can leave a hedge before trimming or reshaping it, the more flowers and berries it will bear – three years is good, seven or even fifteen better still. Secondly, cutting hard back every other year is more or less pointless if done in the early autumn, say September or October: while there may be some small increase in flowers, any increase in the number of berries available for wildlife

OPPOSITE FROM TOP:
The longer that hedges like this Blackthorn one are left uncut, the more blossom and berries they will bear.

Blackberries and rose hips on a hedge that was cut the previous season but not hard back, retaining 10cm of new growth – incremental growth trimming.

Blackthorn responds particularly well to incremental growth trimming, producing good numbers of sloes on the short lengths of twig that remain.

during the winter is likely to be negligible – they will have been cut off too soon. For Hawthorn, even late-winter cutting of hedges managed on a two-year rotation is unlikely to boost the number of berries available compared to annual cutting, although it will increase sloe, blackberry and hip numbers. On the other hand, managing on a three-year rotation, even if cutting in the autumn, significantly increases the number of flowers and berries of all the species studied, that is Hawthorn, Blackthorn, bramble and Dog-rose; doubtless this holds true for other native shrubs and trees as well. Altering the frequency and timing of hedge cutting to increase the abundance of blackberries will be especially beneficial, since they have a high energy content and are a preferred food of many birds, small mammals and invertebrates. Their seeds, which often remain available in the flower receptacles long after the soft parts have rotted, are especially valuable (Evans *et al.* 2011).

Many farmers have a deep dislike of cutting on a three-year rotation, as much for cultural as practical reasons. Not only do the hedges look unkempt and uncared for, but when three-year-old growth is flailed all the shattered and splintered ends can look like carnage, resulting in negative press from local communities. So it has been exceptionally useful that the UKCEH study has proven that incremental growth trimming is a good alternative to cutting on a three-year rotation. Even when done annually in the autumn, the increase in number of flowers and berries is not far behind that delivered by cutting once every three years. For Blackthorn hedges it is particularly advantageous. Knowing that incremental trimming delivers substantial gains for wildlife is a game changer.

UKCEH's research into the impact of flail cutting on flower and berry production: Results

In their first experiment, which ran from 2006 to 2014, the team from UKCEH led by Jo Staley looked at the impacts of the frequency and timing of flail trimming on flower and berry production in the Hawthorn hedges planted in 1961 at the Monks Wood research station in Cambridgeshire (Staley *et al.* 2018a). In a second experiment, the research was expanded to cover a further five sites across lowland England, over the period 2010 to 2016. The hedges on two of these new sites were dominated by Hawthorn (Marsh Gibbon in Oxfordshire and Woburn in Buckinghamshire), on one by Blackthorn (Waddesdon in Oxfordshire), and on the remaining two by a mixture of woody species (Waddesdon again, and Yarcombe in Devon). Thus, the three most frequent types of hedge found in the UK were covered by the study.

At the Monks Wood Hawthorn hedges, over the entire eight-year course of the experiment, those cut every two years had 1.7 times more flowers than those cut every year, averaged across both years, and those cut every three years 1.9 times more, averaged across the three years. For berries, cutting every two years produced on average 1.7 times more than cutting every year, and cutting every three years 3.5 times more. It's as well that the study was carried out over a good number of years, since there was considerable variation in flower and berry production between years, a circumstance with which many naturalists will be familiar.

Unexpectedly, early-autumn (September) cutting at Monks Wood consistently resulted in more Hawthorn flowers the following spring than cutting in late winter, in January or February. Presumably, this is because Hawthorn bushes can replace some of the flower buds lost in September but none of those lost in late winter. For berries, however, cutting in the early autumn had a drastic impact on numbers available in the winter, as would be expected, removing nearly all. Indeed, cutting in the early autumn on a two-year rotation resulted in no increase in the number of berries available for wildlife compared to cutting every year. This is an important result, since on many farms across Britain and Ireland it is impractical to cut hedges after the end of October at the latest: the ground becomes too soft to bear the weight of a tractor without causing severe rutting and soil damage.

The second experiment, the one where the impacts of cutting on two-year or three-year rotations as opposed to annually were investigated over a six-year period across five sites, looked not just at flower and berry production on Hawthorn but also on Blackthorn and bramble, with additional data on Dog-rose. It also investigated the impact of incremental growth trimming.

Here, for Hawthorn, Jo Staley and her team found results similar to those from the Monks Wood experiment. One notable difference was that hedges cut on a two-year rotation in late winter showed no significant increase in either flower or berry production compared to annual cutting. This difference may be because the Hawthorns in the Monks Wood hedges were much younger, having been planted in 1961, while those at Woburn and Marsh Gibbon were planted in the 1790s and 1840s respectively.

The results for Blackthorn flowering were much the same as for Hawthorn, except that it was only cutting on a three-year rotation that had a significant benefit in terms of increased flowers, producing 1.8 times as many as annually cut hedges. The impact of cut timing on flower production was less marked than for Hawthorn, although late-winter cutting resulted in more, not fewer, flowers than autumn cutting. Turning to berries, winter-cut hedges managed on a three-year rotation produced five times the number

Hawthorn trees that have been allowed to grow freely for many years can produce huge crops of berries.

of sloes compared to those cut annually; unlike for Hawthorn, there was a small but significant increase in berry yields on two-year-rotation winter-cut hedges.

For brambles, there was a lot of variation between sites and years, but on average there were about one and a half times more blackberries on plots cut once every two or three years compared to annually, and two and a half times more on plots cut in the winter than in the autumn. For Dog-rose too, the number of fruits (hips) was greater on plots cut every two or three years than on those cut every year, but only if the cut was a winter one.

The impacts of incremental growth trimming had never been researched before, so here the results were eagerly awaited and did not disappoint. For Hawthorn, allowing hedges to become about 10cm wider and higher each time they were trimmed rather than removing all new growth produced 1.9 times more flowers and 1.7 times more berries. Intriguingly, it did not seem to matter whether the hedges were cut annually, biennially or triennially – the proportionate increase in flowers and berries remained much the same. For Blackthorn, the equivalent figures were 1.8 times more flowers and 2.6 times more berries. The latter figure, being higher than for Hawthorn, suggests that Blackthorn is particularly vulnerable to hard cutting. Even more promisingly, for both Hawthorn and Blackthorn the benefits of incremental growth trimming appeared to become stronger over time. With bramble and Dog-rose, while there was a trend for more flowers and berries with incremental trimming, it was not significant in either case.

In this second, five-site, experiment, the number of flowers and weight of berries on control lengths of hedge, those cut only at the start of the work and not during the seven years that followed, were assessed. Tellingly, by the seventh year, these Hawthorns produced five times as many flowers as hedges cut annually in the autumn, and the Blackthorn nine times the number. For berries, the uncut sections produced seven times the weight of haws and an impressive 28 times the weight of sloes. At Monks Wood, over the 8 years of the first experiment, stretches of Hawthorn hedges which had not been cut for 15 or more years produced 20 times as many flowers as those cut every year, and 31 times the weight of berries. If you want as many flowers and berries as possible, cut your hedges as infrequently as possible.

An additional concern expressed by farmers is that trimming just once every three years hastens the rate at which hedges become gappy compared to annual trimming. So this too was examined by UKCEH in their research programme. After six years of the various cutting treatments, in late winter, after all the leaves had fallen, they placed white sheets behind the hedges to maximise contrast, before taking high-resolution digital images of the bottom 90cm. These were analysed using specialised image-processing software to determine hedge density and gappiness. Comparing hedges cut hard back on two- or three-year rotations to those trimmed hard back annually, the results showed no decrease in hedge density, just a hint of this where the cutting was done in winter, nor any increase in the number of gaps more than 20cm wide, taken as the minimum width through which a lamb can easily pass. When trimmed incrementally, the trend was for hedges to become thicker at the base, and for gaps to become smaller. Although this experiment would need to be carried out over more than three consecutive two-year rotations or two three-year ones to test long-term changes in structure, it seems likely that farmer concerns about changes to hedge structure following triennial trimming have no basis in fact. If each time hedges are cut they are allowed to grow just a little taller and a little wider, biennial and triennial cutting may even result in them becoming denser and less gappy.

How flail cutting affects butterflies and moths

For most wildlife, cutting late in the winter but before the bird breeding season, usually taken to begin on 1 March, is best. But for the Brown Hairstreak butterfly, a species of high conservation concern strongly associated with hedges (see box on page 163), it is different. Each year the flail cutter destroys many of its eggs. But need this be so? To find out, Jo Staley and her colleagues (2018b) looked into it as part of their wider study into the impacts of trimming. Choosing the farm at Yarcombe in East Devon for this research, they found, as others had suspected, that hedge cutting by flail in the latter half of September is actually better for the butterfly than cutting in late winter. This is because the females are late flying and often still on the wing in late September and early October, searching for fresh Blackthorn twigs on which to lay their eggs. Cutting early, before the butterfly has finished flying, means more eggs survive. The research also showed that cutting frequency affects populations: hedges cut once every three years in September had 1.3 times the number of eggs found on hedges cut annually in that month. But the best result came from incremental growth trimming, which led to 2.3 times the number of eggs surviving compared to cutting back hard each time.

ABOVE: Brown Hairstreak caterpillar on its foodplant, Blackthorn.

A larger UKCEH study, across five experimental sites, confirmed that the Brown Hairstreak is exceptional in favouring September trimming, and that for most Lepidoptera cutting in late winter is better (Staley *et al.* 2016). The sampling technique adopted seems somewhat crude, involving as it did whacking foliage with a stick to dislodge caterpillars and pupae into pieces of guttering inserted into the hedge. But it worked. Over three years, 1,100 caterpillars and pupae were collected, belonging to at least 61 species, all moths except for the Brown Hairstreak. The abundance of caterpillars and pupae was most strongly affected by the timing of cutting, doing so in late winter resulting in 16% more individuals than in September. Incremental cutting once again proved beneficial, resulting in an 18% increase in the number of species present compared to cutting hard back. A previous study, part of the same research programme, had found similar results for those moths that have leaf-mining larvae or that construct cases within which to live, like *Coleophora* species (Facey *et al.* 2014). These moths were more numerous on hedges cut once every two or three years than on those cut annually, and the number of species was greater on hedges cut in winter than those cut in September.

In general, we would expect hedges that have not been trimmed for a few years to be associated with more insects than those just cut, and this has been confirmed by a study in south-west England (Froidevaux *et al.* 2019a, 2019b). Light traps placed beside hedges that had not been trimmed for between three and ten years caught more insects, notably flies but also moths with shrub and tree foliage-eating caterpillars, than those placed alongside hedges that had just been trimmed. Correspondingly, more Greater Horseshoe Bats, Lesser Horseshoe Bats and Brown Long-eared Bats were recorded along untrimmed hedges, the response of the two horseshoe bat species being mainly correlated to increased insect abundance, that of the long-eared bats to increased hedge height. Indeed, Greater Horseshoe Bat activity was almost four times greater alongside untrimmed hedges than trimmed ones. Somewhat unexpectedly though, frequency of trimming was not found to have any impact on commoner bats, specifically pipistrelle and *Myotis* species.

The shape of hedges – what's best for both carbon and wildlife?

The trimming of hedges into an A shape is often advocated, although it's rare to see it done. Proponents argue that this shape has several advantages – a thicker base to the hedge, easier avoidance of emerging standard trees when cutting, and better shedding of snow. Personally, I'm not convinced by any of these arguments and have not come across any good evidence to support them, although it's possible the last may be true. Hedges cut to a box profile can have perfectly good thick bases. It is worth bearing in mind that if a hedge cut to an A profile is to have the same volume as that of a square-cut one, then unless it is allowed to double in height its base will have to be twice as wide, putting at risk herbaceous margins. Nor is it obvious to me why it should be any easier for a tractor driver with flail attachment to spot and avoid promising stems growing out of the apex of an A-shaped hedge than out of the top of a square-cut one. Nevertheless, I do accept that if a finger bar cutter or shaping saw is being used, an A shape may make sense because the cut branches will fall more easily to the ground for collection – but, as noted above, use of either piece of machinery in Britain and Ireland is unusual.

On the other hand, hedges that have had their top edges bevelled or chamfered to roughly a 45-degree angle are common. The bevels on each side may meet to form a point at the top but often don't, a final pass with the flail truncating the apex. Whether cutting to this chamfered shape has any particular advantages to hedge health or wildlife, I don't know, but hedges cut in this way do look good. Cutting to an A shape may be a carry-over from times before flails, when slashers, long-handled bladed tools, were used to trim hedges, work best done by cutting upwards. There's a wonderful short film available on the internet, dated 1942, produced by the Ministry of Information for the Ministry of Agriculture that's

LEFT: Chamfered hedges along a Sussex lane.

well worth watching. It is called 'Hedging' and shows a skilled Northamptonshire hedger, Dighton, laying a hedge in the Midland style, assisted by a land girl. At the end Dighton demonstrates the technique of trimming a hedge with a slasher.

To my mind, the most topical question relating to hedge shape is whether there's an effective compromise between a shape that is optimal for carbon storage and one that is best for wildlife. As a general rule, line-of-trees hedges contain much more biomass and therefore carbon, while those that are kept short, compact and dense by regular trimming are best for wildlife, including the flora and fauna of associated margins and ditches. One compromise is for a good number of standard trees to be encouraged to emerge from a classic bushy hedge, but as we have seen these cannot be too close together or they will risk shading out the bushes, margins and ditches below. Another is to manage hedges on a coppice rotation: provided they are harvested for wood fuel, such hedges are even better for the climate than unmanaged line-of-trees hedges, as explained in Chapter 13. Coppiced hedges can be side-trimmed if necessary, both to reduce their spread onto cropping land and to help keep them dense at the bottom, while retaining a few stems to lay will help with habitat continuity. A minor issue with side-trimming is that side branches tend to have more flowers and berries than those growing vertically out of the top of hedges.

A new, neat, solution to the dilemma of managing for both carbon and wildlife, one that is likely to be especially attractive to farmers, has recently been advocated by hedge management specialist Colin Clutterbuck and Devon farmer George Steer. Colin is extremely proficient at hedge laying, having twice been announced Supreme Champion at the National Hedgelaying Association's annual competition, while George is keen to trial innovative ways of managing hedges on his land. The idea is delightfully simple: the sides and tops of hedges are flail trimmed in the customary way, except that a narrow band of growth down the centre line is left to grow on. The top growth on either side can be bevelled if wanted, as long as there remains an uncut central strip. The beauty of this idea is the ease with which it can be achieved from hedges trimmed to a square profile, currently the most common form of management. All that needs to be done is to leave the central strip untouched – the rest of the hedge can continue to be trimmed. The hedge will sequester carbon as it grows and, being in part bushy but with top growth unconstrained and so free to flower, wildlife will thrive. If the central growth gets too big and starts to shade out the hedge, margins, ditches or crops, then it can be cut back using tree shears, a firewood crop being taken if desired. It remains to be seen whether the idea will catch on, but I think it has strong legs.

RIGHT: The central strip of this hedge has been left uncut by farmer George Steer, both to provide good wildlife habitat and to increase carbon capture and storage.

Hedge laying

Hedge laying, or hedging as many refer to it, is the practice of rejuvenating hedges by partially cutting through stems at the base and then bending them over to create a dense barrier. It often used to be called 'plashing' and still is in parts of England, while in my home county of Devon it is 'steeping'. In Ireland it is often 'layering' or 'stretching'.

As a technique to thicken up hedges so they become an effective barrier, it has been known for a very long time. As far back as around 55 BC Julius Caesar wrote that the Nervii, a tribe living in Flanders, obstructed invading cavalry with defences made by cutting into slender trees and bending them over, intertwining thorns and brambles. By doing this, according to Caesar, they created barriers so dense it was almost impossible to see through let alone penetrate them (Rackham 1986). My own theory, having visited a wet woodland in Devon into which Eurasian Beavers had been introduced a few years beforehand, is that ancient humans would have got the idea from these large industrious rodents. I saw willow stems lying near horizontal sending up fresh shoots along their lengths – the beavers had felled them by gnawing their bases but not right the way through, leaving a living hinge. The parallels with hedge laying were obvious.

There are many ways that hedge stems and branches can be manipulated to form living fences. Georg Müller (2013) describes and beautifully illustrates a wide range of these from across Europe. On mainland Europe, traditional methods have been largely based on bending and weaving stems, assisted sometimes by partially snapping or notching them. The intricate patterns that can result are extraordinary, as ornamental as any fine trellis work. Frequently, stems are bent over at varying heights to create a stepped appearance. Occasionally they are crisscrossed, stems going in both directions. Others are fashioned so they have a sturdy top rail. Many techniques involve the use of living stakes, upright stems that are retained to add strength to the hedge but cut, bent or folded at the desired hedge height.

BELOW: A tree 'laid' by a beaver – the inspiration for ancient humans?

It is only in Britain that hedge laying, as opposed to bending or weaving for example, has evolved to become the dominant means of rejuvenating hedges. Here we have developed a rich variety of regional and local styles – you can tell where you are by the way the hedges are laid. These styles are described in Murray Maclean's book *Hedges and Hedgelaying* (2006) and in Valerie Greaves' booklet *Hedgelaying Explained* (2002). They are also outlined on the National Hedgelaying Society's website. Georg Müller provides superb line drawing of many in Volume 1 of *Europe's Field Boundaries* (pages 131 to 137). For the hedge enthusiast these are a must-see, and his wife and son have kindly agreed to me reproducing an example plate (see opposite). Across much of Ireland, Neil Foulkes tells me there's never been a strong tradition of laying to rejuvenate hedges, stems being laid only to 'bush a gap' – that is, to mend holes.

British hedge laying styles vary from the Midland, also known as the Midland Bullock, to the Devon. In the Midland style, designed primarily to keep cattle in pastures and out of crops, the stems, often called pleachers, are carefully cut part-way through at the base with billhook or axe, and laid over at an angle of 30–45 degrees. Vertical stakes are driven between the stems into the ground at regular intervals of 30–60cm. These in turn are held in place by thin Hazel, willow or Ash rods called heathers or binders, woven around each other and the stakes to hold the pleachers in place. Side branches and twigs are trimmed away from the side facing crops or ditches to encourage good regrowth but retained on the other as a barrier against the cattle.

RIGHT: Nigel Adams having a well-earned beer after demonstrating hedge laying in the Midland style, at a competition in the Netherlands.

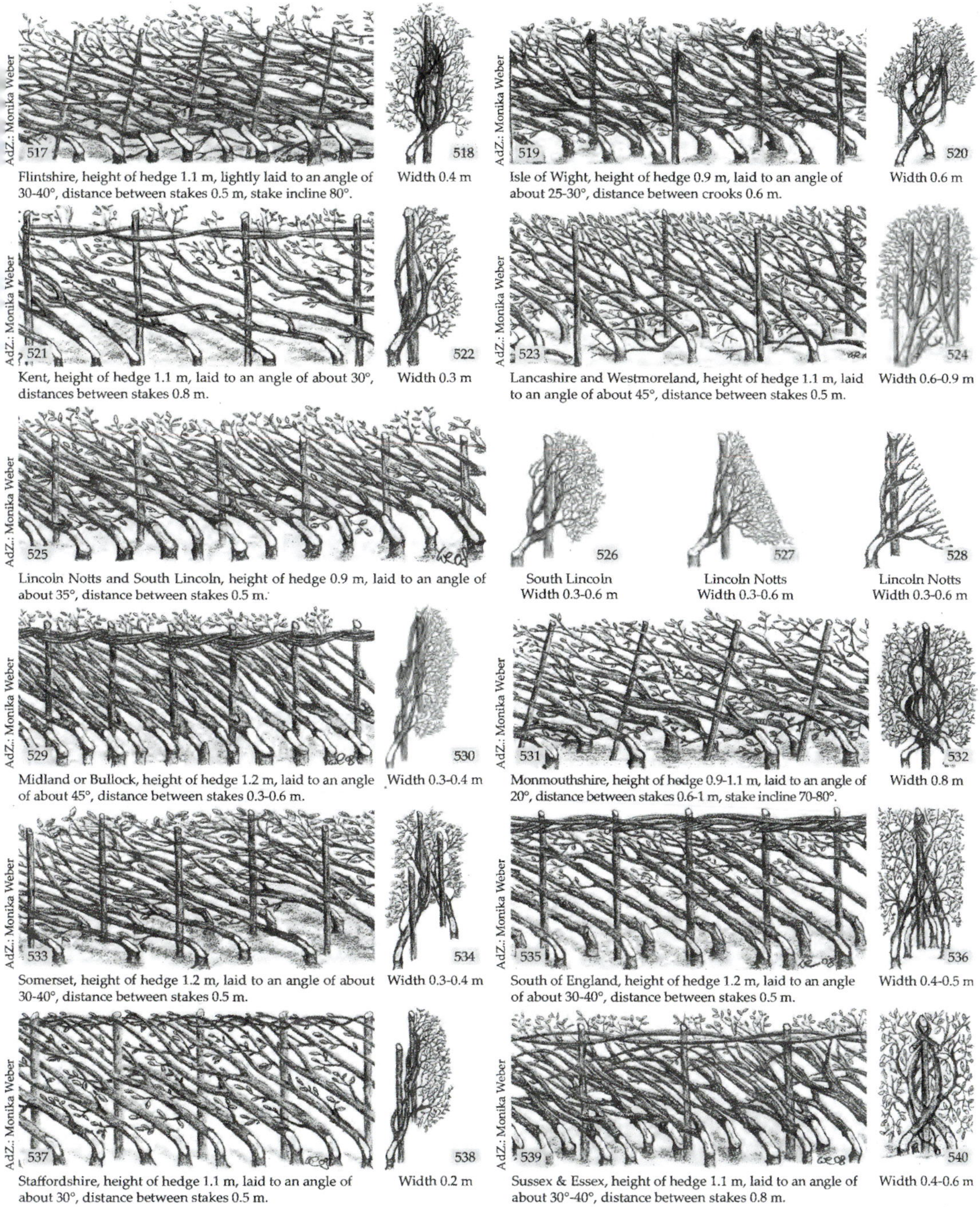

ABOVE: An example plate from Georg Müller's *Europe's Field Boundaries* (2013) showing a selection of British hedge laying styles.

RIGHT: Hedge superbly laid in the Devon style by Mike Reed and Russell Woodham (shown). The lines of the laid stems follow the top edges or combs of the bank.

In the Devon style, by contrast, where the shrubs and trees grow on top of banks, the pleachers, known locally as steepers (but stretchers in west Somerset), are laid flat along the top of the bank and held in place by crooks driven into the ground. There's no need for either stakes or top binding. Between these two styles there are many variations and intermediates. For example, the South of England style is similar to the Midland one but among other differences the side branches are retained on both sides, so the regrowth is not vulnerable to grazing on either. This is known as a 'double brush' style. The various Welsh styles focus on creating hedges that are dense at the base so that they are sheep-proof – in some, inserting dead wood into thin parts is acceptable practice, whereas in most English styles it is frowned upon. The Yorkshire style is highly distinctive because sawn stakes and top rails are used, reflecting it is said the lack of Hazel and other rods available for top binding, although in East Yorkshire a hedger would say when cutting thorns, 'If that weean't mak a steak, it'll mak a yether' (J. B. Smith, personal communication). The end result is of a hybrid between a post-and-rail fence and a laid hedge.

This rich variety of traditional ways to convert upright hedge growth into effective barriers against livestock has been lost over much of Europe. In the Netherlands, over the last 20 years or so, hedgers have effectively had to reinvent their country style, based on clues in old paintings and drawings and remnant artefacts in the few ancient hedges that remain. Meanwhile they have imported the English Midland style and become proficient at it. Only in a few parts of France and in Britain have traditional skills been continuously maintained, and even here they were in danger of dying out in the latter half of the 20th century. Fortunately, in Britain there has been a considerable revival in hedge laying, led by the formation of the National Hedgelaying Society in 1978. Ireland followed in 2004 with the founding of the Hedge Laying Association of Ireland, an organisation which covers both the Republic and Northern Ireland. Now across our islands there are many local groups actively encouraging people to learn and practise the craft. It's become so popular that training courses are frequently oversubscribed.

To see a hedge laid by a master of the craft is to see a creation of great structural beauty. Elevated to these standards, hedging is much more than simply a practical means of creating a living barrier, it is also an art form, a living sculpture. The way that an unruly hedge, often full of holes and straggly, can be tamed and combed into a living fence appears little short of miraculous. The survival and revival of this traditional craft is a cause for celebration.

There's no better place to see the conversion from chaos into order than at one of the many hedge laying competitions that take place today. There are classes for beginners and intermediates as well as

BELOW: Pairs compete in a hedge laying competition in Devon, organised by the Blackdown Hills Hedge Association.

open ones for experts. Both men and women take part, young and old – it's an inclusive craft, one which keeps practitioners fit and healthy. It's truly remarkable how many hedgers keep going well into their eighth decade.

New ways of hedge laying

It could be said that traditional hedge laying techniques are somewhat anachronistic, since they were developed for a time when the primary function of a hedge was to control cattle and sheep. In modern times this function is typically performed by wire fencing. As if to underline such redundancy, most hedges in livestock areas are now fenced immediately after being laid – indeed, if the work is grant-aided, fencing is almost an automatic condition. At times, it seems as if traditional hedge laying has no place on the modern farm, except perhaps for ornamental purposes.

Happily, however, laying very much remains relevant, as it is an effective technique for rejuvenation, giving hedges a new lease of life. As we have seen, without hedge laying or coppicing, most hedges face becoming derelict, mere relics, eventually to disappear altogether. As this happens, farmland wildlife disappears, precious soil is lost, water quality falls and carbon is released back into the atmosphere. Nevertheless, done to championship standards, traditional hedge laying, while very pleasing to the eye, is costly and time-consuming and, as we have seen, not necessarily particularly good for wildlife. Nature likes something rougher, scruffier. To address this there are many who now use less refined styles, ones that are not only quicker and cheaper but that also result in a thicker more natural hedge from year one.

When I lay a hedge on our farm, in my own informal Devon style, I fill gaps with cut brushwood, retain as much dead wood as I can, trim off protruding twigs only where strictly necessary, keep marginal brambles, roses and suckering Blackthorn growth wherever feasible, and don't use crooks, intertwining the stems instead. All these things would be frowned on if I was in a competition, yet the end result is better for wildlife, the job done faster and regrowth good. Many others I know take the same pragmatic approach.

Nigel Adams, ever a leader in such matters and himself a champion hedge layer, has developed a simplified style suitable for use with unbanked hedges. Fewer branches are removed in comparison to say

ABOVE: Hedge laid by Nigel Adams in his 'conservation style'. Thick and good for wildlife when laid, and regrowth the following year is excellent.

the Midland or South of England styles, stakes are used sparingly, and there's no top binding. The result is a good strong thick hedge, well suited to much wildlife. He has labelled this style 'conservation hedge laying'. While this term makes it clear it is better for wildlife than traditional laying, I think it might well be used more generally, to cover all hedge laying done with wildlife in mind, including the mechanical hedge laying I shall describe a little later.

In 2010, Jo Staley, working with UKCEH colleagues and Nigel, started an investigation into the speed and costs of different hedge rejuvenation techniques, and their impact on regrowth and berry production (Staley *et al.* 2018a). They selected five study sites, spread across Oxfordshire, Buckinghamshire, Cambridgeshire and Northamptonshire. Four of these had Hawthorn-dominated hedges and the fifth mixed-species hedges. Nigel's conservation hedge laying proved to be nearly three times as fast as the Midland style, taking on average 12 minutes per metre length, compared to 33 minutes for Midland hedging. The cost was half that of Midland hedging. Satisfyingly, three years after laying, regrowth and structural condition were just as good. And, since more stems were retained, in the first winter following laying there were slightly more berries; it was not until the fourth winter that the Midland style-laid hedges had fully caught up with conservation-laid ones in terms of berry production. So, just as Nigel had predicted, his pragmatic style was faster, less costly and better for wildlife than the pukka Midland style, and the resulting regrowth and structure just as good.

Going a step further, another new way of hedge laying is truly radical. It is 'mechanical hedge laying', sometimes but confusingly called 'wildlife hedge laying'. This technique was developed in the year 2000 by a Buckinghamshire farmer, John Morris of Long Crendon (Dodds 2005). Disappointed with the results of traditional hedge laying, finding it costly and time-consuming and the result a thin sparse hedge with no berries, he decided to try another approach. This was to cut part-way through the stems of a standing hedge with a chainsaw before pushing them over, lock, stock and barrel, a few metres at a time, without removing any brash. To do this he used a tractor-mounted telescopic handler. The result was a thick broad hedge that retained its autumn berry crop and flowered well next spring. Morris noted too that it contained much more dead wood than traditionally laid hedges, to the benefit of invertebrates and much other wildlife. Better still, the cost was a fraction of that incurred by traditional hedge laying; two men, one cutting the stems and the other operating the telehandler, could lay up to 250m in a day. By contrast, a traditional hedge layer can only manage about 20m a day. The local Aylesbury Vale Countryside Service was so impressed with the technique that they started to offer grants to local farmers to adopt it.

Hearing about this novel approach, traditional hedgers voiced considerable concern, verging on horror. It went completely against accepted wisdom. They worried that the dense thick layer of branches would smother the stools, severely repressing the growth of new shoots, so the hedge would die. And how would the hedge be managed next time it needed rejuvenating?

To address these doubts, the UKCEH team included mechanical hedging in their research. Firstly, they found it was indeed cheaper than traditional laying in the Midland style, just a third of the cost. Astonishingly, it was also about 30 times faster, taking just 1 minute per metre. When they looked at regrowth, they found that while in the first year there were fewer shoots from stools in mechanically laid hedges than in hedges laid in the Midland style, by the second year there was no difference – nor was there any in the third year. Overall, the volume of basal shoots per stool was the same, regardless of treatment, and Hawthorn shoots growing from mechanically laid stools were 1.4 times taller, as would be expected given the increased shade deep within mechanically laid hedges. The team noted that during mechanical laying some stems were either inadvertently cut right through or broken off entirely when pushed over, explaining

ABOVE: Mechanical hedge laying.

no doubt the high proportions of dead foliage found in the summer following laying.

Four years after laying, the density and gappiness of the hedges were assessed using the same white sheet and photo method described earlier to evaluate the impact of trimming frequency on hedge structure. This revealed the mechanically laid hedges to still be denser at the base and to have smaller gaps than the traditionally laid ones. Finally, Jo's team looked at the weight of berries produced. In sharp contrast to traditionally laid hedges, the mechanically laid ones had just as many Hawthorn berries over the three winters following laying as hedges which had not be laid at all. It took until the fourth winter for traditionally laid hedges to catch up with mechanically laid ones in terms of berry yield.

So concerns about poor regrowth on mechanically laid hedges can be allayed. Rather, their increased size, higher density and fewer basal gaps, together with the retention of dead wood and continuity of berry production, all argue in favour of promoting the technique, quite apart from it being cheaper and faster. The only concern is that being so wide, the laid shrub layer may cover ditches and flower-rich margins and shade them out.

Others have now adopted mechanical hedge laying; indeed it seems likely that some were doing it even before John Morris in Oxfordshire brought it to the attention of the conservation community. I have seen

ABOVE: A mechanically laid hedge the following autumn. Note the many berries, far more than would have been present if the hedge had been laid traditionally by hand.

mixed-species banked hedges in Devon laid in this way by skilled excavator operators: after a few years you would never know they had not been laid by hand. The stems are pushed down with the bucket and kept in place with sods of turf lifted from the field close to the base of the bank. With light material, such as when there's a high proportion of Blackthorn, there's not even a need to part cut through stems; they can simply be bent over.

The evidence is clear: both mechanical hedge laying and quick and ready manual techniques can result in just as good regrowth as traditional hedge laying executed to a high standard. And they are better for wildlife. In the early years following laying more flowers and berries are produced, while the wider and denser mass of laid stems, branches and twigs is better for nesting birds and small mammals like dormice than the typically sparse, sometimes skeletal, outcomes of traditional styles. For invertebrates too, research reveals that these new techniques are normally beneficial (Amy *et al.* 2015). Their greatest advantage, though, is that they are considerably faster and cheaper – this more than anything else makes them attractive alternatives. Like incremental growth cutting, they too are game changers.

I am certainly not advocating that these new techniques should replace traditional hedge laying styles, for these styles are culturally important rural crafts worthy of our respect and support. There are more than enough hedges in need of laying for both traditional and new techniques to be practised alongside one another. Given the magnitude of the task before us, we should take advantage of every approach open to us.

Still, would it not be good to see wildlife-friendly classes included in hedge laying competitions? Participants could be marked up for continuity of habitat, for the retention of dead and decaying wood and of fresh twigs (for the coming season's flowers and berries), and for keeping a good cover of plants and leaf litter on the ground. All classes would be encouraged to retain promising young trees and stems wherever more standard trees are required. And would it not be good too if organisers of hedge laying competitions could ask for brash taken out of the hedge line to be either placed in heaps to rot down, for wildlife and carbon reasons, or used as a source of renewable energy, not burnt in bonfires? This would be leading by example.

Coppicing

The other way of rejuvenating a hedge is to coppice it. It's a simpler and less skilled process than laying: all it entails is cutting down stems to within a few centimetres of the ground and removing them. The stools send up new shoots the following spring and the cycle of hedge growth is restarted. It's said that trees are traditionally coppiced when the trunk is 'as thick as a man's thigh at breast height' (Menneer 1994) – for laying I guess the equivalent would be as thick as a man's upper arm. My impression is that coppicing is becoming more popular, partly because more and more hedges have become too big to lay easily and partly because of the arrival of tree shears.

In some parts of Britain coppicing has long been the method of choice for rejuvenating hedges, notably in Essex, Suffolk and the clay lands of Norfolk. Rebecca Inman, hugely experienced farm adviser working in these parts of East Anglia, tells me that coppicing was integrated with rotational farming, the hedges being coppiced just before the fields were put to the plough; after the hedges had grown up again and become stock-proof the land was returned to pasture until the next coppicing. But, Rebecca says, not even the landscape historians really know why it is that hedges over much of East Anglia have traditionally been coppiced, not laid.

Coppicing is also the usual practice over most of Cornwall, wherever salt-laden gales do not keep woody growth pruned close and thick. This may, perhaps, have been because hedge firewood or charcoal

ABOVE: Hedge being coppiced at the Organic Research Centre's Elm Farm in Berkshire. The work in the foreground was done by chainsaw, while that further down the hedge is being carried out by tree shears attached to the excavator.

was in great demand in the county, woods being few and far between. Letting hedges grow a few years beyond their optimal state for laying, to the point where they are ideal for coppicing, results in much heavier crops of logs (see Chapter 14). Perhaps too there was a high demand for the likes of hurdles, thatching spars, rafters or mine supports, materials only readily sourced from hedges with more mature growth. Another possible explanation is that local farmers may not have thought it mattered too much if hedges were thin at the base, as the job of keeping stock in or out of fields was done by the high, usually stone-faced, banks on which the trees and shrubs grew. Yet in Devon, where the tradition is to lay hedges, high banks are the norm too.

In Ireland, where there has never been a strong tradition of hedge laying, coppicing has become the main means of rejuvenation, largely by default rather than intent. During the Emergency declared during the Second World War, many hedges were fortuitously rejuvenated by coppicing when cut for wood fuel: Neil Foulkes tells me this was the saving grace of many. Subsequently, now that agri-environment schemes encourage rejuvenation, hedges tend to be coppiced rather than laid because it's easier and cheaper, and in any case many hedges are too big to be laid successfully.

Coppicing has its advantages and drawbacks. On the positive side, not only does it yield more wood fuel than laying, it's also much quicker – if done by chainsaw it is some 16 times faster than Midland-style hedge laying (Staley *et al.* 2018a), if by tree shears many times faster again. It's also better suited to thicker and taller stems that can be physically challenging to lay by hand and are more likely to break off. On the negative side, initially there's a complete loss of habitat continuity and, because it looks like hedge removal, it can send shock waves around local communities. Many a time unannounced and unexplained hedge coppicing has led to negative press and bad feeling – I have been asked to become involved as a mediator in several such cases. There are ways, though, to lessen both visual and wildlife impact. For a start, it's good practice not to coppice more than about one in 20 hedges on a farm in any one year, and to ensure that these coppiced hedges are well separated. Furthermore, the differences between laying and coppicing need not be so stark – it's usually perfectly feasible to retain a few stems to lay. For sure this takes a little more time, but not much.

Another downside is that the regrowth from coppiced stools is more vulnerable than that in laid hedges to browsing by livestock, deer, hares and Rabbits simply because there's no protective overgrowth. If fences have to be erected, this adds to the expense. Partly it is a problem of our own making – the almost innate desire of many farmers and hedgers to do as neat and tidy a job as possible. We are driven to remove as much growth as possible, far beyond that strictly necessary for rejuvenation or safe working. Often only bare ground is left behind from which nothing but the cut stools protrude, their new shoots temptingly obvious to all passing browsers. Yet, as less tidy-minded individuals have found, leaving some growth including brambles and roses can be an effective deterrent, at least for deer, hares and Rabbits. If that fails, then why not lay a few cut branches along the hedge line, to cover the cut stumps? As we have seen with mechanically laid hedges, new shoots are perfectly capable of growing up through vegetation, even dense living stems. In past times, using cut branches to protect stools was common practice – it still is in woodland coppices. The option we now have of using tree shears to coppice makes retaining some scruffy growth in hedges easier, since stems can be lifted out of the brambles, as they can be over fence lines.

If precautions are taken to minimise these risks, the results of coppicing can be excellent, leading to thick gapless hedges after just a few years. I described earlier how hedges managed on a 20-year coppice rotation by Ross Dickinson in the Marshwood Vale in Dorset are wonderfully thick and bushy. I have seen other such examples too. On a 2011 Hedgelink visit to Dennington Hall Farms in Suffolk we saw fine, healthy, coppiced hedges that you would never know had not been rejuvenated by laying. Here mixed hedges with much Hawthorn, Field Maple and elm were being coppiced when 40–50 years old. The same good results were apparent when we visited Jonathan Boaz's Mill Farm in Worcestershire: the mainly Blackthorn hedges he had coppiced looked as good as any laid hedges. At both Dennington Hall Farms and Mill Farm the hedges were being regularly trimmed, starting within a few years of coppicing, to keep them thick and bushy and to lengthen the management cycle: unlike at Ross's farms, firewood production was not a primary objective.

ABOVE: This mixed hedge at Mill Farm in Worcestershire has been rejuvenated by coppicing rather than laying and is in fine condition.

Planting

Everywhere I go in England these days I see freshly planted hedges, or so it seems. I find this most encouraging, since new hedges are much needed. The urgency is such that setting aside strips of land and waiting for natural colonisation is rarely, if ever, appropriate, even though spontaneous hedges growing up along fence lines are surprisingly frequent. Pragmatically, planting is the way forward. Besides delivering faster results, it allows much greater control over the species mix. For most of us, it is also psychologically more satisfying than just letting nature run its course, and it is an excellent way to engage people with nature. Volunteers like planting hedges.

Where best to plant? This will depend on the intended roles for the new hedge, but generally in rural locations hedges are best placed where they fill in gaps in the existing network, join up loose ends, make huge fields smaller, and connect woodlands and other habitats. Often this means reinstating lost hedges, thereby restoring historic field patterns and strengthening local landscape character. Planting alongside existing hedges to create green lanes is also a good idea, since these lanes are especially useful for wildlife. So too is planting along the contours of slopes, or besides ditches and other watercourses, to capture soil and polluting chemicals, even to reduce flooding. In addition, hedges may be planted in places where they

ABOVE: The new hedge on the left has been planted parallel to an existing one to create a green lane at the Worcestershire Wildlife Trust's Lower Smite Farm reserve. The emergent tree covered in white blossom is a pear.

LEFT: A welcome new hedge. The use of plastic tree protectors like these rabbit spirals is now discouraged.

screen ugly buildings, increase privacy or deter trespass. They may even be planted as linear fruit and nut orchards, or for wood fuel – the list goes on. In built-up areas new hedges can reduce traffic noise and, if carefully sited, keep vehicle pollutants away from pedestrians. In this context hedges are nearly always in the wrong place – they should lie between road and pavement!

ABOVE: Steve Jefferies planting a hedge on a new turf-faced bank at Brook Manor on Dartmoor. The same hedge, four years later, can be seen on page 223. The growth has been remarkably good.

But new hedges are not welcome everywhere – not across wildflower meadows or heathlands, for example. Some birds like Skylarks and waders require expansive tree-free open spaces as found on moorland and grazing marsh, so these are seldom the right places for new hedges. Planting on deep organic soils like peat is also to be avoided, as the soil disturbance and chemical changes involved are likely to lead to a net loss of carbon to the atmosphere for decades. These exceptions apart, new hedges are to be welcomed nearly everywhere in the lowlands, whether on farms, in gardens or in parks. The ultimate goal must be to achieve the average 10 kilometres of hedge per square kilometre suggested to be optimal for wildlife (Staley *et al.* 2020, 2023).

Just as with hedge trimming and laying, our thinking about the best ways to plant new hedges has changed radically over the last decade or two and will doubtless evolve further over the coming years. Within the conservation community it is no longer considered appropriate to follow the customary practice of planting predominantly Hawthorn in a double row at close spacing, eight plants per metre, the rows just 50cm apart. This may have been justified in the time of the Enclosure Acts, but not now. Likewise, the use of herbicides or polythene sheeting to suppress competition from other plants is now frowned upon, as indeed is the use of any non-biodegradable substance: plastic tree guards are being phased out as viable alternatives become available.

Even in landscapes where Hawthorn hedges are characteristic, multi-species native mixes are now being encouraged. The imperatives of the biodiversity and climate crises demand change. We need hedges to be as good for wildlife as possible, and to have high levels of resilience to pests, diseases and environmental stresses such as summer droughts and winter flooding. The priority is to plant for the present and the future, not for the past. Preserving traditional practices and cultural landscape character is surely less important.

After all, producing a stock-proof barrier as rapidly as possible is now very rarely the objective. The continued popularity of Hawthorn as a major part of any planting mix perplexes me, as it is a species that will not throw up new shoots from the base unless cut down to ground level and is, management-wise, demanding. Other shrubs and trees like Blackthorn and Hazel are constantly sending up new shoots, while the likes of Beech and Holly have lower branches that persist for a long time. Still, there's a worthy place for Hawthorn in hedge mixes because its abundant blossom and berries attract much wildlife, but at a proportion limited to a quarter or less, not a half or more.

Nor do we need to plant so many shrubs per metre: six or even four will do just fine for most purposes, including making good-quality wildlife habitat. And why just have two rows, not three or even four? The wider a new hedge the better, with space being allocated not just for the line of bushes but also for decent margins together with any bank or ditch. On a visit to Flanders a few years ago I saw hedges where the ground on each side was ploughed so close to the stems of the bushes that when trimmed they assumed an inverted A shape. This, I was told, was a cultural legacy of the starvation years that followed occupation during the First World War, when every last scrap of land was needed for food production That is not the case now in the UK or Ireland: here, while recognising the need for food security, our national food strategies stress that we must make space for nature (Rialtas na hÉireann 2021, Defra 2022b): we can spare enough farmland for more and wider hedges without risking food shortages. After all, in the UK 85% of farmland is used either to graze livestock or to produce feed for animals even though we know we ought to consume less meat and dairy for climate and health reasons; plants grown over a much smaller area can provide the calories and nutrients we need (Dimbleby 2021). The massive contribution hedges

RIGHT: Planting trees grown from seeds collected in more southerly climatic zones is likely to be unnecessary and risky. Wind-pollinated species like this Pedunculate Oak are, in any event, capable of exchanging genetic material over substantial distances.

make to the ecological health of the farmed environment, and to our wellbeing, should trump all else.

The debate about whether we should plant non-native trees, or natives grown from seeds sourced from more southerly climatic zones, has yet to be fully resolved. However, the emerging consensus outside forestry circles where timber production is the primary objective is that we already have an adequate variety of native species and that these have sufficient variability, both genetically and in terms of phenotypic expression, to make it unnecessary to run the considerable risks associated with introducing foreign material. These risks include importing further diseases, pests and aggressive invasives, and a mismatch between the seasonality (phenology) of the trees and the wildlife reliant upon them. Still, provided plants are UK and Ireland sourced and grown (UKISG, as defined by the Woodland Trust) we have little need to be fussy about more local provenance. It's worth bearing in mind that many of our trees are wind-pollinated, allowing genetic exchange to occur naturally over considerable distances, including across the Channel.

It does, though, make good sense to tailor species mixes to local conditions both as they are now and as it is anticipated they will be towards the end of the 21st century. For example, Beech, a shallow-rooted tree, should probably not be planted in drought-prone soils or areas with low rainfall. Only in urban situations can a reasonable case be made for planting non-natives, both because the heat-island effect adds stresses that natives may be unable to survive and because many hedges are planted for ornamental purposes.

When planting new hedges there's a tendency only to think about trees and shrubs, but creepers and ramblers like wild roses, bramble, Honeysuckle, Hop, Bittersweet (Woody Nightshade), Black Bryony and White Bryony are all excellent plants for wildlife and will enhance any hedge. Traveller's-joy is also good for wildlife, but in places with lime-rich soils it can at times smother hedges. Although these plants may colonise naturally with time, why not propagate and plant them or broadcast their seeds?

ABOVE: Terry Coombe, professional hedger and champion hedge layer, planting a new hedge on a low earth bank in Dorset.

Standard trees

By far the easiest way to get more standard trees in hedges is to identify promising stems and saplings and just let them grow. The most convenient time to do this is when a hedge is being laid or coppiced; once found, the young trees can simply be left untouched. Being already prominent, they should be in no need of marking, and given a fair wind they should survive unaided to grow into old age. Otherwise, if hidden within surrounding vegetation, marking and perhaps halo pruning will be necessary to ensure they don't fall foul of the flail trimmer.

That may be the easiest way to increase numbers, but the opportunity to plant future standards in a new hedge should never by missed. Any number of species can be chosen for this purpose, not just those that can grow large and stately such as oaks and Beech, but also smaller ones like Rowan, Holly and Crab Apple. These are often the best for flowers and berries. Even Hawthorn will make a fine tree if allowed to grow freely. Tim Sparks, who has carried out much research on the Monks Wood hedges (e.g. Sparks and Martin 1999, Croxton and Sparks 2002, Sparks and Croxton 2007), tells me that a single mature Hawthorn tree can yield as many berries as 200m of

RIGHT: A promising young stem marked so the tractor driver can see and avoid it.

regularly cut hedge. The choice of tree species will depend a lot on location – in a garden setting smaller ones are likely to be appropriate, and consideration must always be given to overhead lines.

It's normal to plant a standard about once every 10 metres, or even more frequently in new hedges. This is, of course, far too close a spacing for large mature specimens: they will close canopies and shade out the bushes beneath, excepting a few shade-tolerant species like Holly and Beech. But this close spacing is of no matter since the trees can always be thinned out as they grow, favouring more vigorous or characterful individuals to leave a final spacing of one every 30–50 metres. This may require a certain amount of gritting of teeth, though – it is difficult to fell a tree you have planted yourself!

As a tree grows, to reduce shading on the hedge below or to allow the passage of tractors and other farm machinery, it may well be advisable to cut off lower limbs while it is still young, lifting the crown. Other trees may be pollarded, that is lopped off at a height beyond the reach of grazing animals, to prolong their lives or to ensure a legacy of ancient veterans.

Otherwise, hedgerow standards should require little management provided they remain healthy and safe. Just take care not to compact, cultivate or otherwise damage the ground beneath them, ideally to a distance twice the radius of the canopy, since the roots of large trees typically spread this far. As for Ivy, as I've mentioned before (in

ABOVE: A single Hawthorn tree allowed to mature can produce as many flowers and berries as 200m of annually close-cut hedge.

LEFT: A willow being pollarded. The job had last been done 17 years beforehand.

Chapter 7), it's sad to see its stems cut in the mistaken belief that the creeper is somehow damaging the tree. Such action is only justified in places of high lichen value, as Ivy can smother rare species growing on trunks (Sanderson and Lamacraft 2022).

It's also distressing to see large sweeping boughs of hedgerow trees cut off without good reason. There was a line of old oaks with such boughs on a farm near ours where one summer I watched Purple Hairstreak butterflies laying eggs and next spring found their caterpillars. A year or two later I passed by to find that these limbs had been cut off, for no reason that I could fathom. A fine wildlife feature had been needlessly removed, and with it a haven of dense shade for cattle anxious to get out of the summer sun.

The management and restoration of basal and marginal flora and fauna

For a hedge to be in fine fettle for wildlife, the base and margins must be cared for too, although this is far from straightforward. In many ways managing a hedge's shrub layer is the easy part, explaining no doubt the near single-minded focus upon it. Hedge bottoms and margins should be free from excessive nutrients, unaffected by pesticides, whether herbicides, fungicides or insecticides, and either lightly grazed or mown occasionally. Challenging requirements, to say the least.

The greatest of all these challenges is the replacement of the blankets of Nettles, Goosegrass and Bracken that have come to dominate hedge floras over vast tracts of farmland with more natural and diverse plant communities. The spread of these aggressive plants is the direct result of the application of manures and fertilisers on adjacent land or the deposition of atmospheric nitrogen. The only recourse is either to wait a long time, probably decades, for the unwanted nutrients to leach away, or to take the drastic measure of stripping off the topsoil together with its unwanted plants, followed by reseeding with desirable ones. Either way, action must be taken to ensure no more nutrients reach the hedge from adjacent land, such as by placing it under organic management. Using herbicides to kill off the unwanted plants is likely to be only a temporary remedy: as long as soil conditions remain favourable for them they will rapidly recolonise. In many instances the best that can be achieved is the establishment of a margin of perennial grasses – but even this is a

ABOVE: Dense Bracken stands are becoming more frequent in hedges: the reasons for this are unclear but are likely to include atmospheric nitrogen deposition.

major achievement. We need more case studies to inform good practice (Critchley *et al.* 2010).

Pesticides reach hedges through spray or dust drift, through the leaching of soil drenches or seed dressings, or sometimes through direct application (Botías *et al.* 2019). In arable areas, hedgerow herbs may be exposed to herbicides multiple times during the crop-growing season and over many years, and cumulatively this is bound to have a profound impact on plant communities. Neonicotinoids, systemic insecticides used as seed dressings, have been found to contaminate the foliage, pollen and nectar of flowers growing in hedge margins, occurring at levels similar to those in the target crop. Consequently there is grave concern about their impacts, both lethal and sublethal, on insects including pollinators (Goulson 2021). Numbers of moth and butterfly caterpillars, and of hoverfly larvae, have all been shown to fall when neonic residues are present in the soil. Vertebrates are not immune either: the ingestion of just a few treated seeds following an accidental spill would be lethal to grain-eating birds, voles and mice. More worrying still, neonics are just one of many types of pesticide applied to crops; cocktails of them are suspected to have even deeper perverse impacts.

RIGHT: Tussocky grass margins like this are of great value to wildlife and should be welcomed.

The situation is so serious that the continued use of pesticides of ever-increasing potency risks changing hedges from being refuges and lifelines for farmland wildlife into death-traps, depressing populations even further. Radical changes in farming practice are urgently needed to address this major threat. Those of us concerned with the state of the environment and nature must continue to press for the widescale adoption of appropriate organic and regenerative practices and the banning of toxic chemicals with proven or probable harmful effects on wildlife.

In pastoral areas high stocking rates with cattle and sheep are the norm these days, leading to a great many hedge bottoms and edges being eaten out or becoming dominated by just a few coarse unpalatable grasses. Banks become eroded, the undergrowth is lost and the wind whistles through the base of the hedge unhindered. Sooner or later the hedge becomes derelict. The obvious solution, and one adopted by many farmers, is to erect a wire fence. But this too has its problems. The total exclusion of livestock and probably deer, together the difficulty of machine-cutting growth behind fences, leads to fundamental changes in plant communities. Typically, low-growing and annual herbs are lost, to be replaced by tall perennials like Common Nettle and docks, by coarse grasses like Cock's-foot

and Yorkshire-fog, and by bramble and Blackthorn suckers. The vegetation tends to become more uniform, losing the diversity of structure so essential to much wildlife. Patches of bare soil, vital to solitary bees and wasps for nesting, disappear. Thus permanent fencing is a poor solution. It is far better to reduce stocking levels, but such a course of action is seldom attractive to farmers needing to make a living. An alternative is to use temporary electric fencing, but this is maintenance-heavy. For all that, it's better to have a rank hedge margin between shrub and fence, even if species-poor, than to have only close-grazed grass or mainly bare, eroding ground.

ABOVE: Flower-rich hedge margins next to pastures are rather rare but can be achieved with low stocking levels.

Bank management

Where hedges have substantial banks, these too need looking after. Inevitably the forces of gravity and erosion mean they slump and diminish over time. Trampling by cattle and sheep, loss of ground cover through dense overhead shade or heavy grazing, and the windblow of trees all hasten this decline; it does not take long for tall banks to become mere mounds, eventually little more than a trace. This is a real loss not only to wildlife, but also, since so many are of medieval origin or even older, to our cultural heritage.

RIGHT: The stones facing this Cornish hedge have been placed in the traditional local herringbone pattern. The core of the hedge is an earth bank.

Ideally, every time a hedge is rejuvenated by laying or coppicing, soil that has slumped to the base should be put back on top, a process known as casting up. In Devon this used to be done by men with strong backs and a Devon shovel, the handle made from hedgerow Ash and with just the right curve to serve as a form for the concave batter to which the sides of a bank need to be shaped for long-term stability. Now the work is nearly always done with an excavator, effective but crude by comparison.

If the banks have deteriorated further, then more drastic measures are necessary, restoration rather than maintenance. Rebuilding with an excavator is usually feasible given an experienced operator unless the soils are sandy or friable and won't compact. But if you want to rebuild faces of sufficient height and stability to be effective barriers to livestock, that's a different story. It's difficult! In Cornwall and Devon, where many banks are distinctively faced with stones, these are often placed vertically or in a beautiful herringbone pattern, something that can only be done by hand. Other West Country banks are faced with turf, sods being taken from adjacent fields and carefully laid one on top of the other. Either way, these are costly, time-consuming and skilled activities. Far better to keep on top of bank repair, casting up and patching faces regularly.

To reduce costs and manual labour, engineers have developed new ways of stabilising earth banks. Between 2006 and 2008,

LEFT: Rebuilt to the original dimensions following pipeline laying, this massive section of banked hedge in Devon has been stabilised with soil-filled hessian bags and sackcloth, the end stonework being built in the traditional way, by hand.

National Grid laid a new gas pipeline through the south Devon countryside that required over 400 banked hedges to be breached. These were reinstated using a range of techniques, some innovative. One of these involved covering the surface of the compacted earth with hessian sheeting secured by biodegradable pegs. Another approach was to build up the faces with pre-seeded soil-filled hessian sacks. The resulting banks proved stable and rapidly blended back into the surrounding landscape despite looking very out of place initially (Wolton 2014a). National Grid commissioned the Devon Rural Skills Trust to carry out fine stone facing to some sections, so traditional methods were blended with new ones.

Ditch management

Just as banks slump and erode over time, so ditches fill with sediment and debris. Only where the water at times flows fast and furious do they self-cleanse, even deepen. There are hedge ditches on our farm that have become deep gullies, but this is not the norm: most farm hedges become shallower and choked with vegetation with the passage of time, requiring periodic cleaning to retain their land-drainage function. The flora of hedge ditches, like that of banks and margins, also declines with heavy shading, as happens when the shrubs and trees above are allowed to grow unchecked. Shallow, dry

ABOVE: A string of Black Bryony berries. These last long into the winter, being distasteful and poisonous until late in the season. The seeds require a six-month wet period to germinate (Pollard *et al.* 1974).

or heavily shaded ditches are not the effective powerhouses of invertebrate biomass discussed earlier (Chapter 9), and generally support fewer and less interesting species. From a wildlife perspective, therefore, as well as from a land-drainage one, ditches need management to maintain their value.

That is, unless more radical approaches are being taken. There's increasing interest in rewetting farmland, to counter the impacts of excessive drainage in the past and more frequent droughts and heatwaves as the climate changes. Leaky dams are being installed, to retain sediment in watercourses, as an anti-pollution measure, and to reduce flood risk, and so ditch bottoms remains at least moist during the summer. We've just put in a series of these dams across a minor stream on our farm. In a few places, such as the National Trust's Holnicote Estate in Somerset, ditches are being filled in completely, allowing the water to find its own way across the land, spreading out to create natural channels and boggy patches – a process of ecological reset termed 'Stage Zero'. On top of that, reintroduced Eurasian Beavers are rapidly spreading and, in places, are already having a dramatic impact on watercourses of all sizes, including ditches and drains, much to the benefit of wildlife. So, rather than seeing them cleaned we may well soon see some hedge ditches being deliberately allowed to become choked to make farmland wetter, perhaps with the assistance of wild beavers. If so, there will inevitably be conflicts with food production that need to be carefully managed.

Concluding remarks

We have rediscovered some things that should not have been forgotten. A symposium paper I've just read, published in 1992, has a series of illustrations of hedges in various stages of growth. There's one showing a closely trimmed hedge, shoulder high, simply captioned 'hedge trimmed'. Since the other five illustrations have negative captions, I assume the authors considered this to be the only satisfactory state. The one showing a taller, uneven yet continuous line of bushes, looking just right for laying, is entitled 'hedge derelict'. It may seem obvious to us now, but this example shows that just 30 years ago many were unaware that hedges need to be taken through a cycle of management to remain healthy – it took Nigel Adams to remind us. This period of temporary amnesia no doubt explains why we now have an unfortunate legacy of thinking that hedges that have grown into lines of tall bushes or trees are not hedges at all. Forgive me for keeping returning to this matter, but it's been the source of so much confusion and frustration, hindering our understanding of changes in hedge fortunes. Amongst other things, it has resulted in the unnecessary inflation of hedgerow loss figures; we have enough to be despondent about without that.

Hedge cutting is without doubt a vexed issue, causing great anxiety among the public, but we now have good scientific evidence to show that it can be done in nature-friendly ways. The emergence of incremental growth trimming, as a practical means of keeping hedges thick and bushy while allowing farmers to demonstrate good husbandry, is probably the single new technique that has the greatest potential to improve the way hedges are managed. Alongside that we now know that hedges can be laid in ways that are cheaper and faster than traditional ones, with excellent results in terms of regrowth and wildlife, another big step forward. The arrival of tree shears means that coppicing too can now be done more efficiently and quickly, a further important advance. Like it or not, most hedges can be rejuvenated perfectly well by machine, whether by laying or by coppicing. Now the challenges are around communication and persuasion. And solving the most intractable problem of all – how to replace swathes of Nettle and Goosegrass with a diversity of more desirable flowering plants.

I have alluded several times in this chapter to a powerful, almost innate, drive that so many of us have, namely to carry out, above all else, a neat and tidy job. Regrettably, this frequently overrides our desire to do what is best for hedge health and wildlife. Nature likes things scruffy! Oliver Rackham (1986) recognised the problem when he wrote that hedges full of spring blossom can still be seen in places 'that have escaped the Vandal hand of tidiness'. More recently, Benedict Macdonald (2019) urges us to 'be messy', coining the term 'ecological tidiness disorder (ETD)' for the cultural obsession we have in the UK for neatness. It seems we need to exert control over natural growth at any cost.

ABOVE: An intense desire for tidiness is reflected in the way British hedges, wherever they are, are so often managed.

In the hedge world, ETD manifests itself in the way hedges are trimmed closely to precise shapes and straight lines, at least once a year, often more frequently in gardens. Topiary is held to be perfection. Rough irregular edges or outgrowths of brambles or wild roses are intolerable. Margins must be cut close to remove unsightly dead stems and to make them clean and even. It is apparent too in the way we are told to lay our hedges – every last bit of dead wood must be removed, every protruding twig cut off, every bramble or Ivy plant cut back if not dug out, every last piece of brash burnt. Macdonald says that most other European countries do not suffer from ETD to nearly such a degree, and ascribes the comparative wealth of wildlife in their towns, villages and countryside to this. Even in the Republic of Ireland, in rural areas at least, it is less prevalent – north of the border hedges are much more likely to be pared back to the bone.

ETD is, it seems, a peculiarly British affliction. One that we need to be constantly on guard against, taking remedial action wherever it's encountered. Usually this means doing less, so theoretically it should be easy enough, but the drive to be excessively tidy is deeply embedded. Attitudes are slowly changing, particularly among younger generations, but it will take considerable time yet, and conscious effort, before a more nature-friendly culture prevails.

On the other hand, as I write, there's a growing movement in favour of just letting hedges grow, unhindered by any management.

ABOVE: A wonderfully scruffy hedge.

Partly this is a simple result of neglect, but increasingly it reflects the growing popularity of rewilding. In its purest form rewilding is all about letting natural processes such as vegetational succession and large herbivore grazing take place with as little interference as possible. It works well only at scale, the larger the area the better. Standing back and letting nature take its course at the scale of Yellowstone National Park or the area depopulated because of the Chernobyl nuclear disaster works well. It can even work at the scale of the famous and pioneering Knepp Wildland project in West Sussex, which covers some 1,400 hectares, although here there remains much human intervention, largely in the form of herbivore management.

ABOVE: A hedge spreads into a field at the Knepp Rewilding Project, creating excellent habitat for Nightingales and other songbirds. For how much longer will it be recognisable as a hedge?

But rewilding is not a concept easily applied to hedges, at least not in a pure form. Not only is even the longest far too small in area, but hedges are by definition artificial features within heavily managed landscapes, either agricultural or urban. By their very nature and functions hedges are dependent on management intervention to remain hedges. In 2019 I visited Knepp and saw first-hand what had happened to the hedges there since the rewilding project started in 2001. Many had spread out sideways, largely due to the growth of suckering Blackthorn, to form wide bands with a central row of oak trees. Some were merging seamlessly with thorn and other scrub colonising previously open fields. Without doubt they provided magnificent wildlife habitat, much appreciated by Turtle Doves and Nightingales, but some were scarcely recognisable any more as hedges, and were becoming subsumed within scrub and woodland. There's no harm in this, for hedges are not sacrosanct. However, it is not a land-use model fit for most of our hedged landscapes, at least not as long as we rely on our fields for food or wish to be able to walk unhindered through gardens and parks. Pragmatically, though, hedges can be taken part-way down the rewilding path by allowing them to increase in width and height, within limits determined by management.

Management for healthy hedges: 12 top tips

1. **Think holistically** – across all structural parts, not just the woody bits but the hedge bottom, margins and any banks and ditches too.
2. **Promote connectivity** – retain, mend and plant hedges to join up the landscape.
3. **Generate structural diversity** – at the farm level have hedges at all stages of the management cycle present, favouring those that are thick and bushy.
4. **Foster a wide range of shrubs and trees** – for wildlife, and to increase resilience to diseases, pests, climate change and other stressors.
5. **Keep the shrub layer dense** – this will often mean careful trimming but can be achieved by regular laying and coppicing alone.
6. **Allow shrubs and trees to blossom and fruit** – retain at least some of the previous season's growth. Each time hedges are trimmed, raise the cutting height a little so they gradually expand, and try to trim on a three-year rotation or longer, not every year. Cut as late in the autumn or winter as ground conditions allow.
7. **Look after standard trees and encourage new ones** – not just tall ones like oaks and Beech, but smaller ones such as Crab Apple, Rowan and Holly, even Hawthorn.
8. **Encourage outgrowths** – the very best hedges for wildlife have wavy soft edges formed by clumps of brambles, roses, Blackthorn suckers or similar billowing out to the side.
9. **Allow dense ground cover to develop** – mainly by cutting basal and marginal growth only once every five years or so, and by controlling grazing. This cover is vital for much wildlife such as nesting Grey Partridges, Grass Snakes and Hedgehogs, and especially for overwintering invertebrates. Welcome tall perennials like Hogweed, Wild Angelica, Hemp-agrimony, Common Valerian and Great Willowherb.
10. **Encourage flower-rich margins and bank sides** – the wider the better, by careful mowing and grazing management and reducing overhead shade. Moving further out into the field, carefully cultivated wildflower strips will complement the hedge, providing further resources for pollinators, seed-eating birds and other wildlife.
11. **Manage ditches** – so that they remain at least moist during the summer, have gently sloping and well-vegetated sides, and are not heavily shaded by overhead growth.
12. **Prevent contamination with either fertilizers (artificial compounds and organic manures) or pesticides (including herbicides and insecticides).** Even tiny amounts of these substances may be very harmful to wildlife, either individually or in combination. When applying them to adjacent fields, keep them well clear of all hedge parts including margins and ditches, and take care to avoid spray drift or rainfall wash-off.

Looking ahead

chapter sixteen

I am optimistic about British and Irish hedges. Thirty years ago, when they piqued my interest, few people thought about hedges as being anything other than human-made features of limited function, of little importance in the grand scheme of things. Scant attention was paid to them in comparison to native woodlands, heathlands or flower-rich grasslands. How different things are today! Is it too much to say that hedges have a higher profile now than they've had for well over a century? I don't believe so. Politicians, farmers, environmentalists are all taking about them in positive terms: these are encouraging times. To be a hedge enthusiast is no longer to beat a lonely path.

For a start, there is much greater awareness of the importance of hedges, both culturally and for a healthy environment, far beyond simple appreciation of their aesthetic value or obvious agricultural functionality. Just a decade ago, it would have been astonishing if a government minister had said 'we know that hedgerows are probably the single most important ecological building block that we have in the farm landscape, and how we manage them matters'. But this is exactly what George Eustice, Secretary of State for the Environment, said when addressing the Oxford Farming Conference in January 2022.

True, there's a lot of neglect and not nearly enough hedge laying or coppicing, and in Ireland it seems hedges are still being removed, but my bet is that in England at any rate the length of hedges is currently increasing, new planting outweighing losses from dereliction and from housing, road and rail development, although we must await the results of the new round of Countryside Survey to be sure. As well as new hedge planting, I also see fewer hedges being flailed back to their core – the message that this is good neither for wildlife nor for hedge health seems to be percolating through. Last year a farm near to us

OPPOSITE PAGE: What will our hedges landscapes look like in 50 years' time? The Usk Valley, Powys.

was sold to an intensive dairy farmer, and this year, for the first time, the hedges have not been given the customary short back and sides, but instead have been allowed to keep several centimetres' growth, just as those concerned for nature would wish. Incremental growth trimming is gaining traction. I am heartened by all this.

There are a number of reasons for the change in attitude. Foremost has been the availability of agri-environment schemes under which farmers can enter into agreements with government to get paid for maintaining or enhancing features of environmental value. Often called stewardship schemes, they have been through repeated iterations on both sides of the Irish Sea. In truth, keeping track of it all is quite bewildering. In England for example, starting in the late 1980s we have had, in chronological order: the Wildlife Enhancement Scheme, Environmentally Sensitive Areas, the Hedgerow Incentive Scheme, Countryside Stewardship, Environmental Stewardship, Countryside Stewardship (again) and now Environmental Land Management schemes that include the Sustainable Farming Incentive. A memory test, especially since the last three are presently operating concurrently and each has multiple tiers. Other countries have had, have and are developing such schemes too. Whatever the complexity, the main point is that collectively they have been successful at improving hedge management, albeit some with more success than others. Uptake of hedge options has generally been high because they are a quick win for most farmers – a relatively easy way to receive public payments for delivering environmental improvement.

Surveys of farmer attitudes towards hedges confirm that agri-environment schemes and guidance from the conservation community have had a major impact. When asked in 1999, almost 80% of farmers in England said they trimmed most of their hedges annually. Astonishingly, by 2009, just 10 years later, this figure had nearly halved, to 45% (Britt *et al.* 2011a, 2011b). It would have been even lower, but the 2009 survey encompassed Powys in Wales, where farmers were more likely to trim annually than in England. The number of farms where hedges had been recently laid increased from 42% in 1999 to 50% in 2009, where hedges had been coppiced from 15% to 35%, and where new hedges had been planted from 30% to 43%. Just as encouragingly, the proportion of farmers expressing the view that hedges are very important for wildlife rose from 33% to 40%.

In 2022, when readers of the magazine *Farmers Weekly* were asked to complete a questionnaire, 86% of the 1,160 farmers who responded

said that wildlife habitat and nature corridors were key benefits of hedges (CPRE 2022). Shelter and shade for crops or livestock came second, at 64%. The same survey reported that 86% of farmers believed hedges to be important to them and their businesses. Good news indeed!

ABOVE: Content cattle, shaded, sheltered and fed by the hedges on our farm in Devon, at sunrise.

Recognition that hedges have a big role to play in helping to address both climate and biodiversity crises also helps to explain why so many are thinking positively about them. The Climate Change Committee's 2019 recommendation that the extent of hedges in Britain should be increased by 40% has been enormously helpful. From a biodiversity point of view, we now know that even a half-decent network of hedges can make a huge difference to the richness and abundance of farmland wildlife, especially so where the farming is intensive. If we could achieve an average density of 10 kilometres of well-managed hedge per square kilometre, the suggested optimum, then just imagine how much more wildlife there would be. It would be a huge step forward towards achieving national goals for nature's recovery.

Another reason to be optimistic is our newfound technical capacity to coppice or even lay hedges by machine, saving huge amounts of manual effort and money. With the best will in the world we will never succeed in laying or coppicing the vast numbers of hedges currently in need of such management if we do so only by hand, using traditional skills. That's not to say that we should abandon these skills – it's essential that we don't, because they are important culturally and because there will always be hedges that are unsuited, for one reason or another, to be managed by machine.

Not just one, but several countrywide organisations are currently taking a strong lead on hedge conservation. Operating across the UK, the Tree Council is doing much to raise interest and awareness in hedges, in both urban and rural environments. It has recently introduced and managed a successful annual National Hedge Week and has taken on the leadership of Hedgelink, a group that brings together organisations and individuals interested in hedges, to exchange knowledge and ideas, to engender collaboration and to take forward conservation priorities. Initially called the UK Hedgerow Habitat Action Plan Group, this has been running now for close on 30 years and has had many successes, influencing national policy and initiating research, as well as being a thinktank. I have been privileged to be a member from the very start, and was chair for a few years.

Next there's CPRE, the Countryside Charity, an organisation that has been campaigning for hedges in England for very nearly a hundred years. It led the call for legal protection that culminated in the 1997 *Hedgerows Regulations*, giving protection to most rural hedges in England and Wales. Despite being complex and difficult to enforce, they have been a major deterrent to hedge removal. Recently CPRE launched a '40by50' campaign, to extend the hedge network by 40% by 2050, in line with the Climate Change Committee recommendation.

Meanwhile, the People's Trust for Endangered Species is providing a strong lead on hedge recording, establishing the Great British Hedgerow Survey as well taking a leading role in Hazel Dormouse and Hedgehog conservation. The Trust has even developed a Healthy Hedgerows app which provides immediate feedback to surveyors and land managers on where within the 10-point management cycle a hedge lies and gives management options.

Other UK organisations too are actively engaging with hedges: the National Trust, the Woodland Trust, the Wildlife Trusts and the RSPB among others. Then there's the long-running and much valued National Hedgelaying Society, whose patron from 2001 was the Prince of Wales – until his succession to the throne as King Charles III in 2022. He is himself a hedge layer and enthusiastic proponent of the skill.

ABOVE: Countryside at its best. Dartmoor National Park.

In Ireland too there is now a strong movement for hedges. Alan Moore, leading member, tells me that the group Hedgerows Ireland was formed in 2019 in response to the unnecessary removal by Irish Rail of kilometres of trackside hedges in County Tipperary, along the Limerick Junction to Waterford line. The group has since rapidly become a national organisation actively campaigning for better hedge conservation and protection, forming wide-ranging partnerships with the likes of beekeepers, fishing groups, rough shooters and the hunting fraternity, as well as with environmental and farming bodies. Furthermore, Birdwatch Ireland, the Irish Wildlife Trust, An Taisce and Woodlands of Ireland all have a keen interest in hedges, while the Hedge Laying Association of Ireland receives core funding from the Irish government.

Everyone wants to be involved with ensuring a good future for hedges – how excellent is that?

Yet another promising development is the increasing interest in agroforestry, where trees are grown amongst crops and pasture to improve food production and the environment (Raskin and Osborn

2019). About 20 years ago I was invited to join an agroforestry conference in France to talk about hedges. My immediate thought was that hedges have nothing to do with agroforestry, but of course I was entirely wrong. In mainland Europe and the United States hedged field systems have long been seen to be as much agroforestry as are orchards, wood-pastures and alley cropping (where lines of trees and crops alternate). In these islands, however, we have only just started to break down the false dichotomy that arose in the latter half of the last century between agriculture and forestry: we are beginning to think again of hedges as an integral part of good farm business management. After all, as we have seen, hedges provide shelter and shade, scarce nutrients for livestock, windbreaks for arable and horticultural crops, and conserve soil, all of which improve production. Likewise, they can improve crop pollination and reduce pest damage to crops. They even have the capacity to increase farm revenue by producing wood fuel, and perhaps also through the sale of carbon credits. In short, there are many good reasons why farmers are now thinking more positively about their hedges as a business asset and so looking after them.

There are, it is true, major challenges ahead. All is not rosy. In Ireland, on both sides of the border, hedges continue to be removed faster than they are being planted: some level of legal protection is urgently required. In the Republic far better enforcement is needed of compliance rules linked to EU farm support payments and of environmental impact assessment regulations. In Britain, on the other hand, increasing numbers of hedges are being allowed to develop into lines of trees, not through design but through lack of management. There's a worry too that Brexit will result in less protection. Already we are starting to see fields once more being ploughed right up to the very edge – although the government has clearly signalled its intent not to let this happen (Defra 2023b).

Throughout Britain and Ireland, only a low proportion of our hedges are in as good a condition for wildlife as they should be. Large numbers are either too thin or too gappy, or both. Far too many have their bottoms and margins smothered in Nettles and Goosegrass because the soil has been enriched by nutrients originating from adjacent land or the air – this is the most troublesome problem of all. Then there's the likelihood that many are contaminated by a cocktail of pesticides to such an extent that their wildlife is much depleted. There is much more to do.

ABOVE: Autumn hedge colours. Locks Park Farm, Devon.

Perhaps most of all, we need to learn to enjoy and appreciate scruffiness. Our hedges, whether in gardens, parks or fields, need to look for most of the time as if they have been dragged through a hedge backwards.

Much has changed in the 50 years since the publication of Ernie Pollard, Max Hooper and Norman Moore's authoritative book *Hedges*. Public money is now available to plant rather than remove hedges, and they even have their own protective legislation. Across farms, villages, towns and cities there is far greater appreciation of their true value, whether it be for wildlife, climate or business viability. As I write, great concern exists for their future; many are their supporters and advocates. What, I ask, will the next 50 years bring for hedges?

OVERLEAF: Our idyllic ancient countryside. Sourton, Dartmoor.

Key results from 33 local hedge surveys in England, 2006–2010

appendix A

These surveys were all sponsored by the Department for Environment, Food and Rural Affairs (Defra) and carried out using the standard Defra/Hedgelink methodology (Defra 2007). All bar one (the Moss Valley survey in Derbyshire) either surveyed every hedge in their areas (full censuses) or selected hedges using the stratified random system set out in the standard methodology. Further information is available in Wolton 2010 and 2011 and linked spreadsheets, and in the individual survey reports.

Overall, 5,824 hedges were surveyed with a total length of 1,100km, and 10,605 standard hedgerow trees recorded. The areas sampled covered about 6% (just over 76,000ha) of the country's land area. For the 28 surveys where the information was provided, on average 41% of hedges were in favourable condition for all these attributes: height, width, gappiness of shrub layer, width of margins and introduced species, while 63% were in favourable condition for nutrient status, their herbaceous flora not being dominated by Nettle or Goosegrass.

Species-rich hedges are defined as those with five or more species per 30m length, except in northern England (Lancashire, South Yorkshire, north-east Lincolnshire and more northerly counties), where four suffice, unless otherwise specified.

Bedfordshire 2006 (Ward 2006)

Survey area: Studham Parish
Survey type: full census
Area surveyed: 971ha
Hedge length surveyed: 48.23km
Number of hedges surveyed: 212
Proportion of hedges species-rich: 68% of hedges with four or more species per 30m
Number of standard trees: 1,380 (1 per 35m on average)
Condition: 39% in favourable condition for height, width, gappiness of shrub layer, width of margins and introduced species; 57% in favourable condition for nutrient enrichment. 47% of hedges were in favourable condition for all attributes including nutrient enrichment.
Comments: the most frequent shrubby species found were Hawthorn (in 86% of hedges), Blackthorn (60%), Elder (60%) and Hazel (50%), and the most frequent standard trees were Pedunculate Oak (41% of the total) and Ash (23%).

Bedfordshire 2007 (Webb 2007)

Survey area: Maulden Parish
Survey type: full census
Area surveyed:1,053ha
Hedge length surveyed: 34.75km
Number of hedges surveyed: 154
Proportion of hedges species-rich: 43% of hedges with four or more species per 30m
Number of standard trees: 602 (1 per 58m on average)
Condition: 17% in favourable condition for height, width, gappiness of shrub layer, width of margins and introduced species; 60% in favourable condition for nutrient enrichment. 8% of hedges were in

favourable condition for all attributes including nutrient enrichment.

Comments: 35% of hedges were less species-rich and 13% more species-rich than in 1975.

Bedfordshire 2008 (Webb 2009)

Survey area: Sundon-Sharpenhoe–Smithcombe–Barton Hills

Survey type: full census

Area surveyed: 962ha

Hedge length surveyed: 30.9km

Number of hedges surveyed: 120

Proportion of hedges species-rich: 63% of hedges with four or more species per 30m

Number of standard trees: 847 (1 per 36m on average)

Condition: 43% in favourable condition for height, width, gappiness of shrub layer, width of margins and introduced species; 69% in favourable condition for nutrient enrichment. 34% of hedges were in favourable condition for all attributes including nutrient enrichment.

Comments: much of the land was under National Trust management and in an SSSI. Hedge extent and connectivity found to be similar to that after the Enclosure Acts and after the Second World War, and the number of trees greater than after the war. In built-up areas, 50% of hedges were next to roads, tracks and other public rights of way. 13% of hedges were bordered by uncropped field margins.

Bedfordshire and Hertfordshire 2009 (Banham 2010)

Survey area: area east of Barton Hills SSSI including Deacon Hill SSSI and the farmland to the north around Knocking Hoe SSSI on the border of Bedfordshire and Hertfordshire

Survey type: full census

Area surveyed: 1,092ha

Hedge length surveyed: 44.56km

Number of hedges surveyed: 161

Proportion of hedges species-rich: 70% of hedges with four or more species per 30m

Number of standard trees: 709 (1 per 63m on average)

Condition: 25% in favourable condition for height, width, gappiness of shrub layer, width of margins and introduced species; 88% in favourable condition for nutrient enrichment.

Comments: study area was dominated by arable land. Since many nature reserves and SSSIs were present in the area, uptake of agri-environment schemes had been encouraged. The survey found an average density of 40.8m of hedge per hectare, higher than the area immediately to the west along the chalk hills (32.1m/ha), surveyed in 2008. Analysis of historic aerial photographs found that the density of hedges had not varied significantly since 1946. Hawthorn, Dog-rose, Elder and Blackthorn were the primary hedge shrubs. The most frequent hedgerow trees were Ash (40%) and Field Maple (15%). Just 7 standard trees (1%) had a diameter at breast height greater than 1m.

Cheshire 2008 (Atkinson and Lowe 2009)

Survey area: Shacklock Oviatt and Church Shocklach parishes

Survey type: random sample

Area surveyed: not specified

Hedge length surveyed: 18.18km

Number of hedges surveyed: 104

Proportion of hedges species-rich: 29%

Number of standard trees: 169

Condition: 46% in favourable condition for height, width, gappiness of shrub layer, width of margins and introduced species; 32% in favourable condition for nutrient enrichment.

Comments: builds on previous Cheshire Hedgerow Biodiversity Action Plan surveys carried out by volunteers. By 2007 these had covered 108 parishes (53% of those in Cheshire). Although different methodology used, some results are comparable. The average length of hedge was 177m, compared to 251m for the rest of Cheshire, height 2.44m (cf. 2m), width 2.71m (cf. 1m), and 29% of hedges were species-rich (cf. 28%). There was an average of 1.6 trees per 30m section, mainly oaks (43%) but with fair numbers of Ash (17%), Field Maple (15%) and Hawthorn (9%). 21 veterans were recorded: 18 oaks, 2 Black Poplars and 1 Crab Apple.

Chilterns AONB (Oxfordshire, Buckinghamshire, Hertfordshire, and Bedfordshire) 2006 (Mortimer and Adams 2007)

Survey area: Chilterns Area of Outstanding Natural Beauty

Survey type: random sample

Area surveyed: 3,200ha

Hedge length surveyed: 56.88km

Number of hedges surveyed: 197

Proportion of hedges species-rich: 46%

Number of standard trees: 55 (1 per 1,030m on average)

Condition: 52% in favourable condition for height, width, gappiness of shrub layer, width of margins and introduced species; 77% in favourable condition for nutrient enrichment.

Comments: 44% of hedges were in favourable

condition for all attributes including nutrient enrichment. Estimated to be 4,010km of hedgerow and 44,000 hedgerow trees in the AONB (833km²). A further 83 hedges were surveyed in 2007, bringing the total to 280.

Cornwall 2009 (Bennett and Dale 2010)

Survey area: St Dennis and Tregony parishes
Survey type: random sample
Area surveyed: 1,691ha
Hedge length surveyed: 24km
Number of hedges surveyed: 134
Proportion of hedges species-rich: 20%
Number of standard trees: 65 (1 per 370m on average)
Condition: 48% in favourable condition for height, width, gappiness of shrub layer, width of margins and introduced species; 58% in favourable condition for nutrient enrichment.
Comments: the two parishes were considered representative of the Cornish Killas Joint Character Area. Hawthorn was the main hedgerow tree in one parish and oaks in the other. All farms within which hedges were surveyed were under an agri-environment agreement.

Derbyshire 2006–2008 (Evans 2009)

Survey area: Part of Chesterfield Borough
Survey type: random sample
Area surveyed: 2,000ha
Hedge length surveyed: 33km
Number of hedges surveyed: 322
Proportion of hedges species-rich: 47%
Number of standard trees: 991 (1 per 33m on average)
Condition: 53% in favourable condition for height, width, gappiness of shrub layer, width of margins and introduced species; 69% in favourable condition for nutrient enrichment.
Comments: hedgerow trees were present in 64% of hedges, but young trees with a diameter of less than 5cm appeared very rare. The five most frequent woody plants were Hawthorn, Elder, Ash, Dog-rose and Sycamore. 29 of the 991 hedgerow trees were classified as veterans. Just 9% of hedges had an uncropped margin between hedge and arable crop. 32% of hedgerows showed signs of management, including trimming, within the last 2 years, 61% within the past 10 years.

Derbyshire 2008 (Gregory 2009)

Survey area: Moss Valley
Survey type: non-random, focusing on hedges known to be of high quality
Area surveyed: 200ha
Hedge length surveyed: 9.3km
Number of hedges surveyed: 57
Proportion of hedges species-rich: 46%
Number of standard trees: 122 (1 per 76m on average)
Condition: 36% in favourable condition for height, width, gappiness of shrub layer, width of margins and introduced species; 37% in favourable condition for nutrient enrichment. A higher proportion (35%) of species-rich hedges were in a favourable condition than for all hedges (21%).
Comments: 18% of hedges contained 7 or more species in a 30m stretch and so were 'important' under the *Hedgerows Regulations*. On average, hedges contained 4.5 woody species per 30m. 28% of hedges would have benefited from coppicing/laying/planting to reduce gappiness, and 21% from a change in flail cutting practice to allow an increase in overall size.

Devon 2007 A (Hodgson 2010)

Survey area: parishes of Brayford, Monkleigh, Sourton, Hennock, Brixton and Burlescombe
Survey type: random sample
Area surveyed: 14,651ha
Hedge length surveyed: 150km
Number of hedges surveyed: 773
Proportion of hedges species-rich: 63%
Number of standard trees: 1,063 (1 per 141m on average)
Condition: 17% in favourable condition for height, width, gappiness of shrub layer, width of margins and introduced species; 60% in favourable condition for nutrient enrichment. 19% of hedges were favourable for all attributes including nutrient enrichment. More species-rich hedges were in favourable condition than all hedges together (25% vs 19%).
Comments: 26% of hedges contained 7 or more species so were 'important' under the *Hedgerows Regulations*. The most frequent woody species were Blackthorn, Hazel and Hawthorn, these species dominating 19%, 16% and 13% of hedges respectively. Ash dominated a further 16%, and Beech 11%. The percentage of hedges in favourable condition varied considerably between parishes, for example just 3.7% in Monkleigh but 32% in Burlescombe and Sourton. 31% of hedges would have benefited from laying or coppicing to reduce gappiness, and 35% from changes in flail practice to increase overall size. 70% of banks were in good condition.

Devon 2007 B (Hodgson 2010)

Survey area: Bradninch parish
Survey type: random sample
Area surveyed: 1,777ha

Hedge length surveyed: 24.4km
Number of hedges surveyed: 120
Proportion of hedges species-rich: 76%
Number of standard trees: 155 (1 per 157m on average)
Condition: 51% in favourable condition for height, width, gappiness of shrub layer, width of margins and introduced species; 56% in favourable condition for nutrient enrichment. 30% of hedges were favourable for all attributes including nutrient enrichment.
Comments: the proportion of hedges in Bradninch parish that were species-rich was higher than the Devon average (76% vs 63%), more hedges were in favourable condition (30% vs 19%), and more banks were in good condition (77% vs 70%).

Devon 2008 (Hodgson 2010)

Survey area: parishes of Farway and Southleigh
Survey type: random sample (but some farmer selection)
Area surveyed: 1,117ha
Hedge length surveyed: 27.4km
Number of hedges surveyed: 148
Proportion of hedges species-rich: 82%
Number of standard trees: 1,323 (1 per 21m on average)
Condition: in favourable condition for height, width, gappiness of shrub layer, width of margins and introduced species; in favourable condition for nutrient enrichment. 38% of hedges were favourable for all attributes including nutrient enrichment.
Comments: the two parishes that were surveyed had a higher than Devon average number of species-rich hedges (83% vs 63%), a higher proportion of hedges in favourable condition (37% vs 19%), and a higher proportion of banks in good condition (73% vs 70%). More hedges had received some form of traditional management in the last 2 years than for other parishes surveyed in Devon (7% vs 1.5%), reflecting past high uptake of agri-environment schemes. 71% of hedges were, or had recently been, under agri-environment agreements. Results suggest that fencing reduces floral diversity of hedge banks, but that wire-strand fencing is better than wire netting because stock can graze under and through the fence.

Devon 2009 (Hodgson 2010)

Survey area: Spreyton Parish
Survey type: random sample
Area surveyed: 1,691ha
Hedge length surveyed: 24.0km
Number of hedges surveyed: 134
Proportion of hedges species-rich: 92%
Number of standard trees: 170 (1 per 141m on average)
Condition: 69% in favourable condition for height, width, gappiness of shrub layer, width of margins and introduced species; 98% in favourable condition for nutrient enrichment.
Comments: no separate report was produced for this parish. Instead, Devon FWAG produced a report summarising results for all 10 parishes surveyed in Devon between 2007 and 2009 (9 by Devon FWAG, 1 by Exmoor National Park) (Hodgson 2010).

Dorset 2007–2008 (Edwards and Newbould 2009)

Survey area: Kingcombe and Powerstock Strategic Nature Area, west Dorset
Survey type: random sample
Area surveyed: 3,405ha
Hedge length surveyed: 56.3km
Number of hedges surveyed: 200
Proportion of hedges species-rich: not specified
Number of standard trees: not specified
Condition: not specified, other than that 27% of hedges had gaps more than 5m wide.
Comments: the survey area is one of three with the highest concentrations of ancient and species-rich hedges in the county. On average there were 5 woody species per 30m. Nearly all the hedges were on banks, 21.5% of hedges were fenced on both sides, and 27% had associated ditches. Hedgerow trees were present in 31.5% of hedges, excluding those frequent hedges with large Ash or Field Maple stools: many hedges had old laid or coppiced stools hundreds of years old. 25% of hedges were tall and leggy. In addition to the 3% that had been recently laid and 2% recently coppiced, 2% had been recently replanted. 47% of hedges showed signs of recent management, 26% older management, and 27% no management at all. The most frequent shrubs were Hazel (in 94% of hedges), Blackthorn (92%), Hawthorn (79.5%), Ash (78.5%), Field Maple (70%), Elder (65.5%), Dog-rose (62.5%) and Holly (60%). Four plants other than shrubs were found in more than 50% of ground flora samples: Nettle, Ivy, bramble and Red Campion. 28 ancient woodland indicator plants were recorded. 26% of the hedges had a cover of more than 25% Nettles.

Dorset 2010 (Phillips 2010)

Survey area: parishes of Hammoon, Manston and Hinton St Mary, together with small areas of Marnhull, Sturminster Newton and Okeford Fitzpaine, in north Dorset
Survey type: random sample

Area surveyed: 1,535ha
Hedge length surveyed: 30.25km
Number of hedges surveyed: 135
Proportion of hedges species-rich: 61%
Number of standard trees: 60 (1 per 504m on average)
Condition: 19% in favourable condition for height, width, gappiness of shrub layer, width of margins and introduced species; 76% in favourable condition for nutrient enrichment. The 64 hedge sides next to grassland were in better condition that the 96 next to arable, with 25% being in favourable condition for grassland sides and 10% for arable sides.
Comments: survey area was a People's Trust for Endangered Species Hazel Dormouse hotspot. 19% of hedges had 7 or more woody species per 30m, so were 'important' under the *Hedgerows Regulations*. 78% of hedges surveyed had been flailed within the past 2 years. The most common shrubs found were Blackthorn, Hawthorn, Field Maple, Dog-rose, bramble, Elder and Hazel. Most hedges were dominated by Hawthorn or Blackthorn, with Hawthorn being found in 73% of hedges and Blackthorn in 84%. The most frequent species of hedgerow tree recorded were oaks (56%), followed by Ash (23%), Field Maple (10%) and willows (7%). The average number of veteran trees and those with a diameter at breast height greater than 100cm was 0.4 per 1km of hedge.

Durham 2006 (Beeston 2007)

Survey area: Durham Biodiversity Action Plan Area
Survey type: random sample
Area surveyed: 263,700ha
Hedge length surveyed: 36km
Number of hedges surveyed: 193
Proportion of hedges species-rich: 35%
Number of standard trees: 270 (1 per 133m on average)
Condition: 17% in favourable condition for height, width, gappiness of shrub layer, width of margins and introduced species; 55% in favourable condition for nutrient enrichment.
Comments: there were estimated to be 9,100 hedges in the Durham Biodiversity Action Plan area, and 68,000 standard hedgerow trees. Since 1979 there had been a loss of about 21% in hedge length, but a large increase in average height and width. While in 1979 72% of hedges had been recently trimmed, in 2006 only 36% had been recently managed. On average there were 3.0 species of woody plant per 30m length. Hawthorn was the most frequent shrub, covering on average 55–75% of each hedge in which it was found, followed by Blackthorn, Ash and Elder. Ash was by far the most frequent hedgerow tree.

Durham 2010 (Harrison 2011)

Survey area: the Borough of Hartlepool
Survey type: random sample
Area surveyed: 9,859ha
Hedge length surveyed: 41.53km
Number of hedges surveyed: 183
Proportion of hedges species-rich: 27%
Number of standard trees: 76 (1 per 546m on average)
Condition: 5% in favourable condition for height, width, gappiness of shrub layer, width of margins and introduced species; 95% in favourable condition for nutrient enrichment. The 8 hedges in favourable condition were dense and trimmed, and just 2 of these were adjacent wholly or in part to arable land.
Comments: 40% of the hedges contained more than 80% Hawthorn; the rest, with rare exceptions, were mixed species. After Hawthorn the next most frequent species were Dog-rose, bramble, Elder and Blackthorn. Ash was the main hedgerow tree, followed by Sycamore. 50 young trees were tagged during the survey. 1,675m of new hedge was planted in the Borough of Hartlepool in 2010, along with 83 trees.

Essex 2010 A (Banks 2010)

Survey area: Hazeleigh Woods Living Landscape area, Maldon District. The area stretches from the parish of Woodham Mortimer eastwards to the Blackwater estuary.
Survey type: random sample
Area surveyed: 2,124ha
Hedge length surveyed: 28.83km
Number of hedges surveyed: 104
Proportion of hedges species-rich: 23%
Number of standard trees: 25 (1 per 1,153m on average)
Condition: 41% in favourable condition for height, width, gappiness of shrub layer, width of margins and introduced species; 28% in favourable condition for nutrient enrichment. 17% of hedge sides next to arable were in favourable condition for all attributes including nutrient enrichment, but just 5% of those next to grassland. Grazing was believed to be the main reason for poor condition, with herbicides possibly a contributing factor. Gaps along the length of hedges may have been under-recorded.
Comments: the survey area is a national stronghold for Grey Partridge. 18 (15%) woody linear features selected for survey were found no longer to fall within the definition of a hedge, 10 because they were too wide. Three woody linear features had been lost altogether, but these were offset by three new hedges. 47% of hedges were single-species. The modal number of species per 30m was 3, and 4%

of hedges were 'important' under the *Hedgerows Regulations*, having 7 or more species per 30m stretch. Blackthorn, Hawthorn and elms were dominant in the shrub layer. 87% of hedgerows had no standard hedgerow trees. Where such trees were present, the most common species were oaks (59%), followed by elms (14%). On average, there were 0.4 trees over 100cm diameter at breast height or in veteran condition per 1km. 68% of hedges had been flail cut within the last 2 years, and just 1% either laid or coppiced within the last 10 years. 31% of hedges had received no management in the last 10 years. 66% of hedges were within holdings with agri-environment agreements. Overall, lack of management and inappropriate management were considered the greatest threats. Recruitment of new hedgerow trees, and reconnection of isolated hedges, were additional major challenges.

Essex 2010 B (Moore 2010)

Survey area: parishes of Moreton, Fyfield and The Lavers in Epping Forest District
Survey type: random sample, although a large proportion of farms refused access permission, necessitating changes in sample areas
Area surveyed: not specified
Hedge length surveyed: 26.1km
Number of hedges surveyed: 99
Proportion of hedges species-rich: 44%
Number of standard trees: 301 (1 per 87m on average)
Condition: 51% in favourable condition for height, width, gappiness of shrub layer, width of margins and introduced species; 40% in favourable condition for nutrient enrichment. Nearly every hedge on conventional farms failed to be in favourable condition because of nutrient enrichment. However, on the two organic farms surveyed, most passed this test. One of these farms had been under organic management for 11 years, the other for about 5 years.
Comments: the 11 farms surveyed were within an area of intensive arable farming earmarked for urban expansion, along the M11 corridor. The most frequent shrubs were Hawthorn, Blackthorn and Field Maple. The main tree species were oaks, Ash and Field Maple. Most veterans were pollard oaks, with a few Ashes. The number of young trees may have been underestimated because these were difficult to identify. Most hedges were managed by side-trimming to a height of 3.5–4m, the top being left untrimmed. Although the traditional manner of rejuvenating hedges in the area is through coppicing, there was no indication of any such management (or laying) in the last 10 years.

Exmoor National Park 2008 (Davis and Hawkins 2008)

Survey area: Exford parish in west Somerset and Parracombe parish in north Devon
Survey type: random sample
Area surveyed: 4,199ha
Hedge length surveyed: 54.4km
Number of hedges surveyed: 289
Proportion of hedges species-rich: 15%
Number of standard trees: 41 (1 per 1,327m on average). The number of hedgerow trees less than 5cm in diameter is likely to have been underestimated.
Condition: 33% in favourable condition for height, width, gappiness of shrub layer, width of margins and introduced species; 84% in favourable condition for nutrient enrichment.
Comments: most hedges in each parish were shrubby hedges, and virtually all were associated with banks. In Exford over 80% of hedges showed signs of being managed traditionally at some stage, while over half had clearly been laid, coppiced or planted up in the last 10 years. Likewise, in Parracombe many hedges (37%) had recently been laid, coppiced or planted up. This demonstrated the success of the Exmoor Environmentally Sensitive Area agri-environment scheme. The ground flora under Beech hedges was considerably less species-rich than under mixed hedges.

Gloucestershire 2007–2008 (Pattison 2008)

Survey area: parishes of Siston and Pucklechurch, South Gloucestershire
Survey type: full census
Area surveyed: 1,710ha
Hedge length surveyed: 37km
Number of hedges surveyed: 219
Proportion of hedges species-rich: 75%
Number of standard trees: 664 mature trees with a diameter of more than 30cm found (1 per 56m on average).
Condition: not specified.
Comments: the survey had a strong and successful focus on publicity and awareness raising, including in local schools. 44% of hedges had no mature hedgerow trees.

Gloucestershire 2009 (Stroud Valleys Project 2009)

Survey area: parishes of Longney, Frampton, Elmore, Slimbridge and Frocester
Survey type: random sample
Area surveyed: 4,800ha

Hedge length surveyed: 9.2km
Number of hedges surveyed: 45
Proportion of hedges species-rich: 29%
Number of standard trees: 114 (1 per 81m on average)
Condition: 31% in favourable condition for height, width, gappiness of shrub layer, width of margins and introduced species; 51% in favourable condition for nutrient enrichment.
Comments: the low sample size means that the results may not be representative of either the parishes or landscape types covered. Average woody species-richness was 4 species per 30m. The dominant shrubs were Blackthorn (38%) and Hawthorn (28%). 60% of hedges had standard hedgerow trees, with an overall average of 2 trees per hedge. Following the loss of elms, the most frequent standard hedgerow trees were Crack-willow (23%) and Ash (22%). Grasses dominated 54% of herbaceous quadrats, forbs 46%. The most frequent grass was False Oat-grass, the most frequent herb Nettle. 13% of hedges were in favourable condition for all attributes, including nutrient enrichment. 68% of hedges had ditches alongside them. No hedges had been recently laid or coppiced, but two had new planting.

Gloucestershire (Severn Vale) 2010 (Greensweig and Lewis 2011)

Survey area: Severn Vale parishes: parts of Arlingham, Cam, Coaley, Eastington, Elmore, Frampton on Severn, Frocester, Fretherne with Saul, Longney, Slimbridge and Whitminster
Survey type: random sample (but difficulty in obtaining access permission required some changes to samples)
Area surveyed: 500ha
Hedge length surveyed: 22.8km
Number of hedges surveyed: 90
Proportion of hedges species-rich: 47%
Number of standard trees: 453 (1 per 50m on average)
Condition: 69% in favourable condition for height, width, gappiness of shrub layer, width of margins and introduced species; 21% in favourable condition for nutrient enrichment. 21% of hedges were in favourable condition for all attributes, including nutrient enrichment.
Comments: on average 32% of hedges were sampled within each 1km square. Just one hedge out of 90 contained 7 or more woody species, the threshold for a hedge to be 'important' under the *Hedgerows Regulations*. The most frequently recorded shrubs were Hawthorn and Blackthorn. 14 grass species and 114 herbs were recorded: on average 3 grasses and 17 herbs were found along each hedge. Non-native species were rare – Snowberry was found in one hedge, and Himalayan Balsam in two. 72% of hedges had standard trees, Ash (185) and oaks (75) being most frequent. Of the 453 trees found, 107 were considered saplings, but this included 47 elms not expected to reach even semi-maturity. 2% of hedges surveyed had been recently coppiced, but none had been laid. The stage in the Adams 10-point management cycle was recorded: 62% of hedges were at stage 5, 23% at stage 6, 14% at stage 7 and 1% at stage 9. There were no hedges in any of the other stages.

Hampshire 2010 (Leggett 2010)

Survey area: Kingsclere parish, north-west Hampshire. Half of the parish is in the North Wessex Downs AONB
Survey type: random sample
Area surveyed: 2,167ha
Hedge length surveyed: 17.76km
Number of hedges surveyed: 76
Proportion of hedges species-rich: 66%
Number of standard trees: 27 (1 per 658m on average)
Condition: 32% in favourable condition for height, width, gappiness of shrub layer, width of margins and introduced species; 78% in favourable condition for nutrient enrichment. 25% of hedges were in favourable condition for all attributes, including nutrient enrichment.
Comments: overall, Kingsclere parish had 702 hedgerows totalling 110km in length, with a mean hedgerow length of 157m. 11% of these hedges were sampled. 33% of the hedges had 7 or more species per 30m so were 'important' under the *Hedgerows Regulations*.

Kent 2006 (Buell 2006)

Survey area: Lamberhurst Parish
Survey type: random sample
Area surveyed: 5,700ha
Hedge length surveyed: 22.21km
Number of hedges surveyed: 148
Proportion of hedges species-rich: 57%
Number of standard trees: not specified (but see comments)
Condition: 47% in favourable condition for height, width, gappiness of shrub layer, width of margins and introduced species.
Comments: about 90% of hedges were present either on 1840 Tithe maps or on 1860 OS maps. Comparison between 1995 and 2006 data suggested there had been an increase in the species-richness of hedges, and in the number of hedgerow trees. There were estimated to be 5,417 hedgerow trees in the parish. 22% of hedges surveyed were under agri-environment agreements: these were in slightly better condition and more species-rich than those not under agreement.

Kent 2008 (Phillips 2008)

Survey area: Canterbury City Council area
Survey type: random sample
Area surveyed: 30,884ha
Hedge length surveyed: 15.41km
Number of hedges surveyed: 92
Proportion of hedges species-rich: 42%
Number of standard trees: not specified
Condition: 51% in favourable condition for height, width, gappiness of shrub layer, width of margins and introduced species; 62% in favourable condition for nutrient enrichment.
Comments: The report highlights some inconsistencies in data collection, reducing confidence in the results.

London Borough of Bromley 2008 (Hayes 2008)

Survey area: Darwin's Landscape Laboratory
Survey type: Full census
Area surveyed: 996ha
Hedge length surveyed: 26.80km
Number of hedges surveyed: 184
Proportion of hedges species-rich: 67%
Number of standard trees: 823 (1 per 33m on average)
Condition: 58% in favourable condition for height, width, gappiness of shrub layer, width of margins and introduced species; 81% in favourable condition for nutrient enrichment.
Comments: this survey was carried out in a Green Belt, in the Downlands landscape where Charles Darwin studied hedgerows in *c.* 1870, and comparisons were made with his findings wherever possible. Just 2% of hedges present in 1868 were missing in 2008. There were 7% more hedges present in 2008 than in 1868, most of these being gained after 1940. Overall, 91% of hedges present in 2008 had been present 140 years beforehand, in Darwin's time. A further 5km of wide woody linear featured were surveyed, although they were considered to fall outside the definition of a hedge. Another hedge survey covering the Downland landscape was undertaken in 1995, using a different methodology. Comparison between survey results suggest that the hedge network remained stable between 1995 and 2008. Both reports show multi-species hedges to be prevalent, a high degree of connectivity, and that more than half of the hedges contained standard trees (68.5% in the 2008 survey). Ash was the most frequent hedgerow tree.

Malvern Hills AONB, lying within Worcestershire, Herefordshire and Gloucestershire, 2009 (Lashley 2009)

Survey area: Malvern Hills Area of Outstanding Natural Beauty
Survey type: random sample
Area surveyed: not specified
Hedge length surveyed: 13.05km
Number of hedges surveyed: 63
Proportion of hedges species-rich: 54%
Number of standard trees: 195 (1 per 67m on average)
Condition: 46% in favourable condition for height, width, gappiness of shrub layer, width of margins and introduced species; 32% in favourable condition for nutrient enrichment.
Comments: hedges were sampled within one or two 1km grid squares in each of six Landscape Character Types where hedges are a dominant character. 99.2% of hedges had a 2m uncultivated width from the base of the hedge. 30 of the 195 hedgerow trees recorded were veterans; only one young standard hedgerow tree was found. 50% of landowners reported via a questionnaire that they were still flail cutting their hedges every year, despite many being aware that this was not best practice for maintaining and enhancing biodiversity.

South Yorkshire 2006–2008 (Barber 2009)

Survey area: Rotherham
Survey type: random sample
Area surveyed: 28,650ha
Hedge length surveyed: 42.8km
Number of hedges surveyed: 175
Proportion of hedges species-rich: 61%
Number of standard trees: 43 (1 per 995m on average)
Condition: 27% in favourable condition for height, width, gappiness of shrub layer, width of margins and introduced species; 59% in favourable condition for nutrient enrichment.
Comments: about 70% of the hedges were dominated by Hawthorn - these were less species-rich than others. Where Hawthorn did not dominate, then Blackthorn, Field Maple and Holly were likely to do so. However, Elder was the second most frequent species after Hawthorn.

Sussex 2007 (Challis 2009)

Survey area: Low Weald, High Weald, Wealden Greensand, South Downs and Coastal Plains Natural Areas
Survey type: random sample
Area surveyed: 370,000ha
Hedge length surveyed: 48km

Number of hedges surveyed: 361
Proportion of hedges species-rich: 51%
Number of standard trees: 273 (1 per 176m on average)
Condition: 44% in favourable condition for nutrient enrichment
Comments: a continuation of a Hedgerow Inventory Project started in 2001/02, using the newly available standard methodology developed by Defra/Hedgelink (Defra 2007). Hedges for survey were selected within 0.25km² randomly selected squares within each Natural Area: by autumn 2007, 54 such squares had been surveyed.

West Yorkshire 2006 (Duke *et al.* 2007)

Survey area: Calderdale
Survey type: partial census: volunteers asked to visit every hedge in 1km squares selected centrally
Area surveyed: 1,600ha (estimated from report)
Hedge length surveyed: 15.2km
Number of hedges surveyed: 139
Proportion of hedges species-rich: 47%
Number of standard trees: 519 (1 per 29m on average)
Condition: 11% in favourable condition for height, width, gappiness of shrub layer, width of margins and introduced species; 45% in favourable condition for nutrient enrichment. Only two out of the 139 hedges were in favourable condition for all attributes, including nutrient enrichment.
Comments: 99 of the hedges surveyed were in the Coal Measures Natural Area, 40 in the South Pennines Natural Area. The mean hedge density was 0.95km per km². 9% of hedges were dominated by Holly and 63% by Hawthorn. Himalayan Balsam was the most frequently encountered non-native. 40% of hedges had no standard hedgerow trees. Just 4% of trees had diameters less than 10cm, with 27% having diameters between 10cm and 19cm.

West Yorkshire 2009 (Hanson 2010)

Survey area: Harewood, Lotherton Hall and Temple Newsham estates in Leeds District
Survey type: partial census
Area surveyed: 862ha
Hedge length surveyed: 16.86km
Number of hedges surveyed: 94
Proportion of hedges species-rich: 52%
Number of standard trees: 70 (1 per 241m on average)
Condition: 39% in favourable condition for height, width, gappiness of shrub layer, width of margins and introduced species; 79% in favourable condition for nutrient enrichment.
Comments: about 60% of the hedges at Harewood were surveyed, 65% at Lotherton Hall and 20% at Temple Newsham. The results varied considerably between the three estates. For example, 71% of those at Harewood were species-rich, but only 17% at Lotherton Hall and 36% at Temple Newsham. The average length of hedges at Harewood Estate was 212m and at Temple Newsham 224m, but just 76m at Lotherton Hall, where the survey area included formal and bird gardens. 77% of hedges on the Harewood estate were trimmed on a 2–3-year rotation, under an agri-environment agreement. Hawthorn was the main shrub on all three estates, followed by Blackthorn, Hazel and Elder. Species-rich hedges were on average in better condition than others: 55% of such hedges at Harewood were in favourable condition, compared to 41% for all hedges on the estate. Seven of the 70 hedgerow trees recorded were veterans, all on the Harewood estate (4 oaks, 2 Ash, 1 Sycamore).

Worcestershire 2010 (Lashley 2010)

Survey area: north-west of Worcester city, between St John's, Hallow and the villages of Lower Broadheath and Upper Broadheath
Survey type: part census, part random sample
Area surveyed: 322ha
Hedge length surveyed: 14.9km
Number of hedges surveyed: 79
Proportion of hedges species-rich: 78%
Number of standard trees: 305 (1 per 49m on average)
Condition: 40% in favourable condition for height, width, gappiness of shrub layer, width of margins and introduced species; 63% in favourable condition for nutrient enrichment.
Comments: Land use within the survey area was dominated by arable farming and market gardening, with some improved grassland and small blocks of woodland. Of 52 boundaries shown on an 1813 map, just 4 were no longer hedges. 47% of hedges surveyed were associated with a bank. No hedges showed signs of laying, planting or gapping up in last 10 years, although 5 had been coppiced. 39 veteran trees were recorded: 35 Pedunculate Oaks, 3 Ashes and 1 Field Maple. Most veteran hedgerow trees were lapsed pollards. 43% of hedgerow trees had a diameter at breast height of 20cm or less, indicating a good population of young trees, although many of these were elms. Survey timing did not permit full assessment under the *Hedgerows Regulations*, but 19 hedges were thought likely to qualify as 'important' and an additional 24 possibly so.

Trees and shrubs that are native in Britain or probable archaeophytes

appendix B

These are the species that may be used when (a) employing Hooper's rule to age a hedge, (b) determining whether a hedge is species-rich, and (c) deciding whether a hedge is 'important' under Schedule 3 of the 1977 *Hedgerows Regulations* and so may not be removed, in England or Wales, without consent. Archaeophytes are species not considered native but which became established before AD 1500.

Alder	*Alnus glutinosa*
Alder Buckthorn	*Frangula alnus*
*Apple	*Malus domestica*
Ash	*Fraxinus excelsior*
Aspen	*Populus tremula*
*Barberry	*Berberis vulgaris*
Beech	*Fagus sylvatica*
Bird Cherry	*Prunus padus*
*Black Currant	*Ribes nigrum*
Black Poplar	*Populus nigra*
Blackthorn	*Prunus spinosa*
Box	*Buxus sempervirens*
Broom	*Cytisus scoparius*
Buckthorn	*Rhamnus cathartica*
Butcher's-broom	*Ruscus aculeatus*
*Cherry Plum	*Prunus cerasifera*
Common Juniper	*Juniperus communis*
Crab Apple	*Malus sylvestris*
Dogwood	*Cornus sanguinea*
Downy Birch	*Betula pubescens*
Downy Currant	*Ribes spicatum*
*Dwarf Cherry	*Prunus cerasus*
Dwarf Gorse	*Ulex minor*
Elder	*Sambucus nigra*
Elm	*Ulmus* species
Field Maple	*Acer campestre*

**Gooseberry	*Ribes uva-crispa*
Gorse	*Ulex europaeus*
** Grey Poplar	*Populus × canescens*
Guelder-rose	*Viburnum opulus*
Hawthorn	*Crataegus monogyna*
Hazel	*Corylus avellana*
Holly	*Ilex aquilfolium*
Hornbeam	*Carpinus betulus*
Large-leaved Lime	*Tilia platyphyllos*
*Lime	*Tilia × europaea*
*Medlar	*Mespilus germanica*
Mezereon	*Daphne mezereum*
Midland Hawthorn	*Crataegus laevigata*
Mountain Currant	*Ribes alpinum*
Osier	*Salix viminalis*
*Pear	*Pyrus communis*
Pedunculate Oak	*Quercus robur*
Plymouth Pear	*Pyrus cordata*
*Plum, Bullace, etc.	*Prunus domestica*
Rose	*Rosa* species
Rowan	*Sorbus aucuparia*
*Scot's Pine	*Pinus sylvestris*
Sea-buckthorn	*Hippophae rhamnnoides*
Sessile Oak	*Quercus petraea*
Silver Birch	*Betula pendula*
Small-leaved Lime	*Tilia cordata*
Spindle	*Euonymus europaeus*
Spurge-laurel	*Daphne laureola*
*Sweet chestnut	*Castanea sativa*
*Sycamore	*Acer pseudoplatanus*
**Walnut	*Juglans regia*
Wayfaring-tree	*Viburnum lantana*
Western Gorse	*Ulex gallii*
** White Poplar	*Populus alba*
Whitebeam	*Sorbus* species
Wild Cherry	*Prunus avium*
**Wild Cotoneaster	*Cotoneaster cambricus*
Wild Pear	*Pyrus pyraster*
Wild Privet	*Ligustrum vulgare*
Wild Service-tree	*Sorbus torminalis*
Willow	*Salix* species
Yew	*Taxus baccata*

* Species omitted from Schedule 3.
** Included in Schedule 3 but not native and doubtfully archaeophytes.
Names follow Stace (2019).

Priority species significantly associated with hedges in the UK

appendix C

These are species included in statutory lists of threatened species most in need of conservation action. Most are red-listed using criteria developed by the International Union for Conservation of Nature (IUCN), so under threat of extinction at a UK level.

The full list of priority species was drawn up between 1995 and 1999 and revised in 2007 (BRIG 2007). Moths that are priority species yet remain widespread and common despite undergoing substantial population declines, and which are not red-listed, are not included in this table.

The relevant pieces of legislation are Section 41 of the Natural Environment and Rural Communities Act 2006 (England), Section 7 of the Environment (Wales) Act 2016, Section 2(4) of the Nature Conservation (Scotland) Act 2004, and Section 3(1) of the Wildlife and Natural Environment Act (Northern Ireland) 2011.

IUCN threat categories

CR – Critically Endangered
EN – Endangered
VU – Vulnerable
NT – Near Threatened
LC – Least Concern
NA – Not Assessed

	Common name	Scientific name	IUCN threat category
Vascular plants	Bastard Balm	*Melittis melissophyllum*	VU
	Copse-bindweed	*Fallopia dumetorum*	VU
	Crested Cow-wheat	*Melampyrum cristatum*	VU
	Grape-hyacinth	*Muscari neglectum*	VU
	Plymouth Pear	*Pyrus cordata*	VU
	Purple Ramping-fumitory	*Fumaria purpurea*	LC
	Starved Wood-sedge	*Carex depauperata*	EN
Non-vascular plants	Pale Bristle-moss	*Orthotrichum pallens*	NA
	Round-leaved Feather-moss	*Rhynchostegium rotundifolium*	NA
Fungi	Pepper Pot	*Myriostoma coliforme*	NA
	Sandy Stilt Puffball	*Battarrea phalloides*	NA
	Weathered Earthstar	*Geastrum corollinum*	NA
Lichens	a beard lichen	*Usnea florida*	NT

	Common name	Scientific name	IUCN threat category
Lichens cont.	a lichen	*Anaptychia ciliaris* subsp. *ciliaris*	EN
	a lichen	*Bacidia incompta*	VU
	a lichen	*Caloplaca flavorubescens*	EN
	a lichen	*Caloplaca virescens*	EN
	a lichen	*Cryptolechia carneolutea*	EN
	a lichen	*Parmelina quercina* [British material is *P. carporrhizans*]	VU
	a lichen	*Wadeana dendrographa*	NT
	a lichen	*Wadeana minuta*	NT
	Golden Hair-lichen	*Teloschistes flavicans*	VU
	Orange-fruited Elm-lichen	*Caloplaca luteoalba*	EN
	Southern Grey Physcia	*Physcia tribacioides*	VU
	String-of-sausage Lichen	*Usnea articulata*	NT
Beetles	Alder Flea Weevil	*Orchestes testaceus*	VU (pre-1994)
	Hazel Pot Beetle	*Cryptocephalus coryli*	EN
	Red-horned Cardinal Click Beetle	*Ampedus rufipennis*	VU (pre-1994)
	Scarlet Malachite Beetle	*Malachius aeneus*	NT
	Six-spotted Pot Beetle	*Cryptocephalus sexpunctatus*	EN
Butterflies	Brown Hairstreak	*Thecla betulae*	VU
	White-letter Hairstreak	*Satyrium w-album*	EN
Moths	Autumnal Rustic	*Eugnorisma glareosa*	NT
	Barberry Carpet	*Pareulype berberata*	EN
	Beaded Chestnut	*Agrochola lychnidis*	NT
	Broom Moth	*Melanchra pisi*	VU
	Broom-tip	*Chesias rufata*	NT
	Dark-barred Twin-spot Carpet	*Xanthorhoe ferrugata*	NT
	Double Dart	*Graphiphora augur*	NT
	Dusky-lemon Sallow	*Xanthia gilvago*	NT
	Dusky Thorn	*Ennomos fuscantaria*	NT
	Figure of Eight	*Diloba caeruleocephala*	EN
	Flounced Chestnut	*Agrochola helvola*	NT
	Garden Dart	*Euxoa nigricans*	VU
	Garden Tiger	*Arctia caja*	NT
	Heart Moth	*Dicycla oo*	NT
	Hedge Rustic	*Tholera cespitis*	VU
	Lackey	*Malacosoma neustria*	VU
	Large Nutmeg	*Apamea anceps*	NT
	Liquorice Piercer	*Grapholita pallifrontana*	NA (probably LC)
	Minor Shoulder-knot	*Brachylomia viminalis*	NT
	Mouse Moth	*Amphipyra tragopoginis*	VU
	Oak Hook-tip	*Watsonalla binaria*	VU
	Pale Eggar	*Trichiura crataegi*	VU
	Pale Shining Brown	*Polia bombycina*	EN

	Common name	Scientific name	IUCN threat category
Moths cont.	Sallow	*Xanthia icteritia*	NT
	Scarce Vapourer	*Orgyia recens*	EN
	September Thorn	*Ennomos erosaria*	NT
	Sloe Carpet	*Aleucis distinctata*	LC
	Sprawler	*Asteroscopus sphinx*	VU
	White-line Dart	*Euxoa tritici*	VU
	White-spotted Pinion	*Cosmia diffinis*	LC
Herptiles (reptiles and amphibians)	Adder	*Vipera berus*	NA (probably LC)
	Common Lizard	*Zootoca vivipara*	NA (probably LC)
	Common Toad	*Bufo bufo*	NA (probably LC)
	Grass Snake	*Natrix helvetica*	NA (probably LC)
	Great Crested Newt	*Triturus cristatus*	NA (probably LC)
	Slow-worm	*Anguis fragilis*	NA (probably LC)
Birds	Bullfinch	*Pyrrhula pyrrhula*	LC
	Cirl Bunting	*Emberiza cirlus*	LC
	Cuckoo	*Cuculus canorus*	LC
	Dunnock	*Prunella modularis*	LC
	Grey Partridge	*Perdix perdix*	VU
	House Sparrow	*Passer domesticus*	LC
	Lesser Redpoll	*Acanthis flammea cabaret*	LC
	Lesser Spotted Woodpecker	*Dryobates minor*	EN
	Linnet	*Linaria cannabina*	LC
	Marsh Tit	*Poecile palustris*	NT
	Red-backed Shrike	*Lanius collurio*	CR
	Reed Bunting	*Emberiza schoeniclus*	LC
	Song Thrush	*Turdus philomelos*	LC
	Spotted Flycatcher	*Muscicapa striata*	NT
	Starling	*Sturnus vulgaris*	VU
	Tree Pipit	*Anthus trivialis*	LC
	Tree Sparrow	*Passer montanus*	VU
	Turtle Dove	*Streptopelia turtur*	CR
	Willow Tit	*Poecile montana*	EN
	Yellowhammer	*Emberiza citrinella*	LC
Mammals	Barbastelle	*Barbastella barbastellus*	VU
	Bechstein's Bat	*Myotis bechsteinii*	LC
	Brown Long-eared Bat	*Plecotus auritus*	LC
	Greater Horseshoe Bat	*Rhinolophus ferrumequinum*	LC
	Harvest Mouse	*Micromys minutus*	NT
	Hazel Dormouse	*Muscardinus avellanarius*	VU
	Hedgehog	*Erinaceus europaeus*	VU
	Lesser Horseshoe Bat	*Rhinolophus hipposideros*	LC
	Polecat	*Mustela putorius*	LC
	Soprano Pipistrelle	*Pipistrellus pygmaeus*	LC

Species names

appendix D

Where standard English names for plants, animals and other organisms exist, these are used throughout the text. For each species, the list below provides a cross-reference to the scientific name.

Common name	Scientific name
Adder	*Vipera berus*
African Savanna Elephant	*Loxodonta africana*
Alder	*Alnus glutinosa*
Alder Buckthorn	*Frangula alnus*
Alfalfa (Lucerne)	*Medicago sativa*
Alpine Newt	*Ichthyosaura alpestris*
Ash	*Fraxinus excelsior*
Ashy Mining Bee	*Andrena cineraria*
Aspen	*Populus tremula*
Badger	*Meles meles*
Bank Vole	*Myodes glareolus*
Barbastelle	*Barbastella barbastellus*
Barberry	*Berberis vulgaris*
Barberry Carpet	*Pareulype berberata*
Barn Owl	*Tyto alba*
Barren Strawberry	*Potentilla sterilis*
Bastard Balm	*Melittis melissophyllum*
Beautiful Fairyfly	*Mymar pulchellum*
Bechstein's Bat	*Myotis bechsteinii*
Beech	*Fagus sylvatica*
Beefsteak Fungus	*Fistulina hepatica*
Bird Cherry	*Prunus padus*
Bittersweet (Woody Nightshade)	*Solanum dulcamara*
Black Bean Aphid	*Aphis fabae*
Black-bindweed	*Fallopia convolvulus*
Black Bryony	*Tamus communis*
Black-grass (Twitch)	*Alopecurus myosuroides*
Black Poplar	*Populus nigra*

Common name	Scientific name
Black-spotted Longhorn Beetle	*Rhagium mordax*
Black-stem Wheat Rust	*Puccinia graminis*
Black-thighed Flower Fly	*Syrphus vitripennis*
Blackbird	*Turdus merula*
Blackcap	*Sylvia atricapilla*
Blackthorn	*Prunus spinosa*
Blue Tit	*Cyanistes caeruleus*
Bluebell	*Hyacinthoides non-scripta*
Bogbean	*Menyanthes trifoliata*
Borage	*Borago officinalis*
Box	*Buxus sempervirens*
Bracken	*Pteridium aquilinum*
Bramble	*Rubus fruticosus* agg.
Bright-line Brown-eye	*Lacanobia oleracea*
Brimstone butterfly	*Gonepteryx rhamni*
Broad Buckler-fern	*Dryopteris dilatata*
Broad-leaved Dock	*Rumex obtusifolius*
Brown Hairstreak	*Thecla betulae*
Brown Hare	*Lepus europaeus*
Brown Long-eared Bat	*Plecotus auritus*
Buckthorn	*Rhamnus cathartica*
Buddleia (Butterfly-bush)	*Buddleja davidii*
Buff-tailed Bumblebee	*Bombus terrestris*
Bugle	*Ajuga reptans*
Bullace	*Prunus domestica* subsp. *insititia*
Bullfinch	*Pyrrhula pyrrhula*
Bulrush	*Typha latifolia*
Buzzard	*Buteo buteo*

Common name	Scientific name
Cabbage Seed Weevil	*Ceutorhynchus assimilis*
Carrion Crow	*Corvus corone*
Chaffinch	*Fringilla coelebs*
Cherry Laurel	*Prunus laurocerasus*
Cherry Plum	*Prunus cerasifera*
Chiffchaff	*Phylloscopus collybita*
Cirl Bunting	*Emberiza cirlus*
Cleavers (Goosegrass)	*Galium aparine*
Coal Tit	*Periparus ater*
Cobalt Blue Fungus	*Pulcherricium caeruleum*
Cock's-foot	*Dactylis glomerata*
Codling Moth	*Cydia pomonella*
Comma	*Polygonia c-album*
Common Batman Fly	*Myathropa florea*
Common Bird's-foot-trefoil	*Lotus corniculatus*
Common Carder Bee	*Bombus pascuorum*
Common Club-rush	*Schoenoplectus lacustris*
Common Comfrey	*Symphytum officinale*
Common Couch	*Elytrigia repens*
Common Dog-violet	*Viola riviniana*
Common Earwig	*Forficula auricularia*
Common Field-speedwell	*Veronica persica*
Common Figwort	*Scrophularia nodosa*
Common Fleabane	*Pulicaria dysenterica*
Common Juniper	*Juniperus communis*
Common Knapweed	*Centaurea nigra*
Common Lizard	*Zootoca vivipara*
Common Nettle	*Urtica dioica*
Common Pipistrelle	*Pipistrellus pipistrellus*
Common Shrew	*Sorex araneus*
Common Spotted Bristleside	*Parasyrphus punctulatus*
Common Toad	*Bufo bufo*
Common Valerian	*Valeriana officinalis*
Copse-bindweed	*Fallopia dumetorum*
Cotoneaster	*Cotoneaster* species
Cow Parsley	*Anthriscus sylvestris*
Crab Apple	*Malus sylvestris*
Crack-willow	*Salix fragilis*

Common name	Scientific name
Creeping Thistle	*Cirsium arvense*
Creeping Willow	*Salix repens*
Curlew	*Numenius arquata*
Daisy	*Bellis perennis*
Damson	*Prunus domestica* subsp. *insititia*
Dandelion	*Taraxacum* agg.
Dark-edged Bee-fly	*Bombylius major*
Deptford Pink	*Dianthus armeria*
Devon Whitebeam	*Sorbus devoniensis*
Docks	*Rumex* species
Dog-rose	*Rosa canina* agg.
Dog's Mercury	*Mercurialis perennis*
Dogwood	*Cornus sanguinea*
Dotted Bee-fly	*Bombylius discolor*
Duke of Argyll's Teaplant	*Lycium barbarum*
Dunnock (Hedge Sparrow)	*Prunella modularis*
Early-purple Orchid	*Orchis mascula*
Elder	*Sambucus nigra*
Elm	*Ulmus* agg.
Escallonia	*Escallonia macrantha*
Eurasian Beaver	*Castor fiber*
Eurasian Drone Fly	*Eristalis arbustorum*
Evergreen Spindle	*Euonymus japonicus*
Eyed Hawk-moth	*Smerinthus ocellata*
False Oat-grass	*Arrhenatherum elatius*
Field Cow-wheat	*Melampyrum arvense*
Field Horsetail	*Equisetum arvense*
Field Maple	*Acer campestre*
Field-rose	*Rosa arvensis*
Fieldfare	*Turdus pilaris*
Firethorn	*Pyracantha coccinea*
Flame	*Axylia putris*
Flame Shoulder	*Ochropleura plecta*
Fox (Red Fox)	*Vulpes vulpes*
Foxglove	*Digitalis purpurea*
Fuchsia	*Fuchsia magellanica*
Garden Bumblebee	*Bombus hortorum*
Garden Chafer	*Phyllopertha horticola*
Garden Privet	*Ligustrum ovalifolium*

Common name	Scientific name
Garden Warbler	*Sylvia borin*
Garlic Mustard (Hedge Garlic)	*Alliaria petiolata*
Gatekeeper (Hedge Brown)	*Pyronia tithonus*
Germander Speedwell	*Veronica chamaedrys*
Glow-worm	*Lampyris noctiluca*
Goat Willow	*Salix caprea*
Goldfinch	*Carduelis carduelis*
Goosegrass (Cleavers)	*Galium aparine*
Gorse	*Ulex europaeus*
Grape-hyacinth	*Muscari neglectum*
Grass Snake	*Natrix helvetica*
Great Bustard	*Otis tarda*
Great Crested Newt	*Triturus cristatus*
Great Spotted Woodpecker	*Dendrocopos major*
Great Tit	*Parus major*
Great Willowherb	*Epilobium hirsutum*
Greater Horseshoe Bat	*Rhinolophus ferrumequinum*
Greater Pond-sedge	*Carex riparia*
Greater Stitchwort	*Stellaria holostea*
Greenfinch	*Chloris chloris*
Grey Alder	*Alnus incana*
Grey Partridge	*Perdix perdix*
Grey Squirrel	*Sciurus carolinensis*
Grey Willow (Common Sallow)	*Salix cinerea*
Ground-ivy	*Glechoma hederacea*
Guelder-rose	*Viburnum opulus*
Hairy Bitter-cress	*Cardamine hirsuta*
Hart's-tongue	*Asplenium scolopendrium*
Harvest Mouse	*Micromys minutus*
Hawthorn (May)	*Crataegus monogyna*
Hazel Dormouse	*Muscardinus avellanarius*
Heart and Dart	*Agrotis exclamationis*
Heather	*Calluna vulgaris*
Hedge Bedstraw	*Galium album*
Hedge Bindweed	*Calystegia sepium*
Hedge Brown (Gatekeeper)	*Pyronia tithonus*
Hedge Garlic (Garlic Mustard)	*Alliaria petiolata*

Common name	Scientific name
Hedge Mustard	*Sisymbrium officinale*
Hedge Rustic	*Tholera cespitis*
Hedge Snail	*Hygromia limbata*
Hedge Sparrow (Dunnock)	*Prunella modularis*
Hedge Woundwort	*Stachys sylvatica*
Hedgehog	*Erinaceus europaeus*
Hedgerow Crane's-bill	*Geranium pyrenaicum*
Hemlock Water-dropwort	*Oenanthe crocata*
Hemp-agrimony	*Eupatorium cannabinum*
Herb-Robert	*Geranium robertianum*
Himalayan Balsam	*Impatiens glandulifera*
Hogweed	*Heracleum sphondylium*
Holly	*Ilex aquifolium*
Holly Blue	*Celastrina argiolus*
Honeysuckle	*Lonicera periclymenum*
Hop	*Humulus lupulus*
Hornbeam	*Carpinus betulus*
Horse Bot Fly	*Gasterophilus intestinalis*
Horse Chestnut	*Pachycnemia hippocastanaria*
House Martin	*Delichon urbicum*
House Sparrow	*Passer domesticus*
Indian Elephant	*Elephas maximus indicus*
Ivy	*Hedera helix/hibernica*
Jackdaw	*Coloeus monedula*
Jay	*Garrulus glandarius*
Kestrel	*Falco tinnunculus*
Knotted Hedge-parsley	*Torilis nodosa*
Laburnum	*Laburnum anagyroides*, *L. alpinum* & *L. × watereri*
Lapwing	*Vanellus vanellus*
Large-leaved Lime	*Tilia platyphyllos*
Large Skipper	*Ochlodes sylvanus*
Large White	*Pieris brassicae*
Lesser Celandine	*Ficaria verna*
Lesser Redpoll	*Acanthis flammea cabaret*
Lesser Whitethroat	*Curruca curruca*
Leyland Cypress	*× Cuprocyparis leylandii*
Lilac	*Syringa vulgaris*
Linnet	*Linaria cannabina*

Common name	Scientific name
Liquorice Piercer	*Grapholita pallifrontana*
Long-tailed Tit	*Aegithalos caudatus*
Lords-and-ladies	*Arum maculatum*
Magpie	*Pica pica*
Marmalade Hoverfly	*Episyrphus balteatus*
Marsh Thistle	*Cirsium palustre*
Marsh Tit	*Poecile palustris*
May (Hawthorn)	*Crataegus monogyna*
Meadow Brown	*Maniola jurtina*
Meadow Clary	*Salvia pratensis*
Meadowsweet	*Filipendula ulmaria*
Midland Hawthorn	*Crataegus laevigata*
Migrant Aphideater (hoverfly)	*Eupeodes corollae*
Mole	*Talpa europaea*
Moorhen	*Gallinula chloropus*
Muntjac (Chinese Muntjac)	*Muntiacus reevesi*
Napier Grass (Elephant Grass)	*Pennisetum purpureum*
Natterer's Bat	*Myotis nattereri*
Navelwort (Wall Pennywort)	*Umbilicus rupestris*
Nettle (Common Nettle)	*Urtica dioica*
Nightingale	*Luscinia megarhynchos*
Noctule	*Nyctalus noctula*
Nuthatch	*Sitta europaea*
Oak Bush-cricket	*Meconema thalassinum*
Oaks	*Quercus* species
Old Man's Beard (Traveller's-joy)	*Clematis vitalba*
Orange-fruited Elm-lichen	*Caloplaca luteoalba*
Pale Shining Brown	*Polia bombycina*
Palmate Newt	*Lissotriton helveticus*
Peacock (butterfly)	*Aglais io*
Pear Sucker	*Cacopsylla pyri*
Pedunculate Oak	*Quercus robur*
Pittosporum	*Pittosporum tenuifolium*
Plane (London Plane)	*Platanus × hispanica*
Plymouth Pear	*Pyrus cordata*
Prickly Pear	*Opuntia* species
Primrose	*Primula vulgaris*
Purple Hairstreak	*Favonius quercus*

Common name	Scientific name
Pygmy Shrew	*Sorex minutus*
Rabbit (Wild Rabbit)	*Oryctolagus cuniculus*
Radish	*Raphanus sativus*
Ramsons (Wild Garlic)	*Allium ursinum*
Rape Winter Stem Weevil	*Ceutorhynchus picitarsis*
Red Admiral	*Vanessa atalanta*
Red-brown Longhorn Beetle	*Stictoleptura rubra*
Red Campion	*Silene dioica*
Red Clover	*Trifolium pratense*
Red Dead-nettle	*Lamium purpureum*
Red Deer	*Cervus elaphus*
Red Fescue	*Festuca rubra*
Red-tailed Bumblebee	*Bombus lapidarius*
Redshank	*Tringa totanus*
Redstart	*Phoenicurus phoenicurus*
Redwing	*Turdus iliacus*
Reedmace (Bulrush)	*Typha latifolia*
Rhododendron	*Rhododendron ponticum*
Ringlet	*Aphantopus hyperantus*
Robin	*Erithacus rubecula*
Roe Deer	*Capreolus capreolus*
Rook	*Corvus frugilegus*
Rosebay Willowherb	*Chamerion angustifolium*
Rosy Apple Aphid	*Dysaphis plantaginea*
Rough Chervil	*Chaerophyllum temulum*
Rowan	*Sorbus aucuparia*
Scarlet Elfcup	*Sarcoscypha austriaca*
Scarlet Malachite Beetle	*Malachius aeneus*
Scots Pine	*Pinus sylvestris*
Serotine	*Eptesicus serotinus*
Skylark	*Alauda arvensis*
Sloe Carpet	*Aleucis distinctata*
Slow-worm	*Anguis fragilis*
Small-leaved Lime	*Tilia cordata*
Small-spotted Spineleg	*Myolepta potens*
Small Skipper	*Thymelicus sylvestris*
Small Tortoiseshell	*Aglais urticae*
Snipe	*Gallinago gallinago*
Snowberry	*Symphoricarpos albus*

Common name	Scientific name
Soapwort	*Saponaria officinalis*
Soft-rush	*Juncus effusus*
Song Thrush	*Turdus philomelos*
Soprano Pipistrelle	*Pipistrellus pygmaeus*
Southern Bracket	*Ganoderma australe*
Speckled Bush-cricket	*Leptophyes punctatissima*
Speckled Wood	*Pararge aegeria*
Spike-rushes	*Eleocharis* species
Spiked Star-of-Bethlehem	*Ornithogalum pyrenaicum*
Spindle	*Euonymus europaeus*
Spotted Flycatcher	*Muscicapa striata*
Spreading Hedge-parsley	*Torilis arvensis*
Starling	*Sturnus vulgaris*
Starved Wood-sedge	*Carex depauperata*
Stoat	*Mustela erminea*
Stock Dove	*Columba oenas*
Straight-tusked Elephant	*Palaeoloxodon antiquus*
String-of-sausage Lichen	*Usnea articlulata*
Stripe-legged Robberfly	*Dioctria baumhaueri*
Swallow	*Hirundo rustica*
Sweet Vernal-grass	*Anthoxanthum odoratum*
Swift	*Apus apus*
Tamarisk	*Tamarix gallica*
Tapered Drone Fly	*Eristalis pertinax*
Tawny Owl	*Strix aluco*
Thrift	*Armeria maritima*
Tiger Marsh Fly	*Helophilus pendulus*
Traveller's-joy (Old Man's Beard)	*Clematis vitalba*
Tree Sparrow	*Passer montanus*
Tufted Vetch	*Vicia cracca*
Upright Hedge-parsley	*Torilis japonica*
Vetiver	*Chrysopogon zizanioides*
Violet Ground Beetle	*Carabus violaceus*
Wall butterfly	*Lasiommata megera*
Wasp Beetle	*Clytus arietis*
Water Mint	*Mentha aquatica*
Water Shrew	*Neomys fodiens*
Water Vole	*Arvicola amphibius*

Common name	Scientific name
Wavy Hair-grass	*Deschampsia flexuosa*
Wayfaring-tree	*Viburnum lantana*
Weasel	*Mustela nivalis*
Western Gorse	*Ulex gallii*
Western Honey Bee	*Apis mellifera*
Western Ramping-fumitory	*Fumaria occidentalis*
White Bryony	*Bryonia dioica*
White-letter Hairstreak	*Satyrium w-album*
White-tailed Bumblebee	*Bombus lucorum* agg.
White Willow	*Salix alba*
Whitethroat	*Curruca communis*
Wild Angelica	*Angelica sylvestris*
Wild Carrot	*Daucus carota* subsp. *carota*
Wild Cherry	*Prunus avium*
Wild Garlic (Ramsons)	*Allium ursinum*
Wild Plum	*Prunus domestica*
Wild Privet	*Ligustrum vulgare*
Wild Service-tree	*Sorbus torminalis*
Wild Strawberry	*Fragaria vesca*
Willow Tit	*Poecile montanus*
Willow Warbler	*Phylloscopus trochilus*
Willows	*Salix* species
Winter Moth	*Operophtera brumata*
Wood Anemone	*Anemone nemorosa*
Wood Crane's-bill	*Geranium sylvaticum*
Wood Mouse	*Apodemus sylvaticus*
Wood-sorrel	*Oxalis acetosella*
Woodpigeon	*Columba palumbus*
Woody Nightshade (Bittersweet)	*Solanum dulcamara*
Wren	*Troglodytes troglodytes*
Wych Elm	*Ulmus glabra*
Yarrow	*Achillea millefolium*
Yellow Dung Fly	*Scathophaga stercoraria*
Yellow-barred Pond Fly	*Sericomyia silentis*
Yellowhammer	*Emberiza citrinella*
Yew	*Taxus baccata*
Yorkshire-fog	*Holcus lanatus*

Further reading

General

Barker, H. 2012. *Hedge Britannia: a Curious History of a British Obsession*. London: Bloomsbury.

Dover, J. W., ed. 2019. *The Ecology of Hedgerows and Field Margins*. Abingdon: Routledge.

Müller, G. 2013. *Europe's Field Boundaries*. Stuttgart: Neuer Kunstverlag. Two volumes.

Pollard, E., Hooper, M. D. and Moore, N. W. 1974. *Hedges*. New Naturalist 58. London: William Collins and Sons.

Staley, J. T., Wolton, R. and Norton, L. 2020. *Definition of Favourable Conservation Status of Hedgerows*. Natural England Report RP2943.

Warwick, H. 2017. *Linescapes: Remapping and Reconnecting Britain's Fragmented Wildlife*. London: Square Peg.

Wolton, R. J., ed. 2014. *Devon Hedges*. Devon Hedge Group and Devon County Council.

Wright, J. 2016. *A Natural History of the Hedgerow, and Ditches, Dykes and Dry Stone Walls*. London: Profile Books.

History

Barnes, G. and Williamson, T. 2006. *Hedgerow History: Ecology, History and Landscape Character*. Macclesfield: Windgather Press.

Muir, R. and Muir, N. 1987. *Hedgerows: Their History and Wildlife*. London: Michael Joseph.

Rackham, O. 1986. *The History of the Countryside*. London: J. M. Dent & Sons.

Rackham O. 1994. *The Illustrated History of the Countryside*. London: Weidenfeld and Nicolson.

Williamson, T. 2002. *Hedges and Walls*. London: National Trust.

Hedgerow trees

Feber, R. 2017. The role of trees outside woods in contributing to the ecological connectivity and functioning of landscapes. Woodland Trust Research Report.

Reid, C., Hornigold, K., McHenry, E. *et al.* 2021. *State of the UK's Woods and Trees 2021*. Grantham: Woodland Trust.

Stokes, J. and Hand, K. 2004. *The Hedge Tree Handbook*. The Tree Council.

Hedges and wood fuel

Chambers, M., Crossland, M., Westaway, S. and Smith, J. 2019. *Harvesting Woodfuel From Hedges: Guidance on bringing England's hedges back into the farm business by managing them for woodfuel*, revised edition. Newbury: Organic Research Centre.

Westaway, S. and Smith, J. 2019. *Productive Hedges: Guidance on Bringing Britain's Hedges Back into the Farm Business*. Cirencester: Organic Research Centre.

Wolton, R. J. 2014. *Wood Fuel From Hedges: How to Manage and Crop Hedges in South-west England for Fuel*. Devon County Council, Tamar Valley AONB and the Devon Hedge Group.

Management

Barr, C. J. Britt, C. P., Sparks, T. H. and Churchward, J. M. 2005. Hedgerow management and wildlife. A review of research on the effects of hedgerow management and adjacent land on biodiversity. Report to Defra.

Botías, C., Basley, K., Nicholls, E. and Goulson, D. 2019. Impact of pesticide use on flora and fauna of field margins and hedgerows. In *The Ecology of Hedgerows and Field Margins*, edited by J. W. Dover. Abingdon: Routledge, pp. 90–109.

Graham, L., Gaulton, R., Gerard, F. and Staley, J. T. 2018. The influence of hedgerow structural condition on wildlife habitat provision in farmed landscapes. *Biological Conservation* 220: 122–131.

Greaves, V. 2002. *Hedgelaying Explained*, revised edition. National Hedgelaying Society.

Maclean, M. 2000. *Resource Management: Hedges*. Tonbridge: Farming Press.

Maclean, M. 2006. *Hedges and Hedgelaying: a Guide to Planting, Management and Conservation*. Marlborough: Crowood Press.

Wolton, R. J. 2022. *New Devon Hedges: How to Make Them*. Devon Hedge Group.

References

Abhijith, K. V. and Kumar, P. 2019. Field investigations for evaluating green infrastructure effects on air quality in open-road conditions. *Atmospheric Environment* 201: 132–147.

Abhijith, K. V., Kumar, P., Gallagher, J., McNabola, A. *et al.* 2017. Air pollution abatement performances of green infrastructure in open road and built-up street canyon environments: a review. *Atmospheric Environment* 162: 71–86.

Adams, N. 2012. The hedgerow management cycle. In *Hedgerow Futures: Proceedings of the First International Hedgelink Conference*, edited by J. W. Dover, held at Stoke-on-Trent, pp. 107–110.

Albrecht, M., Kleijn, D., Williams, N. M. *et al.* 2020. The effectiveness of flower strips and hedgerows on pest control, pollination services and crop yield: a quantitative synthesis. *Ecology letters* 23: 1488–1498.

Alison, J., Botham, M., Maskell, L. C. *et al.* 2022. Woodland, cropland and hedgerows promote pollinator abundance in intensive grassland landscapes, with saturating benefits of flower cover. *Journal of Applied Ecology* 59: 342–354.

Amy, S. R., Heard, M. S., Hartley, S. E., George, C. T., Pywell, R. F. and Staley, J. T. 2015. Hedgerow rejuvenation management affects invertebrate communities through changes to habitat structure. *Basic and Applied Ecology* 16: 443–451.

Anon. 1995. *Biodiversity: the UK Steering Group Report. Volume 2: Action Plans.* London: HMSO.

Anon. *c.* 2010. *Knepp Estate Hedgerow Surveys 2009–2010.* Sussex Hedgerow Inventory Project, Sussex Biodiversity Record Centre.

Atkinson, J. and Lowe, K. 2009. *Shocklach Ancient Hedge Survey.* Cheshire Farming and Wildlife Advisory Group (FWAG) and Cheshire Landscape Trust.

Axe, M. S., Grange, I. D. and Conway, J. S. 2017. Carbon storage in hedge biomass: a case study of actively managed hedges in England. *Agriculture, Ecosystems and Environment* 250: 81–88.

Baird, W. W. and Tarrant, J. R. 1972. Vanishing hedgerows. *Geographical Magazine* 44: 545–551.

Banham, K. 2010. *Chalk Hills of Bedfordshire – Part II. Local Hedgerow Survey Report.* Wildlife Trust for Bedfordshire, Cambridgeshire, Northamptonshire and Peterborough.

Banks, R. 2010. *Hedgerows for Hazeleigh Woods Project 2010: Survey Report.* Essex Biodiversity Project.

Barber, C. 2009. *Rotherham Ancient and Species-rich Hedgerow Project: Updated Project Report.* Rotherham Metropolitan Borough Council.

Barker, H. 2012. *Hedge Britannia: a Curious History of a British Obsession.* London: Bloomsbury.

Barnes, G. and Williamson, T. 2006. *Hedgerow History: Ecology, History and Landscape Character.* Macclesfield: Windgather Press.

Barnes, G., Saunders, D. G. O. and Williamson, T. 2020. Banishing barberry: the history of *Berberis vulgaris* prevalence and wheat stem rust incidence across Britain. *Plant Pathology* 69: 1193–1202.

Barr, C. 2009. Hedgerow survey review. Spreadsheet. Report by Barr Ecology for Defra. www.hedgelink.org.uk.

Barr, C. and Norton, L. 2008 (amended 2009). Reporting on UK hedgerow HAP targets using Countryside Survey 2007 data. Unpublished paper for Hedgelink.

Barr, C. and Petit, S., eds. 2001. *Hedgerows of the World: Their Ecological Functions in Different Landscapes.* Proceedings of the 2001 Annual IALE (UK) conference, held at the University of Birmingham. International Association for Landscape Ecology.

Barr, C. J., Stuart, R. C., Smart, S. M. and Firbank, L. G. 1999. Results from MAFF-funded work in the CS2000 programme. Unpublished report for the UK Steering Group for the Habitat Action Plan for Ancient and/or Species-rich Hedgerows, funded by the Ministry of Agriculture, Food and Farming.

Baudry, J. and Burel, F. 2019. Multi-scale control of carabid assemblages in hedgerow network landscapes. In *The Ecology of Hedgerows and Field Margins*, edited by J. W. Dover. Abingdon: Routledge, pp. 147–162.

Beeston, R. 2007. *Durham Hedgerow Survey 2006.* Durham County Council and Durham Biodiversity Partnership.

Bellamy, P. E. and Hinsley, S. A. 2005. The role of hedgerows in linking woodland bird populations. In *Planning, People and Practice: the Landscape Ecology of Sustainable Landscapes.* Proceedings of the 13th Annual IALE (UK) Conference, pp. 99–106.

Benfield. B. 2021. Lockdown Lichen GR ST00, East Devon. Unpublished note.

Bennett, S. and Dale, J. 2010. *Cornwall Hedgerow Survey Project 2009.* Cornwall Farming and Wildlife Advisory Group (FWAG).

Besnard, A. G. and Secondi, J. 2014. Hedgerows diminish the value of meadows for grassland birds: Potential conflicts for agri-environment schemes. *Agriculture, Ecosystems and Environment* 189: 21–27.

Biber, J.-P. 1988. *Hedges*. Steering Committee for the Conservation and management of the environment and natural habitats, Council of Europe. Planning and Management Series No.1. Strasbourg.

Biffi, S., Chapman, P. J., Grayson, R. P. and Ziv, G. 2022. Soil carbon sequestration potential of planting hedgerows in agricultural landscapes. *Journal of Environmental Management* 307: 114484.

Biffi, S., Chapman, P. J., Grayson, R. P. and Ziv, G. 2023. Planting hedgerows: biomass carbon sequestration and contribution towards net-zero targets. *Science of the Total Environment* 892: 164482

Billington, G. 2005. *Radio Tracking Study of Greater Horseshoe Bats at Buckfastleigh Caves Site of Special Scientific Interest, 2003*. English Nature Research Report 573. Peterborough: English Nature.

Black, K., Lanigan, G., Ward, M., Kavanagh, I., Ó hUallacháin, D. and O'Sullivan, L. 2023. Biomass carbon stocks and stock changes in managed hedgerows. *Science of the Total Environment* 871: 162073.

Blaikie, F. 1821. *A Treatise on the Management of Hedges and Hedge-row Timber*. London: John Harding.

Borin, M., Passoni, M., Thiene, M. and Tempesta, T. 2010. Multiple functions of buffer strips in farming areas. *European Journal of Agronomy* 32: 103–111.

Botías, C., Basley, K., Nicholls, E. and Goulson, D. 2019. Impact of pesticide use on flora and fauna of field margins and hedgerows. In *The Ecology of Hedgerows and Field Margins*, edited by J. W. Dover. Abingdon: Routledge, pp. 90–109.

Boughey, K. L., Lake, I. R., Haysom, K. A. and Dolman, P. M. 2011. Improving the biodiversity benefits of hedgerows: How physical characteristics and the proximity of foraging habitat affect the use of linear features by bats. *Biological Conservation* 144: 1790–1798.

Boyes, D. H., Evans, D. M., Fox, R., Parsons, M. S. and Pocock, M. J. 2021. Street lighting has detrimental impacts on local insect populations. *Science Advances* 7: eabi8322.

Breed, M. F., Ottewell, K. M., Gardner, M. G. and Lowe, A. J. 2011. Clarifying climate change adaptation responses for scattered trees in modified landscapes. *Journal of Applied Ecology* 48: 637–641.

Brewer, A., Ditchburn, B., Cross, D., Whitton E. and Ward, A. 2019. *Tree Cover Outside Woodland in Great Britain*. National Forest Inventory, Statistical Report. Forestry Commission.

BRIG. 2007. *Report on the Species and Habitat Review: Report by the Biodiversity Reporting and Information Group (BRIG) to the UK Standing Committee*. Peterborough: JNCC.

Bright, P. W. 1998. Behaviour of specialist species in habitat corridors: arboreal dormice avoid corridor gaps. *Animal Behaviour* 56: 1485–1490.

Bright, P. W. and MacPherson, D. 2002. *Hedgerow Management, Dormice and Biodiversity*. English Nature Research Report 454. Peterborough: English Nature.

Britt, C., Sparks, T., Roberts, A. and Kirkham, F. 2011a. Hedgerow management in England and Wales: current practice and factors influencing farmers' decisions. *Aspects of Applied Biology* 108: 133–143.

Britt, C., Tuffnell, N., Kirkham, F., Roberts, A. and Sparks, T. 2011b. Hedgerow management: a survey of land managers' and contractors' practices and attitudes. Research report BD2117 for the Department for Environment, Food and Rural Affairs. Britt Vegetation Management and Risk and Policy Analysts Ltd.

Broughton, R. K. and Hinsley, S. A. 2015. The ecology and conservation of the Marsh Tit in Britain. *British Birds* 108: 12–29.

Browne, S. J., Aebischer, N. J., Yfantis, G., and Marchant, J. H. 2004. Habitat availability and use by turtle doves *Streptopelia turtur* between 1965 and 1995: An analysis of common birds census data. *Bird Study* 51: 1–11.

Buell, S. 2006. *A Pilot Assessment of Biological and Historical Hedgerow Features in Kent at Lamberhurst Parish*. Kent and Medway Biological Records Centre.

Burel, F. and Baudry, J. 2012. Hedgerow connectivity. In *Hedgerow Futures: Proceedings of the First International Hedgelink Conference*, edited by J. W. Dover, held at Stoke-on-Trent, pp. 75–86.

Cann, D. 2016. *Sorbus* subgenus *Torminaria* (*Sorbus latifolia* agg.) in Devon newspapers. *BSBI News* 133: 12–17.

Carey, P. D., Wallis, S., Chamberlain, P. M. *et al*. 2008. *Countryside Survey: UK Results from 2007*. NERC/UK Centre for Ecology and Hydrology.

Carroll, Z. L., Bird, S. B., Emmett, B. A., Reynolds, B. and Sinclair, F. L. 2004. Can tree shelterbelts on agricultural land reduce flood risk? *Soil Use and Management:* 20, 357–359.

Carson, R. 1962. *Silent Spring*. Boston, MA: Houghton Mifflin.

Carter, S. 2008 (with 2019 postscript). The life and death of a flailed Cornish hedge. Cornish Hedges Library. www.cornishhedges.co.uk.

Castle, D., Grass, I. and Westphal, C. 2019. Fruit quantity and quality of strawberries benefit from enhanced pollinator abundance at hedgerows in agricultural landscapes. *Agriculture, Ecosystems and Environment* 275: 14–22.

Caubel, V., Grimaldi, C., Merot, P. and Grimaldi, M. 2003. Influence of a hedge surrounding bottomland on seasonal soil-water movement. *Hydrological Processes* 17: 1811–1821.

Challis, P. 2009. *Sussex Hedgerow Inventory Project*. Sussex Biodiversity Record Centre.

Chambers, M., Crossland, M., Westaway, S. and Smith, J. 2019. *Harvesting Woodfuel From Hedges: Guidance on bringing England's hedges back into the farm business by managing them for woodfuel*, revised edition. Newbury: Organic Research Centre.

Chanin, P. and Gubert, L. 2012. Common dormouse (*Muscardinus avellanarius*) movements in a landscape fragmented by roads. *Lutra 55*: 3–15.

Clements, D. K. and Alexander, K. N. A. 2009. A comparative study of the invertebrate faunas of hedgerows of differing ages, with particular reference to indicators of ancient woodland and 'old growth'. *Journal of Practical Ecology and Conservation* 8: 7–27.

Climate Change Committee. 2019. *Net Zero: The UK's Contribution to Stopping Global Warming*. London: Climate Change Committee.

Coombes, D. S. and Sotherton, N. W. 1986. The dispersal and distribution of polyphagous predatory Coleoptera in cereals. *Annals of Applied Biology* 108: 461–474.

Coppins, A. M. 2001. Wayside trees, hedgerows and shrubs. In *Lichen Habitat Management*, edited by A. Fletcher. London: British Lichen Society.

Corbet, G. 2011. Life on the links: a perspective on biodiversity. *British Wildlife* 23: 104–109.

Countryside Agency. 2000. *Locally Distinctive Hedgerows*. Research Notes, CRN 3.

CPRE. 2022. *Farming and Hedgerows: Stretching the Boundaries*. London: CPRE.

Cranmer, L., McCollin, D. and Ollerton, J. 2012. Landscape structure influences pollinator movements and directly affects plant reproductive success. *Oikos* 121: 562–568.

Critchley, C. N. R., Wilson, L. A. Mole, A. C, Astbury, S. S. and Bhogal, A. 2010. *Restoration of Herbaceous Hedgerow Flora: Review and Analysis of Ecological Factors and Restoration Techniques. Phase I*. Final report for Defra project BD5301.

Critchley, C. N. R., Wilson, L. A., Mole, A. C., Norton, L. R. and Smart, S. M. 2013. A functional classification of herbaceous hedgerow vegetation for setting restoration objectives. *Biodiversity Conservation* 22: 701–717.

Cross, J., Fitzgerald, J, Fountain, M. *et al.* 2012. Exploiting semiochemicals, conservation biocontrol and selective physical controls in integrated management of pear sucker. Final report, project EMR 32909. East Malling Research.

Crossland, M. 2015. The carbon sequestration potential of hedges managed for woodfuel. The Organic Research Centre.

Croxton, P. J. and Sparks, T. H. 2002. A farm-scale evaluation of the influence of hedgerow cutting frequency on hawthorn (*Crataegus monogyna*) berry yields. *Agriculture, Ecosystems and Environment* 93: 437–439.

Dainese, M., Montecchiari, S., Sitzia, T., Sigura, M. and Marini, L. 2017. High cover of hedgerows in the landscape supports multiple ecosystem services in Mediterranean cereal fields. *Journal of Applied Ecology* 54: 380–388.

Dartmoor Circle. 2012. *Woodfuel from Hedgerows: a Tool-kit for Communities*. https://devonhedges.org/wp-content/uploads/2022/08/Dartmoor-Circle-Woodfuel-Group-Toolkit.pdf.

Darvill, B., Knight, M. E. and Goulson, D. 2004. Use of genetic markers to quantify bumblebee foraging range and nest density. *Oikos* 107: 471–478.

Davies, E. T. and Dunsford, W. J. 1962. *Some Physical and Economic Considerations of Field Enlargement*. University of Exeter, Department of Agricultural Economics. Report No. 133.

Davies, Z. G. and Pullin, A. S. 2007. Are hedgerows effective corridors between fragments of woodland habitat? An evidence-based approach. *Landscape Ecology* 22: 333–351.

Davis, B. and Hawkins, A. 2008. *Exmoor Parish Hedgerow Survey*. Exmoor National Park.

Defra. 2007. *Hedgerow Survey Handbook: a Standard Procedure for Local Surveys in the UK*, 2nd edition. London: Department for Environment, Food and Rural Affairs.

Defra. 2022a. *Official Statistics: Agri-Climate Report 2022*. London: Department for Environment, Food and Rural Affairs.

Defra. 2022b. *Policy Paper: Government Food Strategy (England)*. London: Department for Environment, Food and Rural Affairs.

Defra. 2023a. *Environmental Improvement Plan 2023*. London: Department for Environment, Food and Rural Affairs. www.gov.uk/government/publications/environmental-improvement-plan.

Defra. 2023b. Consultation on protecting hedgerows: ensuring continued protections for hedgerows after the end of cross compliance. London: Department for Environment, Food and Rural Affairs.

Delettre, Y. R. and Morvan, N. 2000. Dispersal of adult aquatic Chironomidae (Diptera) in agricultural landscapes. *Freshwater Biology* 44: 399–411.

Delettre, Y. R., Morvan, N., Trehén, P. and Grootaert, P. 1998. Local biodiversity and multihabitat use in empidoid flies (Insects: Diptera, Empidoidea). *Biodiversity and Conservation* 7: 9–25.

DETR. 1997. *The Hedgerows Regulations 1997: a Guide to the Law and Good Practice*. London: Department of the Environment, Transport and the Regions, now Defra.

Dicks, L. V., Baude, M., Roberts, S. P. M., Phillips, J., Green, M. and Carvell, C. 2015. How much flower-rich habitat is enough for wild pollinators? Answering a key policy question with incomplete knowledge. *Ecological Entomology* 40: 22–35.

Dimbleby, H. 2021. *The National Food Strategy: The Plan*. Independent review for the UK Government.

Dodds, M. J. W. 2005. Mechanical hedge-laying for wildlife, Aylesbury Vale, Buckinghamshire, England. *Conservation Evidence* 2005: 55–56.

Donkersley, P. 2019. Trees for bees. *Agriculture, Ecosystems and Environment* 270–271: 79–83.

Dover, J., Sparks, T., Clarke, S., Gobbett, K. and Glossop, S. 2000. Linear features and butterflies: the importance of green lanes. *Agriculture, Ecosystems and Environment* 80: 227–242.

Dowdeswell, W. H. 1987. *Hedgerows and Verges*. London: Allen and Unwin.

Drake, C. M. 2013. *Bombylius discolor* Mikan and *B. major* Linnaeus (Diptera, Bombylidae) at an *Andrena cinerea* (Linnaeus) colony (Hymenoptera, Andrenidae). *Dipterists Digest* 20: 3–13.

Drexler, S., Gensior, A. and Don, A. 2021. Carbon sequestration in hedgerow biomass and soil in the temperate climate zone. *Regional Environmental Change* 21: 1–14.

Duke, C., Firman, H. A. and Tansey, C. 2007. *The Calderdale Hedge Hunt: a Survey of the Hedgerows of Calderdale 2006*. Calderdale Council.

Dunn, J. C., and Morris, A. J. 2012. Which features of UK farmland are important in retaining territories of the rapidly declining turtle dove *Streptopelia turtur*? *Bird Study* 59: 394–402.

Dunn, J. C., Gruar, D., Stoate, C., Szczur, J. and Peach, W. J. 2016. Can hedgerow management mitigate the impacts of predation on songbird nest survival? *Journal of Environmental Management* 184: 535–544.

Edwards, B. and Newbould, J. A. 2009. *Dorset Hedgerows Survey 2007–08. An Assessment of Hedgerows in Kingcombe and Powerstock Strategic Nature Area*. Dorset Environmental Records Centre.

Emmett, B. E. and the GMEP team. 2017. *Glastir Monitoring and Evaluation Programme*. Final Report to Welsh Government (Contract reference: C147/2010/11). NERC/UK Centre for Ecology and Hydrology.

Evans, A. 2009. *Chesterfield Hedgerow Survey Final Report*. Derbyshire Wildlife Trust.

Evans, D. M., Pocock, M. J. O, Brooks, J. and Memmott, J. 2011. Seeds in farmland food-webs: Resource importance, distribution and the impacts of farm management. *Biological Conservation* 144: 2941–2950.

Evans, D. M., Pocock, M. J. O., Brooks, J. and Memmott, J. 2013. The robustness of a network of ecological networks to habitat loss. *Ecology Letters* 16: 844–852.

Eybert, M. C., Constant, P. and Lefeuvre, J. C. 1995. Effects of changes in agricultural landscape on a breeding population of linnets *Acanthis cannabina* L. living in adjacent heathland. *Biological Conservation* 74: 195–202.

Facey, S. L., Botham, M. S., Heard, M. S., Pywell, R. F. and Staley, J. T. 2014. Lepidoptera communities and agri-environment schemes; examining the effects of hedgerow cutting regime on Lepidoptera diversity, abundance and parasitism. *Insect Conservation and Diversity* 7: 543–552.

Falloon, P., Powlson, D. and Smith, P. 2004. Managing field margins for biodiversity and carbon sequestration: a Great Britain case study. *Soil Use and Management* 20: 240–247.

Feber, R. E., Johnson, P. J., Gelling, M. and Macdonald, D. W. 2019. Ecology and conservation of mammals of hedgerows and field margins. In *The Ecology of Hedgerows and Field Margins*, edited by J. W. Dover. Abingdon: Routledge, pp. 233–249.

Follain, S., Minasny, B., McBratney, A. B. and Walter, C. 2006. Simulation of soil thickness evolution in a complex agricultural landscape at fine spatial and temporal scales. *Geoderma* 133: 71–86.

Forest Research. 2009. *Trends, Long Term Survival and Ecological Values of Hedgerow Trees: Development of Populations Models to Inform Strategy*. Defra Research Report BD2111.

Forest Service. 2017. *Ireland's National Forest Inventory 2017: Results*. Dublin: Department of Agriculture Food and the Marine.

Forster, A. 2010. Urban hedgerows in the Bristol area 2009: local hedgerow survey report. Unpublished report to Defra, BTCV.

Foulkes, N. 2006a. *County Kildare Hedgerow Survey Report*. An Chomhairle Oidhreachta / The Heritage Council.

Foulkes, N. 2006b. *County Leitrim Hedgerow Survey Report*. An Chomhairle Oidhreachta / The Heritage Council.

Fox, R., Dennis, E. B., Harrower, C. A. *et al.* 2021. *The State of Britain's Larger Moths 2021*. Wareham: Butterfly Conservation, Rothamsted Research and UK Centre for Ecology and Hydrology.

French, D. D. and Cummins, R. P. 2001. Classification, composition, richness and diversity of British hedgerows. *Applied Vegetation Science* 4: 213–228.

Froidevaux, J. S. P., Boughey, K. L., Hawkins, C. L., Broyles, M. and Jones, G. 2019a. Managing hedgerows for nocturnal wildlife: do bats and their insect prey benefit from targeted agri-environment schemes? *Journal of Applied Ecology* 56: 1610–1623.

Froidevaux, J. S. P., Broyles, M. and Jones, G. 2019b. Moth responses to sympathetic hedgerow management in temperate farmland. *Agriculture, Ecosystems and Environment* 270: 55–64.

Fuller, R. J., Chamberlain, D. E., Burton, N. H. K. and Gough, S. J. 2001. Distributions of birds in lowland agricultural landscapes of England and Wales: how distinctive are bird communities of hedgerows and woodland? *Agriculture, Ecosystems and Environment* 84: 79–92.

García, A. F., Griffiths, G. J. K. and Thomas, C. F. 2000. Density, distribution and dispersal of the carabid beetle *Nebria brevicollis* in two adjacent cereal fields. *Annals of Applied Biology* 137: 89–97.

Gardiner, T. 2008. Nature reserves and Glow-worms, *Lampyris noctiluca* (Linnaeus, 1758) (Coleoptera: Lampyridae). *Entomologist's Gazette* 59: 131–138.

Gardiner, T. 2010. Hedgerow species richness influences the presence of Orthoptera and Dermaptera along green lanes in Essex, UK. *Entomologist's Gazette* 61: 53–64.

Garratt, M. P., Senapathi, D., Coston, D. J., Mortimer, S. R. and Potts, S. G. 2017. The benefits of hedgerows for pollinators and natural enemies depends on hedge quality and landscape context. *Agriculture, Ecosystems and Environment* 247: 363–370.

Gathmann, A. and Tscharntke, T. 2002. Foraging ranges of solitary bees. *Journal of Animal Ecology* 71: 757–764.

Ghazavi, G., Thomas, Z., Hamon, Y., Marie, J. C., Corson, M. and Merot, P. 2008. Hedgerow impacts on soil-water transfer due to rainfall interception and root-water uptake. *Hydrological Processes* 22: 4723–4735.

Good, R. 1937. An account of a botanical survey of Dorset. *Proceedings of the Linnean Society of London* 149: 114–116.

Goodwin, C. E. D., Hodgson, D. J., Al-Fulaij, N., Bailey, S., Langton, S. and Macdonald, R. A. 2017. Voluntary recording scheme reveals ongoing decline in the United Kingdom hazel dormouse *Muscardinus avellanarius* population. *Mammal Review* 47: 183–197.

Goulson, D. 2021. *Silent Earth: Averting the Insect Apocalypse*. London: Penguin Random House.

Greaves, V. 2002. *Hedgelaying Explained*, revised edition. National Hedgelaying Society.

Green, R. E., Osborne, P. E. and Sears, E. J. 1994. The distribution of passerine birds in hedgerows during the breeding season in relation to characteristics of the hedgerow and adjacent farmland. *Journal of Applied Ecology* 31: 677–692.

Green, S., Martin, S., Gharechelou, S., Cawkwell, F. and Black, K. 2019. *BRIAR: Biomass Retrieval in Ireland Using Active Remote Sensing*. Report No. 305 by Teagasc. Environmental Protection Agency, Ireland.

Green, T. 2016. *Tree Hay: A forgotten fodder*. Agricology blog post. The Organic Research Centre, Cirencester. www.agricology.co.uk/field/blog/tree-hay-forgotten-fodder.

Greensweig, D. and Lewis, R. 2011. *Severn Vale Hedgerow Survey 2010 (Phase 2)*. Stroud: Stroud Valleys Project.

Gregg, R., Elias, J. L., Alonso, I., Crosher, I. E., Muto, P. and Morecroft, M. D. 2021. *Carbon Storage and Sequestration by Habitat: a Review of the Evidence*, 2nd edition. Natural England Research Report NERR094. York: Natural England.

Gregory, G. 2009. *Hedgerows of the Moss Valley Survey Project*. Three Valleys Project, Derbyshire County Council.

Gregory, N. G. 1995. The role of shelterbelts in protecting livestock: a review. *New Zealand Journal of Agricultural Research* 38: 423–450.

Grigson, G. 1975. *The Englishman's Flora*. St Albans: Granada Publishing.

Guyot, G. and Seguin, B. 1976. Influence du bocage sur le climat d'une petite région. In *Les bocages, histoire, écologie, économie*. Inra, CNRS, Ensa et Université Rennes. pp. 121–130.

Haenke, S., Kovács-Hostyánszki, A., Fründ, J. *et al*. 2014. Landscape configuration of crops and hedgerows drives local syrphid fly abundance. *Journal of Applied Ecology* 51: 505–513.

Haigh, A., O'Riordan, R. M. and Butler, F. 2012. Nesting behaviour and seasonal body mass changes in a rural Irish population of the Western hedgehog (*Erinaceus europaeus*). *Acta Theriologica* 57: 321–331.

Hanson, A. 2010. *Historic Parklands and Estates Hedgerow Survey, West Yorkshire*. Yorkshire Farming and Wildlife Advisory Group (FWAG), Leeds City Council and the Harewood Farming Company.

Harrison. J. 2011. *The Hartlepool Hedgerow Survey: a Random Survey of the Hedgerows in the Borough of Hartlepool*. Middlesborough: The Wildflower Ark.

Hayes, A. 2008. *Darwin's Hedgerows*. London: London Borough of Bromley.

The Hedgerows Regulations 1997. UK Statutory Instruments 1997, number 1160.

Henry, T. 2012. The development of hedgerow planting and restoration in Northern Ireland. In *Hedgerow Futures: Proceedings of the First International Hedgelink Conference*, edited by J. W. Dover, held at Stoke-on-Trent, pp. 157–164.

Herbst, M., Roberts, J. M., Rosier, P. T. W. and Gowing, D. J. 2006. Measuring and modelling the rainfall interception loss by hedgerows in southern England. *Agricultural and Forest Meteorology* 141: 244–256.

Hinsley, S. A., Bellamy, P. E., Sparks, T. H. and Rothery, P. 1999. A field comparison of habitat characteristics and diversity of birds, butterflies and plants between game and nongame areas. In *Lowland Game Shooting Study*, edited by L. G. Firbank. ITE final report to the British Association for Shooting and Conservation. Grange over Sands: Institute of Terrestrial Ecology, pp. 69–116.

Hinsley, S. A. and Bellamy, P E. 2000. The influence of hedge structure, management and landscape context on the value of hedgerows to birds: A review. *Journal of Environmental Management* 60: 33–49.

Hodgson, C. 2010. *Devon Hedgerow Surveys 2007–2009*. Devon Farming and Wildlife Advisory Group (FWAG).

Hof, A. R. and Bright, P. W. 2010. The value of agri-environment schemes for macroinvertebrate feeders: hedgehogs on arable farms in Britain. *Animal Conservation* 13: 467–473.

Holden, J., Grayson, R. P., Berdeni, D. *et al*. 2019. The role of hedgerows in soil functioning within agricultural landscapes. *Agriculture, Ecosystems and Environment* 273: 1–12.

Holland, J. M. 2019. Contribution of hedgerows to biological control. In *The Ecology of Hedgerows and Field Margins*, edited by J. W. Dover. Abingdon: Routledge, pp. 123–146.

Hood, A. 2014. An assessment of hedgerow and bank condition in South Devon Area of Outstanding Natural Beauty. MSc thesis, Plymouth University.

Image, M., Gardner, E., Clough, Y. *et al*. 2023. Which interventions contribute most to the net effect of England's agri-environment schemes on pollination services? *Landscape Ecology* 38: 271–291.

Joint Nature Conservation Committee (JNCC). *UK Biodiversity Action Plan*. https://jncc.gov.uk/our-work/uk-bap.

Jauker, F., Diekötter T, Schwarzbach F. and Wolters V. 2009. Pollinator dispersal in an agricultural matrix: opposing responses of wild bees and hoverflies to landscape structure and distance from main habitat. *Landscape Ecology* 24: 547–555.

Jensen, M. 1954. *Shelter Effect: Investigations into the Aerodynamics of Shelter and its Effects on Climate and Crops*. Copenhagen: Danish Technical Press.

Joly, P., Miaud, C., Lehmann, A. and Grolet, O. 2001. Habitat matrix effects on pond occupancy in newts. *Conservation Biology* 15: 239–248.

Kells, A.R. and Goulson, D. 2003. Preferred nesting sites of bumblebee queens (Hymenoptera: Apidae) in agroecosystems in the UK. *Biological Conservation* 109: 165–174.

Kendall, N. R., Smith, J., Whistance, L. K. *et al*. 2019. *Tree Leaves as Supplementary Feed for Ruminant Livestock*. Woodland Trust Research Briefing.

Kendall, N. R., Smith, J., Whistance, L. K. *et al*. 2021. Trace element composition of tree fodder and potential nutritional use for livestock. *Livestock Sciences* 250: 104560.

Kirby, K. 1995. *Rebuilding the English Countryside: Habitat Fragmentation and Wildlife Corridors as Issues in Practical Conservation*. English Nature Science Report 10, Peterborough: English Nature.

Kirby, K. 2020. *Woodland Flowers: Colourful Past, Uncertain Future*. British Wildlife Collection 8. London: Bloomsbury Wildlife.

Kort, J. 1988. 9. Benefits of windbreaks to field and forage crops. *Agriculture, Ecosystems and Environment* 22: 165–190.

Kremen, C., Albrecht, M. and Ponisio, L. 2019. Restoring pollinator communities and pollination services in hedgerows in intensively managed agricultural landscapes. In *The Ecology of Hedgerows and Field Margins*, edited by J. W. Dover. Abingdon: Routledge, pp. 163–185.

Kuemmel, B. 2003. Theoretical investigation of the effects of field margin and hedges on crop yields. *Agriculture, Ecosystems and Environment* 95: 387–392.

Land Use Consultants, in association with GHK Consulting. 2009. *Provision of Ecosystem Services through the Environmental Stewardship Scheme*. Project report. Defra Project Code NR0121. London: Department for Environment, Food and Rural Affairs.

Lashley, R. 2009. *Final Report: Malvern Hills AONB Hedgerow Survey 2009*. Worcestershire Biological Records Centre.

Lashley, R. 2010. *Final Report: West Worcester Tree and Hedgerow Survey*. Worcestershire Biological Records Centre.

Leggett, T. 2010. *Kingsclere Parish Hedgerow Survey 2010*. Hampshire and Isle of Wight Wildlife Trust.

Lewis-Phillips, J., Brooks, S. J., Sayer, C. D. *et al*. 2020. Ponds as insect chimneys: Restoring overgrown farmland ponds benefits birds through elevated productivity of emerging aquatic insects. *Biological Conservation* 241: 108253.

Litza, K., Alignier, A., Closset-Kopp, D. *et al*. 2022. Hedgerows as a habitat for forest plant species in the agricultural landscape of Europe. *Agriculture, Ecosystems and Environment* 326: 107809.

Locke, G. M. 1962. A sample survey of field and other boundaries in Great Britain. *Quarterly Journal of Forestry* 56: 137–144.

Macdonald, B. 2019. *Rebirding: Rewilding Britain and its Birds*. Exeter: Pelagic Publishing.

MacDonald, D. W. and Johnson, P. J. 1995. The relationship between bird distribution and the botanical and structural characteristics of hedges. *Journal of Applied Ecology* 32: 492–505.

MacElwain, L. McAree, J., Douglas, I. and Flynn, B. 2021. *Monaghan Hedgerow Appraisal Survey, 2021*. Flynn Furney Environmental Consultants.

Maclean, M. 2006. *Hedges and Hedgelaying: a Guide to Planting, Management and Conservation*. Marlborough: Crowood Press.

Magurran, A. E. 2004. *Measuring Biological Diversity*. Oxford: Blackwell.

Marshall, M. R., Ballard, C. E., Frogbrook, Z. L. *et al*. 2013. The impact of rural land management changes on soil hydraulic properties and runoff processes: results from experimental plots in upland UK. *Hydrological Processes* 28: 2617–2629.

Marshall, W. 1796. *The Rural Economy of the West of England*. Newton Abbot: David and Charles reprint, 1970. Two volumes.

Maskell, L., Henrys, P., Norton, L., Smart, S., and Wood, C. 2013. Distribution of ash trees (*Fraxinus excelsior*) in Countryside Survey data. UK Centre for Ecology and Hydrology. Unpublished report.

Massingham, H. J., ed. 1939. *The English Countryside: A Survey of its Chief Features*. The Pilgrims' Library. London: B. T. Batsford.

Mathews, F., Lovett, L., Rushton, S. and Macdonald, D. W. 2006. Bovine tuberculosis in cattle: reduced risk on wildlife-friendly farms. *Biology Letters* 2: 271–274.

McCann, T., Rogers, D. and Cooper, A. 2012. *Northern Ireland Countryside Survey NICS2007: Field Boundary – Summary Report 1998–2007*. Northern Ireland Environment Agency Research and Development Series No. 12/13. Belfast: NIEA.

Memmott, J. 1999. The structure of a plant–pollinator food web. *Ecology Letters* 2: 276–280.

Menneer, R. 1994. Wildlife Revival in Cornish Hedges: History, Traditions and Practical Guidance. Redruth, Cornwall: Dyllansow Truran.

Menneer, R. 2007. Prehistoric hedges in Cornwall (5000BC – AD450). Cornish Hedges Library. www.cornishhedges.co.uk.

Merckx, T., Feber, R. E, Riordan, P. *et al*. 2009. Optimizing the biodiversity gain from agri-environment schemes. *Agriculture, Ecosystems and Environment* 130: 177–182.

Merckx, T., Feber, R. E., McLaughlan, C. *et al*. 2010a. Shelter benefits less mobile moth species: the field-scale effect of hedgerow trees. *Agriculture, Ecosystems and Environment* 138: 147–151.

Merckx, T., Feber, R. E., Parsons, M. *et al.* 2010b. Habitat preference and mobility of *Polia bombycina*: are non-tailored agri-environment schemes any good for a rare and localised species? *Journal of Insect Conservation* 14: 499–510.

Merckx, T., Marini, L., Feber, R. E. and Macdonald, D. W. 2012. Hedgerow trees and extended-width field margins enhance macro-moth diversity: implications for management. *Journal of Applied Ecology* 49: 1396–1404.

Merot, P. 1999. The influence of hedgerow systems on the hydrology of agricultural catchments in a temperate climate. *Agronomie* 19: 655–669.

Michelmore, A. P. G. and Proctor, M. C. F. 1994. The hedges of Farley Farm, Chudleigh. *Transactions of the Devonshire Association for the Advancement of Science* 126: 57–84.

Miñarro, M. and Prida, E. 2013. Hedgerows surrounding organic apple orchards in north-west Spain: potential to conserve beneficial insects. *Agricultural and Forest Entomology* 15: 82–390.

Moore, J. 2010. *Hedgerow Survey of Moreton, the Lavers and Fyfield.* Epping Forest District Council Countrycare.

Moore, N. W., Hooper, M. D. and Davies, B. N. K. 1967. Hedges I: introduction and reconnaissance studies. *Journal of Applied Ecology* 4: 201–220.

Moorhouse, T. P., Palmer, S. C. F., Travis, J. M. J. and Macdonald, D. W. 2014. Hugging the hedges: Might agri-environment manipulations affect landscape permeability for hedgehogs? *Biological Conservation* 176: 109–116.

Morandin, L. A. and Kremen, C. 2013. Hedgerow restoration promotes pollinator populations and exports native bees to adjacent fields. *Ecological Applications* 23: 829–839.

Morandin, L. A. and Winston, M. L. 2006. Pollinators provide economic incentive to preserve natural land in agroecosystems. *Agriculture, Ecosystems and Environment* 116: 289–292.

Morandin, L. A., Long, R. F. and Kremen, C., 2014. Hedgerows enhance beneficial insects on adjacent tomato fields in an intensive agricultural landscape. *Agriculture, Ecosystems and Environment* 189: 164–170.

Morandin, L. A., Long, R. F. and Kremen, C. 2016. Pest control and pollination cost–benefit analysis of hedgerow restoration in a simplified agricultural landscape. *Journal of Economic Entomology* 109: 1020–1027.

Mortimer, S. R. and Adams. T. 2007. *Chilterns Hedgerow Survey 2006.* Chilterns Conservation Board.

Moxham, R. 2001. *The Great Hedge of India.* London: Constable.

Muir, R. and Muir, N. 1987. *Hedgerows: Their History and Wildlife.* London: Michael Joseph.

Müller, G. 2013. *Europe's Field Boundaries.* Stuttgart: Neuer Kunstverlag. Two volumes.

Nicholls, C. I. and Altieri, M. A. 2013. Plant biodiversity enhances bees and other insect pollinators in agroecosystems: A review. *Agronomy for Sustainable Development* 33: 257–274.

Nicholls, K. W. 2003. *Gaelic and Gaelicized Ireland in the Middle Ages*, 2nd edition. Dublin: Lilliput Press.

Orford, K. A., Vaughan, I. P. and Memmott, J. 2015. The forgotten flies: the importance of non-syrphid Diptera as pollinators. *Proceedings of the Royal Society B* 282: 20142934.

Osborne, J. L., Martin, A. P., Carreck, N. L. *et al.* 2008. Bumblebee flight distances in relation to the forage landscape. *Journal of Animal Ecology* 77: 406–415.

Pardon, P., Reubens, B., Reheul, D. *et al.* 2017. Trees increase soil organic carbon and nutrient availability in temperate agroforestry systems. *Agriculture, Ecosystems and Environment* 247: 98–111.

Pattison, S. 2008. *Siston and Pucklechurch Hedgerow Project.* South Gloucestershire Council.

Peng, R. K., Sutton, S. L. and Fletcher, C. R. 1992. Spatial and temporal distribution patterns of flying Diptera. *Journal of Zoology* 228: 329–340.

Phillips, A. 2010. *Dorset Hedgerow Surveys: Final Report.* Dorchester: Dorset Farming and Wildlife Advisory Group (FWAG).

Phillips, M. 2008. *Species-rich and Ancient Hedgerows in the Canterbury Area: Final Report.* BTCV.

Pollard, E., Hooper, M. D. and Moore, N. W. 1974. *Hedges.* New Naturalist 58. London: William Collins and Sons.

Précigout, P.-A. and Robert, C. 2022. Effects of hedgerows on the preservation of spontaneous biodiversity and the promotion of biotic regulation services in agriculture: towards a more constructive relationships between agriculture and biodiversity. *Botany Letters* 169(2): 1–29.

Prestt, I. 1971. An ecological study of the viper *Vipera berus* in southern Britain. *Journal of Zoology* 164: 373–418.

Rackham, O. 1986. *The History of the Countryside.* London: J. M. Dent & Sons.

Rackham O. 1994. *The Illustrated History of the Countryside.* London: Weidenfeld and Nicolson.

Rackham, O. 2014. *The Ash Tree.* Dorset: Little Toller Books.

Radford, G. L. 2003. *East Anglesey Hedgerow Survey.* Bangor: North Wales Environment Services.

Raskin, B. and Osborn, S. 2019. *The Agroforestry Handbook.* Bristol: Soil Association.

Reid, C., Hornigold, K., McHenry, E. *et al.* 2021. *State of the UK's Woods and Trees 2021.* Grantham: Woodland Trust.

Rialtas na hÉireann (Government of Ireland). 2021. *Food Vision 2030: A World Leader in Sustainable Food Systems.* Dublin: Department of Agriculture, Food and Marine.

Ricci, B., Franck, P., Bouvier, J. C., Casado, D. and Lavigne, C. 2011. Effects of hedgerow characteristics on intra-orchard distribution of larval codling moth. *Agriculture, Ecosystems and Environment* 140: 395–400.

Robertson, H., Marshall, D., Slingsby, E. and Newman, G., eds. 2012. *Economic, Biodiversity, Resource Protection and Social Values of Orchards: a Study of Six Orchards by the Herefordshire Orchards Community Evaluation Project.* Natural England Commissioned Report NECR090.

Rodwell, J. S., ed. 1991. *British Plant Communities. Volume 1: Woodlands and Scrub.* Cambridge: Cambridge University Press.

Sanderson, N. A. and Lamacraft, D. 2022. *Impacts of Ash Dieback* Hymenoscyphus fraxineus *(Chalara) on Priority Lichens and Potential Mitigation Options.* Natural England Commissioned Report NECR428.

Sardiñas, H. S. and Kremen, C. 2015. Pollination services from field-scale agricultural diversification may be context-dependent. *Agriculture, Ecosystems and Environment* 207: 17–25.

Sarlöv Herlin, I. L. and Fry, G. L. A. 2000. Dispersal of woody plants in forest edges and hedgerows in a southern Swedish agricultural area: The role of site and landscape structure. *Landscape Ecology* 15: 229–242.

Scottish Forestry. *6.1.4 Relative palatability and resilience of native tree seedlings and saplings to browsing.* In Woodland Grazing Toolbox. https://forestry.gov.scot/woodland-grazing-toolbox/grazing-management/foraging/palatability-and-resilience-of-native-trees.

Scottish Natural Heritage. 2003. *Land Cover Change 1940s–1980s: Hedgerow.* Trend Note 009, December 2003.

Slade, E. M., Merckx, T., Riutta, T. *et al.* 2013. Life-history traits and landscape characteristics predict macro-moth responses to forest fragmentation. *Ecology* 94: 1519–1530.

Smith, J., Westaway, S., Mullender, S., Giannitsopoulos, M. and Graves, A. 2023. Making hedgerows pay their way: the economics of harvesting field boundary hedges for bioenergy. *Agroforestry Systems* 97: 291–303.

Smith, S. and Gilbert, J. 2001. *National Inventory of Woodland and Trees: England.* Forestry Commission.

Solagro. 2000. *Trees, Hedges and Water.* Solagro, Rural Development Department of Lower Austria, David Hicknie, An Taisce, European Natural Heritage Fund. Toulouse: Solagro.

Sparks, T. H. and Croxton, P. J. 2007. The influence of timing and frequency of hedgerow cutting on hawthorn flowering and berry yields: preliminary results. *Aspects of Applied Biology* 82: 103–106.

Sparks, T. H. and Martin, T. 1999. Yields of hawthorn *Crataegus monogyna* berries under different hedgerow management. *Agriculture, Ecosystems and Environment* 72: 107–110.

Stace, C. 2019. *New Flora of the British Isles*, 4th edition. Middlewood Green: C&M Floristics.

Staley, J. T., Bullock, J. M., Baldock, K. C. R. *et al.* 2013. Changes in hedgerow floral diversity over 70 years in an English rural landscape, and the impacts of management. *Biological Conservation* 167: 97–105.

Staley, J. T., Botham, M. S., Chapman, R. E. *et al.* 2016. Little and late: how reduced hedgerow cutting can benefit Lepidoptera. *Agriculture, Ecosystems and Environment* 224: 22–28.

Staley, J. T., Adams, N. P., Amy, S. R. *et al.* 2018a. *Effects of hedgerow management and restoration on biodiversity.* Report on Defra research project BD2114.

Staley, J. T., Botham, M. S., Amy, S. R., Hulmes, S. and Pywell, R. F. 2018b. Experimental evidence for optimal hedgerow cutting regimes for brown hairstreak butterflies. *Insect Conservation and Diversity* 11: 213–218.

Staley, J. T., Botham, M. S. and Pywell, R. F. 2019. Hedges for invertebrates and plants: how current and historic hedgerow management alters their structural condition and value as a semi-natural habitat. In *The Ecology of Hedgerows and Field Margins*, edited by J. W. Dover. Abingdon: Routledge, pp. 55–71.

Staley, J. T., Wolton, R. and Norton, L. 2020. *Definition of Favourable Conservation Status of Hedgerows.* Natural England Report RP2943.

Staley, J. T., Wolton, R. and Norton, L. R. 2023. Improving and expanding hedgerows – recommendations for a semi-natural habitat in agricultural landscapes. *Ecological Solutions and Evidence* 4: e12209.

Stanes, R., Jewell, A. and Bass, R. 2008. *The Husbandry of Devon and Cornwall.* Exeter: Stanes.

Stephens, H. 1851. *The Book of the Farm*, 2nd edition. Volume 22. Edinburgh and London: William Blackwood and Sons.

Stephenson, N.L., Das, A.J., Condit, R. *et al.* 2014. Rate of tree carbon accumulation increases continuously with tree size. *Nature* 507: 90–93.

Stoate, C. 2019. Bird and invertebrate ecology in field margins. In *The Ecology of Hedgerows and Field Margins*, edited by J. W. Dover. Abingdon: Routledge, pp. 250–262.

Stroud Valleys Project. 2009. *Severn Vale Hedgerow Survey 2009.* Stroud: Stroud Valleys Project.

Sutter, L., Albrecht, M. and Jeanneret, P. 2018. Landscape greening and local creation of wildflower strips and hedgerows promote multiple ecosystem services. *Journal of Applied Ecology* 55: 612–620.

Thomas, J. and Lewington, R. 2010. *The Butterflies of Britain and Ireland.* Gillingham: British Wildlife Publishing.

Thomas, Z., Ghazavi, R., Merot, P. and Granier, A. 2012. Modelling and observation of hedgerow transpiration effect on water balance components at the hillslope scale in Brittany. *Hydrological Processes* 26: 4001–4014.

Turner, S. 2007. *Ancient Country: the Historic Character of Rural Devon.* Devon Archaeological Society Occasional Paper 20. Exeter: Devon Archaeological Society.

Ulber, B., Williams, I. H., Klukowski, Z., Luik, A. and Nilsson, C. 2010. Parasitoids of oilseed rape pests in Europe: key species for conservation biocontrol.

In *Biocontrol-Based Integrated Management of Oilseed Rape Pests*, edited by I. H. Williams. Dordrecht: Springer, pp. 45–76.

Van Den Berge, S. 2021. Role of hedgerow systems for biodiversity and ecosystem services in agricultural landscapes. PhD thesis, Ghent University, Belgium.

Van Den Berge, S., Baeten, L., Vanhellemont, M. *et al*. 2018. Species diversity, pollinator resource value and edibility potential of woody networks in the countryside in northern Belgium. *Agriculture, Ecosystems and Environment* 259: 119–126.

van Vooren, L., Reubens, B., Ampoorter, E. *et al*. 2018. Monitoring the impact of hedgerows and grass strips on the performance of multiple ecosystem service indicators. *Environmental Management* 62: 241–259.

Vancouver, C. 1808. *General View of the Agriculture of the County of Devon*. Newton Abbot: David and Charles reprint, 1969.

Viaud, V. and Kunnemann, T. 2021. Additional soil organic carbon stocks in hedgerows in crop-livestock areas of western France. *Agriculture, Ecosystems and Environment* 305: 107174.

Wallace, E. E., McShane, G., Tych, W., Kretzschmar, A., McCann, T. and Chappell, N. A. 2021. The effect of hedgerow wild-margins on topsoil hydraulic properties, and overland-flow incidence, magnitude and water-quality. *Hydrological Processes* 35: e14098.

Walter, C., Merot, P., Layer, B. and Dutin, G. 2003. The effect of hedgerows on soil organic carbon storage in hillslopes. *Soil Use and Management* 19: 201–207.

Ward, B. 2006. *Studham Parish Hedgerow Census*. Bedfordshire Farming and Wildlife Advisory Group (FWAG). Bedford: BedsLife.

Webb, H. J. 2007. *Maulden Parish Hedgerow Survey Report*. Bedford: BedsLife.

Webb, H. J. 2009. *Chalk Hills of Bedfordshire Local Hedgerow Survey Report*. Bedford: BedsLife.

Weekers, T., Marshall, L., Leclercq, N. *et al*. 2022. Ecological, environmental, and management data indicate apple production is driven by wild bee diversity and management practices. *Ecological Indicators* 139: 108880.

Westaway, S. 2020. *Ramial Woodchip in Agricultural Production*. WOOFS Technical Guide 2. Cirencester: Organic Research Centre.

Westaway, S., Chambers, M., Crossland, M., Wolton, R. and Smith, J. 2016. Managing traditional hedges for biofuel. In *Social and Technological Transformation of Farming Systems: Diverging and Converging Pathways*. 12th European IFSA symposium. Harper Adams University.

Whistance, L. 2018a. *Browse, Preserved Tree Fodder and Nutrition. How Offering Access to Browse and Feeding Tree Fodder Can Supplement the Diet of Domestic Animals*. AFINET Factsheet 12.

Whistance, L. 2018b. *Trees for Shade, Shelter, Survival and Body Maintenance. How Offering Access to Trees Can Improve the Welfare of Domestic Animals*. AFINET Factsheet 13.

Williamson, H. 1970. The Vanishing Hedgerows: a Cry from the Heart (Fourth Treatment). *The Henry Williamson Society Journal* 53 (September 2017): 58–75.

Williamson, T. 2002. *Hedges and Walls*. London: National Trust.

Wilson, P. 2019. Botanical diversity in the hedges and field margins of lowland Britain. In *The Ecology of Hedgerows and Field Margins*, edited by J. W. Dover. Abingdon: Routledge, pp. 35–54.

Wolton, R. J. 2009. Hazel dormouse *Muscardinus avellanarius* (L.) nest site selection in hedgerows. *Mammalia* 73: 7–12.

Wolton, R. J. 2010. A review of Defra-sponsored surveys carried out in 2006, 2007 and 2008 using the UK Hedgerow Biodiversity Action Plan standard procedure. Report to Defra, on behalf of Hedgelink. www.hedgelink.org.uk.

Wolton, R. J. 2011. Local hedgerow surveys 2006–2010: an overview. Unpublished report to Defra, Hedgelink.

Wolton, R. J. 2012. The yield and cost of harvesting wood fuel from hedges in the Tamar Valley and Blackdown Hills AONBs, South-West England. Unpublished report to the Tamar Valley and Blackdown Hills AONBs. EU CORDIALE project.

Wolton, R. J. 2014a. Devon hedges and development 2: pipelines. In *Devon Hedges*, edited by R. J. Wolton. Devon Hedge Group and Devon County Council.

Wolton, R. J. 2014b. *Wood Fuel From Hedges: How to Manage and Crop Hedges in South-west England for Fuel*. Devon County Council, Tamar Valley AONB and the Devon Hedge Group.

Wolton, R. J., Adams, N., Ledder, E., Stokes, J., Stratton, J. and Wolton, P. 2010. Hedgelink visit to the hedges and orchards of Normandy, France, 8–10 May 2010. Unpublished report, Hedgelink.

Wolton, R. J., Morris, R. K. A., Pollard, K. and Dover J. W. 2013. *Understanding the Combined Biodiversity Benefits of the Component Features of Hedges*. Report of Defra project BD5214.

Wolton, R. J., Bentley, H., Chandler, P. J. *et al*. 2014a. The diversity of Diptera associated with a British hedge. *Dipterists Digest* 21: 1–36.

Wolton, R. J., Pollard, K. A., Goodwin, A. and Norton, L. 2014b. *Regulatory Services Delivered by Hedges: the Evidence Base*. Report of Defra project LM0106.

Wolton, R. J., Westaway, S., Chambers, M., Crossland, M. and Smith, J. 2016. Harvesting fuel from hedges. *Conservation Land Management* 14: 4–8.

Wratten, S. D., Bowie, M. H., Hickman, J. M., Evans, A. M., Sedcole, J. R. and Tylianakis, J. M. 2003. Field boundaries as barriers to movement of hover flies (Diptera: Syrphidae) in cultivated land. *Oecologia* 134: 605–611.

Wright, J. 2016. *A Natural History of the Hedgerow, and Ditches, Dykes and Dry Stone Walls*. London: Profile Books.

Illustration credits

All photographs and figures are copyright © the author or are understood to be out of copyright, except for those listed below.

Bloomsbury Publishing would like to thank those listed below for providing illustrations and for permission to reproduce copyright material within this book. While every effort has been made to trace and acknowledge copyright holders, we would like to apologise for any errors or omissions, and invite readers to inform us so that corrections can be made in any future editions.

Key: T = top; B = bottom; L = left; R = right; BL = bottom left

2 Ernie James / naturepl.com; 8 Guy Edwardes / naturepl.com; 25 Graeme Hart, Perthshire Picture Agency; 30 Lianne Joynson © National Memorial Arboretum; 31 L Map data ©2023 Google, Imagery ©2023 TerraMetrics, R Map data ©2023 Google, Imagery ©2023 CNES / Airbus, Getmapping plc, Infoterra Ltd & Bluesky, Landsat / Copernicus, Maxar Technologies, The GeoInformation Group; 36 BL Neil Tackaberry / Shutterstock; 40 Guy Edwardes / naturepl.com; 46 T Paula Wolton; 59 Tony Seymour; 63 Based on Countryside Survey 2007 data and the former Institute of Terrestrial Ecology's land classification of Great Britain. Published in Staley et al. 2023 and reproduced with kind permission of UKCEH; 65 David Noton / naturepl.com; 87 Marcus Vergette; 89 Erni / Shutterstock; 92 Google Earth ©2013 Google, Image ©2014 Getmapping plc.; 96 L Erni / Shutterstock; 100 Keith Tailby; 102 Nature Photographers Ltd / Alamy Stock Photo; 103 Jon Valters; 104 Steven Falk, 112 Guy Edwardes / naturepl.com; 115 Stephen Dalton / naturepl.com; 116 Jon Stokes; 118 Keith Tailby; 120 Jon Valters; 122 Jack Perks / Alamy Stock Photo; 131 John Walters; 153 Charles E Brown, from *The English Countryside* (B.T. Batsford, 1939); 162 Ray Woods; 165 L Anne Coatesy / Shutterstock, R AGAMI Photo Agency / Alamy Stock Photo; 179 Jean Wynne-Jones; 186 Anne Coatesy / Shutterstock; 187 Andy Sands / naturepl.com; 191 Patrick Fox / Alamy Stock Photo; 196 South West Heritage Trust; 215 T Iorwerth Watkins, West Country Rivers Trust / Catchment Sensitive Farming; 222 B British Geological Survey image P669774 © UKRI 2007. All rights reserved (Permit number CP23/014); 223 Kevin Cox; 224 Steven Falk; 237 R Dawn Painter; 238 B John Walters; 239 L Ian Andrews; 239 R John Walters; 241 Helen J Davies / Shutterstock; 245 John Walters; 255 Matthew Axe; 281 Paula Wolton; 302 Will Field; 307 Lex Roeleveld; 323 Georg Müller, *Europe's Field Boundaries* (2013); 329 Nigel Adams; 335 T Steve Bloomfield/Worcestershire WT; 336 Kevin Cox; 347 Laing O'Rourke; 351 Paula Wolton; 357 Paula Wolton; Guy Edwardes / naturepl.com

195 'Thorny questions' by RH Charles, reproduced with permission of Punch / TopFoto.

Index

Page numbers in **bold** refer to illustrations.
Page numbers in *italics* refer to tables.